Praise for *Applying Design for Six Sigma to Software and Hardware Systems*

"New products are critical for growth and sustainable returns. This book is a comprehensive outline and practical reference that has wide applications to development programs both simple and complex. I am surprised by the number of organizations that rely on 'tribal knowledge' and outdated processes for new product development. *Applying Design for Six Sigma to Software and Hardware Systems* details the tools and the roadmap used by those organizations that are outpacing their competitors."

—*Ralph Quinsey, President and Chief Executive Officer,*
TriQuint Semiconductor

"*Applying Design for Six Sigma to Software and Hardware Systems* is a must-read for anyone involved with product development. The book is very well structured, with systematic tools and guidance. There are several examples and case studies that help illustrate the concepts, and it is easy to read and provides several methods to apply for success. Finally, a great DFSS book covering both hardware and software—the right way."

—*Sam Khoury, Senior Program Manager,*
General Dynamics C4 Systems

"Eric and Patricia generously share their vast technical experience in *Applying Design for Six Sigma to Software and Hardware Systems*. This book is filled with real-world, practical examples that make it interesting to read and easy to apply. I especially appreciate the historical perspective on the genesis of Six Sigma and evolution of DFSS from someone who was part of that process from the beginning. Many detailed flowcharts provide a useful framework for implementation of DFSS. This much-needed book fills a gap in DFSS literature and I highly recommend it. Anyone involved in designing high-tech products, in particular those with significant hardware and software components, will find it to be a very valuable resource."

—*Robert Shemenski, Jr.,*
Achieving Competitive Excellence (ACE) Discipline Chief at Pratt & Whitney and
Adjunct Professor of DFSS at Rensselaer Polytechnic Institute

"As deep as it is clear, the book is a tour-de-force of innovative, practical applications of Design for Six Sigma, and should be on every Master Black Belt's bookshelf."

—*Matthew Barney, Ph.D., Vice President, Leadership Development,*
Infosys Technologies, Ltd.

"Out of many books about Design for Six Sigma, here is a book that is based on extensive empirical evidence and a book that focuses on specific actions that middle managers, institutional leaders, and front-line supervisors can take to make their DFSS deployment as well as actual projects better today. This book challenges beliefs of earlier perception with powerful experience and strong tools used in such a way that the individual stages and maturity of companies are respected and still challenge performance rate."

—*Nina Schwartz, MBB,*
Director, Six Sigma Excellence

"The book provides a comprehensive treatment of DFFS for hardware and software systems. By describing, in detail, common objections and misconceptions about DFSS and by providing the tools required to measure progress and track risks, this book is an invaluable tool to anyone keen on building superior-quality hardware and software systems."

—*Sudhakar Ramakrishna, Corporate Vice President,*
Motorola Wireless Broadband Access Solutions and Software Operations

"It's a must read for anyone who wants to achieve success and cultural institutionalization. I was responsible for deploying DFSS to more than one thousand global technology personnel at Honeywell (for both hardware, software, and HW/SW integrated solutions) and this book would have made my job much easier. Many of the concepts for success and failure are ones that I experienced firsthand. This is the first DFSS book with an approach for both software and hardware integrated systems. It's refreshing to see a book that addresses the aspect of requirements flow down as well as aligning the procurement processes with suppliers. This book is not only theoretical but also practical, allowing the reader access to tools and templates for implementing each step of the DFSS lifecycle. The approach to risk management is the most comprehensive that I've seen that actually characterizes what one would experience in a real development situation and not just from a theoretical perspective. It's clear that the authors have compiled years of practice, experience, and consultation into this excellent book on DFSS."

—*Morris Daniely, Director for Global Quality,*
Honeywell Process Solutions

Applying Design for Six Sigma to Software and Hardware Systems

Applying Design for Six Sigma to Software and Hardware Systems

Eric Maass
Patricia D. McNair

PRENTICE
HALL

Upper Saddle River, NJ • Boston • Indianapolis • San Francisco
New York • Toronto • Montreal • London • Munich • Paris • Madrid
Capetown • Sydney • Tokyo • Singapore • Mexico City

The publisher offers excellent discounts on this book when ordered in quantity for bulk purchases or special sales, which may include electronic versions and/or custom covers and content particular to your business, training goals, marketing focus, and branding interests. For more information, please contact:

U.S. Corporate and Government Sales
(800) 382-3419
corpsales@pearsontechgroup.com

For sales outside the United States please contact:

International Sales
international@pearson.com

Visit us on the Web: informit.com/ph

Library of Congress Cataloging-in-Publication Data is on file.

ISBN-13: 978-0-13-335946-6
ISBN-10:   0-13-335946-8
This product is printed digitally on demand. This book is the paperback version of an original hardcover book.
First printing, August 2009

I'd like to dedicate this book to my wife, Joanelle, whose smiles have brightened my days ever since that summer afternoon when she splashed into my life.

—Eric Maass

This book is dedicated to my husband, Earnie, for twenty-eight wonderful years, and our four children, LaTrisha, Andrew, Eldridge, and William, who have made my life enjoyable and wonderful. Also to my dear parents, Mr. and Mrs. Carl Dean Eatmon, who have shown me incredible unconditional love.

With a special dedication to Dr. Les Lander, at Binghamton University, who shaped my academic life.

—Patricia McNair

Contents

Foreword   xvii

Preface   xxi

Acknowledgments   xxvii

About the Authors   xxix

Chapter 1   Introduction: History and Overview of DFSS   1
              A Brief Historical Perspective on Six Sigma and
                 Design for Six Sigma (DFSS)   1
                    Six Steps to Six Sigma   6
              Historical Perspective on Design for Six Sigma   8
                    DFSS Processes   9
              DFSS Example   14
              Summary   27

Chapter 2   DFSS Deployment   29
              Ideal Scenario for DFSS Deployment   29
              Steps Involved in a Successful DFSS Deployment   30
              DFSS Deployment: Single Project   45
              Minimum Set of Tools, and the "One Tool Syndrome"   47
              Goals for DFSS   48
              "The DFSS Project was a Success, But . . ."   50
              Summary   50

Chapter 3	**Governance, Success Metrics, Risks, and Certification**	**53**
	DFSS Governance	53
	Supportive Project Reviews	54
	Formal Gate Reviews	55
	Success Metrics	57
	Product Development Risks	58
	Risk Management Roles	60
	DFSS Certification	62
	Summary	64
Chapter 4	**Overview of DFSS Phases**	**65**
	DFSS for Projects, Including Software and Hardware	65
	DFSS Process Nomenclatures	69
	Requirements Phase	73
	Architecture Phase	75
	Architecture Phase for the Software Aspects	78
	Design Phase	78
	Integrate Phase	78
	Optimize Phase	78
	Verify Phase	80
	Summary	82
Chapter 5	**Portfolio Decision Making and Business Case Risk**	**83**
	Position within DFSS Flow	83
	Portfolio Decision Making as an Optimization Process	84
	Financial Metric	85
	Portfolio Decisions and Resource Constraints	89
	Goals, Constraints, Considerations, and Distractions	91
	Adjusting Portfolio Decisions Based on Existing Commitments and the Organization's Strategic Direction	92
	Summary: Addressing Business Case Risk	94
Chapter 6	**Project Schedule Risk**	**95**
	Position within DFSS Flow	95
	Project Schedule Model	95
	The "Fuzzy Front End" and Delays Caused by Changing Requirements	97

Time for First Pass: Critical Path versus Critical Chain   98

Critical Chain/Theory of Constraints Project
    Management Behaviors   103

Iterations, Qualification, and Release to Product   105

Summary: Addressing Schedule Risk   106

Chapter 7   **Gathering Voice of the Customer to Prioritize Technical Requirements**   107

Importance and Position within DFSS Flow   107

VOC Purpose and Objectives   110

The VOC Gathering (Interviewing) Team   110

Customer Selection   111

Voices and Images   112

Customer Interview Guide   113

Planning Customer Visits and Interviews   115

Customer Interviews   116

KJ Analysis: Grouping, Structuring and Filtering the VOC   117

Identifying Challenging Customer Requirements (NUDs)   120

Kano Analysis   122

Validation and Prioritization of Customer Requirements   124

Translating Customer Requirements to System
    Requirements: The System-Level House of Quality   124

Constructing a House of Quality   128

Summary: VOC Gathering—Tying It All Together   134

Chapter 8   **Concept Generation and Selection**   137

Position within DFSS Flow   137

Concept Generation Approaches   137

Brainstorming and Mind-Mapping   140

TRIZ   141

Alternative Architecture Generation: Hardware and Software   143

Generation of Robust Design Concepts   146

Consideration of Existing Solutions   147

Feasibility Screening   148

Developing Feasible Concepts to Consistent Levels   148

Concept Selection   149

Summary   152
Appendix: Kansei Engineering   152

Chapter 9   Identification of Critical Parameters and FMEA   153
Position within DFSS Flow   153
Definition of a Critical Parameter   153
Considerations from VOB and Constraints   155
Prioritization and Selection of Critical Parameters   157
FMEA   160
Software FMEA Process (Software Systems, Software
    Subsystems, and Software Components FMEA)   164
Software FMEA Implementation Case Study   169
Considerations of Reliability and Availability   172
Examples of Critical Parameters   174
Summary   176
Appendix: Software FMEA Process Documentation   176

Chapter 10   Requirements Flow-Down   187
Position within DFSS Flow   187
Flow-Down for Hardware and Software Systems   190
Anticipation of Potential Problems: P-Diagrams
    and DFMEA   193
Target and Spec Limits   197
Measurement System Analysis   198
Capability Analysis   202
Flow-Down or Decomposition   203
    Procedure for Critical Parameter Flow-Down
        or Decomposition   204
Flow-Down Examples   206
Initial Tolerance Allocation   208
Summary   210

Chapter 11   Software DFSS and Agile   211
Measuring the Agile Design   218
    Data Collection Plan for ViewHome Prototype   219
Summary   221

Chapter 12	**Software Architecture Decisions**	**223**
	Software Architecture Decision-Making Process	224
	Using Design Heuristics to Make Decisions	227
	Common Design Heuristics and Principles	228
	Using Architecture Tactics to Make Decisions	228
	Using DFSS Design Trade-Off Analysis to Make Decisions	230
	Using Design Patterns, Simulation, Modeling, and Prototyping for Decisions	234
	Summary	235
Chapter 13	**Predictive Engineering: Continuous and Discrete Transfer Functions**	**237**
	Discrete versus Continuous Critical Parameters	238
	Methods for Deriving a Transfer Function for a Discrete Critical Parameter	241
	Logistic Regression for Discrete Parameters	242
	Methods for Deriving a Transfer Function for a Continuous or Ordinal Critical Parameter	244
	Existing or Derived Equation (First Principles Modeling)	245
	Modeling within a Spreadsheet, Mathematical Modeling Software, or Simulation Software	246
	Empirical Modeling using Historical Data: Regression Analysis and General Linear Model	247
	Empirical Modeling using Design of Experiments	251
	Empirical Modeling using Response Surface Methods	256
	DOE with Simulators: Design and Analysis of Computer Experiments (DACE)	259
	Summary	261
Chapter 14	**Predictive Engineering: Optimization and Critical Parameter Flow-Up**	**263**
	Critical Parameter Flow-Up: Monte Carlo Simulation	266
	Critical Parameter Flow-Up: Generation of System Moments (Root Sum of Squares)	267
	Critical Parameter Scorecard	269
	Selecting Critical Parameters for Optimization	270

	Optimization: Mean and/or Variance	271
	Optimization: Robustness through Variance Reduction	273
	Multiple Response Optimization	280
	Cooptimization of Cpk's	282
	Yield Surface Modeling	283
	Case Study: Integrated Alternator Regulator (IAR) IC for Automotive	288
	Summary	290
Chapter 15	**Predictive Engineering: Software Optimization**	293
	Multiple Response Optimization in Software	293
	Use Case Modeling in Optimization	294
	Evaluate the Model	298
	Software Mistake Proofing	299
	Software Stability	303
	Summary	305
Chapter 16	**Verification of Design Capability: Hardware**	307
	Position within DFSS Flow	307
	Measurement System Analysis (MSA)	307
	Improvements for Inadequate Measurement Systems	310
	The Risk of Failures Despite Verification: Test Escapes	313
	Determine the Capability	315
	Summary	316
Chapter 17	**Verification of Reliability and Availability**	319
	Customer Perspective	319
	Availability and Reliability Flow Down	321
	Bathtub Curve and Weibull Model	322
	Software Reliability	325
	Early Life Failures/Infant Mortality	326
	Useful Life/Constant Failure Rate	326
	Wear Out	327
	Detailed Flowchart for Reliability Optimization and Verification	327
	Accelerated Life Testing	328
	WeiBayes: Zero Failures Obtained from ALT	330
	Risk of Failures Despite Verification: Reliability Test Escapes	331

	Methods to Improve Reliability and Availability	332
	Summary	333
	Appendix: Case Studies—Software Reliability, and System Availability (Hardware and Software Availability)	333
	Software Reliability: A Case Study in a Zero Defect Initiative	333
	Case Study: Modeling Availability for a Cellular Base Station	342
Chapter 18	Verification: Software Testing Combined with DFSS Techniques	347
	Software Verification Test Strategy Using Six Sigma	350
	Controlling Software Test Case Development through Design Patterns	354
	Improving Software Verification Testing Using Combinatorial Design Methods	356
	Summary	358
	Bibliography	359
	Glossary of Common Software Testing Terms	359
Chapter 19	Verification of Supply Chain Readiness	363
	Position within DFSS Flow	363
	Verification that Tolerance Expectations Will Be Met	366
	Confidence in Robust Product Assembly (DFMA)	366
	Verification of Appropriate and Acceptable Interface Flows	369
	Confidence in the Product Launch Schedule	369
	Confidence in Meeting On-Time Delivery and Lead-Time Commitments	370
	Case Study: Optoelectronic Multichip Module	380
	Summary	382
Chapter 20	Summary and Future Directions	385
	Future Directions	386
Index		391

Foreword

The challenge of developing and launching a successful product into the marketplace is dependent on the effective resolution of a series of compromises: compromises between design and iteration, research and execution, development and testing, and so on. The ability to quickly and accurately work one's way through this process often spells the difference between a product that is successful in the market and one that is not. The emergence and availability of tools and techniques that can inform these decisions and help improve the "hit rate" of success therefore becomes more and more important.

Product development can be summarized as the process of answering two fairly simple questions: "What is it?" and "How can we tell when we are done?" The ability to clearly and objectively address these questions under significant time and resource pressures distinguishes the top product operations from others.

As one evaluates successes and failures in the product space, it seems that some products have almost a unique "voice" of their own. Whether a phenomenon like the original Motorola RAZR phone, the revolution-causing Apple iPod MP3 player, or the digital video recorder, these industry-changing products are unique in that they specifically address unmet needs. It is notable that only a very few of the actual features of these products form the basis for their success; the Apple iPod wasn't particularly great in audio quality, the RAZR had less talk time than most competitive offerings, but in both cases the excellence and targeting of the anchor attributes outweighed the more minor shortcomings. I once heard a very senior colleague of mine state, without fear of contradiction, that there are no examples of great products that are purely the result of consumer or end-user research. The gist of this comment is that consumers haven't encountered all of the unmet needs that distinguish truly innovative products. This would lead to the need for techniques that integrate the consumer insight process with the potentials for applicable technical innovation in the space. While there is no panacea to this need, the ability to use objective techniques in this space is fundamental to success, particularly in deciding where to focus the time, resources, and costs of the product to get maximum leverage in the marketplace.

The concept of "cost of quality" in its most extended state is a very powerful metaphor for the effectiveness of a development cycle. Simply stated, it is the allocation of all effort into two categories—"value added" and "defect detection and extraction"— and the use of proactive tools and techniques to increase the first at the expense of the second. Let me elaborate. If we hypothesize a perfect product development cycle— crystal clear definition optimally tied to the user target, rendering of this description into the relevant software and hardware sub-elements, and then flawless execution without the introduction of any defects—we arrive at the irreducible minimum cost and time for a product development cycle. Great organizations take on the challenge of identifying their current level of performance, comparing it to their competitors, and then setting out to reduce this cost of error by 15% to 20% per year, using clearly de-fined and communicated tools, methods, and technology improvements.

The third important factor in this discussion is the organizational or human element: how does one deploy new techniques and approaches in a mature organization, over-coming the "not invented here" tendencies of all engineering professionals, and quickly traverse the learning curve phase to achieve results? Here is where the deployment and institutionalization aspects developed in Six Sigma and extended for Design for Six Sigma (DFSS) bring significant value. The combination of formal training, implementa-tion of highly experienced mentors into actual development projects, and gradual development of a "community of practice" has been found to be an extremely effective approach.

Making great products is a combination of art and science, the art being the use of experience, insight, and intuition to decide how much of the available science to employ. DFSS is a structured method for developing new products that are aligned with the customers' needs, using predictive engineering and anticipating and managing potential issues. The authors of this book have developed a unique concept for applying DFSS to hardware and software systems. The collected series of methods, tools, and techniques has been proven in application in leading organizations in a variety of industries. While there is no such thing as a definitive product design "cookbook" that infallibly results in industry leading products, this volume represents a rich collection of techniques and approaches that, if used properly, can identify and address the "sweet spot" aspects of a product definition, proactively identify the high leverage realization challenges, and predict and resolve issues early in the process. The combination of these techniques with talented and trained facilitators in Six Sigma methodologies and change management approaches can and will have major impact in both the effectiveness and efficiency of any product development organization.

—Rey Moré
Former Senior Vice President and Chief Quality Officer
Motorola, Inc.

Preface

PURPOSE AND SCOPE

The goal of this book is to provide a clear roadmap and guidance for developing products—not only simple products but also high-tech, information age products and systems involving both software and hardware development. The intent is to provide clear, practical guidance with real and realistic examples so that the reader will have exactly what he or she needs to successfully apply Design for Six Sigma (DFSS) to products and system development projects involving software, hardware, or both.

The scope of the book encompasses the development project from the development and justification or prioritization of the business case and the associated project schedule through the developing of customer-driven requirements and consequent efforts to fulfill those requirements with high confidence.

DFSS is a structured method for developing robust new products that are aligned with the voice of the customer (VOC), using predictive engineering and anticipating and managing potential issues. Using this proactive process, the development team can:

- Ensure that the team shares a solid understanding of customer needs, and selects a compelling concept that supports and facilitates meeting those needs.
- Define measurable critical parameters that reflect customer needs, and flow them down to quantifiable and verifiable requirements for the subprocesses, subsystems, and components.
- Use predictive engineering and advanced optimization methods to ensure that the product, technology, service, or process is robust to variation in the processing and in

the environment and use conditions, providing confidence that each critical parameter will meet or exceed customer expectations.

- Verify that the new product, technology, and service of process is capable of fulfilling the functional and reliability requirements under normal and stressful conditions.
- Ensure that the supportive organizations and supply chain are aligned and capable of consistently delivering with acceptable cycle times.

WHO CAN BENEFIT FROM THIS BOOK

Although this book is general in approach and could be very helpful for anyone involved in developing almost any product, this book is particularly attuned to meeting the needs of highly skilled people who are motivated to take their new product efforts to the next level.

This book provides the tools and step-by-step guidance for systems engineers, programmers, software engineers, electrical engineers, engineering managers, program managers, and engineering students. It will also be useful for engineers who handle multidisciplinary situations involving software or electronic hardware, such as aerospace, biomedical, and industrial and power engineering. Software and electronics has seeped into many other disciplines as well, so it will also be useful for mechanical engineers, chemical engineers, and civil engineers.

For perhaps the first time, skilled people involved in product development have access to a clear roadmap, clear guidance on what people and teams need to do, step by step, to apply powerful methods—sometimes simple yet elegant, other times more complex—that can enable the product development team to converge quickly on excellent approaches and solutions to deal with even the most complex situations with high-tech and information-age products.

This book addresses a common concern from people who read books and then wonder, "That's great in theory—but what do I need to do?"

Many products involve both software and hardware aspects; this book provides an integrated systems approach that pulls the marketing, software, and hardware communities together to solve hardware and software issues and provide a product that meets customers' expectations, with minimal finger-pointing.

ORGANIZATION AND SUMMARY OF THE CHAPTERS

This book is organized in approximately the same sequence in which the topics are most likely to arise for a new product launch. Although it is hoped that readers will find the engaging literary style grabs them like a fast-paced novel, such that they would read it through in one spellbound sitting, the reader can simply read each topic just as the need arises in the actual product development effort.

The first three chapters set the context and provide the reader with the background and a structure to assist in the challenges involved in Six Sigma deployment in general and DFSS deployment in particular. The first chapter provides a historical perspective followed by a summary of the DFSS process, goals, and an example of a DFSS project. The second chapter provides the deployment perspective, and gives information and tools to assist with the organizational and people challenges involved in DFSS deployment—approaches for engaging management support, obtaining engineering buy-in, overcoming resistance to change, and handling schedule and resource requirements. The third chapter provides support for the reader in handling the ongoing organizational support structure, including suggestions for governance, and continuing support from the management and the people involved in the project. Risks involved in new product development are enumerated, and suggestions provided for success metrics that can enable the team and management to assess progress and, ultimately, success in managing those risks.

The next three chapters delve further into the risks and opportunities involved in the project—topics that might be discussed just before fully launching the new product development effort. Chapter 4 elaborates on the DFSS process and discusses how both the software and hardware development efforts can be aligned. Chapter 5 delves into the business case risk in more detail, and provides a method for assessing the risk and opportunity in the context of a product roadmap that includes a portfolio of potential new products, and how the portfolio can be prioritized in the common situation of resource constraints. Chapter 6 discusses the project schedule with the associated, ever-present schedule risk, and provides some perspective, strategies, and approaches to handle schedule risk. These tools and methods include Monte Carlo simulation for the business case and the project schedule and theory of constraints project management/critical chain as a potential approach for handling schedule risks with respect to the project schedule.

The next several chapters are aligned with a flowchart that provides a step-by-step process for developing systems, software, hardware, or a combination of software and hardware. Chapters 7, 8, and 9 address the approach to gathering the VOC to understand what is important to customers, and the determination of requirements and selection of architecture based on customer and business expectations (VOC and VOB). This sequence of steps includes gathering, understanding, and prioritizing the VOC, making architecture decisions, and selecting measurable critical parameters requiring intense focus. Tools and methods include VOC gathering, KJ analysis, Kano analysis, QFD/House of Quality, concept generation methods, Pugh concept selection, and ATAM for software architecture decisions.

Chapters 10 through 12 discuss the "flow down" of the system-level requirements to requirements for hardware and software subsystems. The alignment of DFSS with Agile processes and software architecture selection are among the key topics involved in the software side, along with the engagement of rest engineering resources to ensure that the requirements can be measured, tested, and evaluated, as well as supply chain resources toward assuring supply chain readiness.

Chapters 13 through 15 discuss the concept of predictive engineering along with the optimization and "flow up" for meeting requirements allocated to both the software and hardware aspects. Methods including FMEA and fault tree analysis (FTA) help to anticipate and prevent problems. For continuous and ordinal requirements, a detailed selection process is provided for determining the transfer function for the critical parameters using a variety of methods relevant to both continuous and discrete variables. These methods include regression, logistic regression, DOE (design of experiments), RSM (response surface methodology), and robust design and stochastic optimization approaches to build high confidence that critical parameters will meet the customer's expectations. Chapter 14 also introduces an approach called Yield Surface Modeling that has had a remarkable success rate in terms of first-pass successes with high yields.

Chapters 16 through 19 correspond to the need to "trust but verify"; models and transfer functions are useful, but there is inherent risk in trusting that the model truly and completely represents reality. These chapters discuss verification and test, in which the capability and reliability of the software or hardware product is assessed on pilot and early production samples. Approaches include accelerated life testing (ALT) for hardware and fault injection testing for software reliability. The supply chain resources anticipate and take preventative action for potential supply chain issues, and verify that the supply chain, including vendors and internal manufacturing and testing facilities, is ready. Chapter 19 also introduces a novel statistical model for supply chains that can be used to attain goals for on-time delivery and quoted lead times with minimal strategic inventory levels.

The final chapter summarizes the topics and challenges discussed in the book, and provides a "look forward" toward future directions for new product development.

SUPPLEMENTARY MATERIAL PROVIDED THROUGH THE WEB SITE

There is a Web site associated with this book, available at http://www.sigmaexperts.com/dfss/. This Web site provides an interactive DFSS flowchart, aligned with the organization of the chapters in the book and with the software and hardware development process, which allow the reader to see a high-level overview of a topic, then click on a specific topic and "drill down" to a more detailed flowchart to aid with decisions on what approaches to consider, and to a summary of each approach.

The Web site also provides templates and Excel spreadsheets that will help the reader apply some of the approaches described in the book. There are examples on both the hardware and software side, including codes. Additionally, exercises are provided, aligned with the related chapters, to reinforce concepts and allow practice for the readers.

If the reader is interested in certification, there are Excel templates that can be used for project charters and for planning and later summarizing the project, and a PowerPoint template for presentations to summarize the project and its impact.

The Web site also provides materials, PowerPoint slides, and Acrobat files that will enable the reader to introduce topics to their organization and assist in selling concepts to management and the people involved in development.

Acknowledgments

The authors would like to thank the many people who have provided support, examples, case studies, and help. We'd like to start by expressing our appreciation to the Prentice Hall team, led by Bernard Goodwin (who seamlessly combined high standards for excellence with strong support for inexperienced authors), with Michelle Housley—a delight to work with. The great team that brought the book into production includes Carolyn Albee and Elizabeth Ryan, and Stephane Nakib, Andrea Bledsoe, and Heather Fox on the marketing and publicity sides. It has been a pleasure to work with each of you!

We'd like to also express our thanks and appreciation for those who reviewed the whole book, or some key chapters, and to those who provided inputs and feedback for the DFSS flowchart that provided a roadmap for the book, including Harry Shah, Bob Shemenski, Lynne Hambleton, Tonda MacLeod, Soo Beng Khoh, Barbara Millet, Peggy Shumway, David Bar-On, Michael Suelzer, Parveen Gupta, Kambiz Motamedi, Shri Gupta, and Dhananjay Joshi. Your help was truly invaluable!

We'd also like to express appreciation for those who enriched the book with the real examples they provided or helped to provide, including Subramanyam Ranganathan, Kevin Doss, Felix Barmoav, Cristina Enomoto, Wendel Assis, Ken Butler, Vivek Vasudeva, Doug Dlesk, Tonda MacLeod, Edilson Silva, Karl Amundson, Alex Goncalves, C.C. Ooi, S.C. Wee, PingPing Lim, Tony Suppelsa, Jason Mooy, Steven Dow, David Feldbaumer and Dirk Jordan. These engineers stepped forward to lead efforts to develop robust and reliable products, and their examples now live on and can hopefully help others to see "how it is done."

Several key people have shared aspects and insights that have increased the clarity and applicability of these methods, including Dr. Brian Ottum, Rick Riemer, Dr. Kevin Otto, Dr. Clyde "Skip" Creveling, Shri Gupta, Keith McConnell, Scott MacGregor, David Karpinia, Dr. John Fowler, Dr. Murat Kulahci, Tania Pinilla, Andy Papademetriou, Maria Thompson, and Carol Adams. We thank you, and the DFSS world is stronger because of your contributions.

Finally, we give our heartfelt thanks and appreciation to those who have provided vital support for Motorola's DFSS deployment efforts, helping to create an environment conducive to new approaches and information sharing, including Bob Epson, Matt Barney, Rick Kriva, Jason Jones, Dan Tegel, Luiz Bernandes, Rosana Fernandes, Steve Lalla, Bob Yacobellis, Andrij Neczwid, Hal Hamilton, Sue Dunmore, Sanjay Chitnis, Nina Fazio, Ruth Soskin, Dawn Start, Kathy Hulina, Zeynel Arslanoglu, Yavuz Goktas, Tom Judd, Jeff Summers, Mike Potosky, and Rey Moré.

About the Authors

Eric Maass has thirty years of experience with Motorola, ranging from research and development through manufacturing, to director of operations for a $160 million business and director of design and systems engineering for Motorola's RF Products Division. Dr. Maass was a cofounder of the Six Sigma methods at Motorola, and was a key advocate for the focus on variance reduction; his article on a "Strategy to Reduce Variance" was published in 1987, the year that Motorola announced Six Sigma. He codeveloped a patented method for multiple response optimization that has resulted in over 60 first-pass successful new products, and most recently has been the lead Master Black Belt for Design for Six Sigma at Motorola.

He coauthored the *Handbook of Fiber Optic Data Communication* and a variety of chapters in books and articles ranging from concept selection to augmentation of design of experiments to multiple response optimization to advanced decision-making methods. Dr. Maass's other accomplishments include driving the turnaround of the Logic Division from "virtual chapter 11" to second-most profitable division (of 22 divisions) in two years, and he also won the contract for Freescale Semiconductor's largest customer, Qualcomm. Dr. Maass has a rather diverse educational background, with a B.A. in biological sciences, an M.S. in chemical and biomedical engineering, a Ph.D. in industrial engineering, and nearly thirty years' experience in electrical engineering. Dr. Maass is currently consulting with and advising several companies and institutions including Motorola, Arizona State University, Oracle, and Eaton.

Patricia McNair is the director of Motorola's software Design for Six Sigma program and a Certified Six Sigma Master Black Belt. She served as cochair of the Software Development Consortium and program director of the Motorola Six Sigma Software Academy. She travels internationally to various countries including France, England, China, Singapore, India, Malaysia, Brazil, and many others for consulting and training of Motorola engineers.

She spent more than twenty-five years in software and systems engineering roles including systems engineering manager, design engineer manager, architect and requirements lead, senior process manager, certified SEI instructor for the introduction to CMMI, certified Six Sigma black belt, and authorized SEI CBA IPI lead assessor for various companies such as Motorola, GE Healthcare, and IBM Federal Systems, where she worked through and managed all phases of a software development life cycle, from requirement gathering, design, development, and implementation, to production and support.

She has served as an adjunct professor at De Paul University in Chicago, the State University of New York at Binghamton, and at the University of Phoenix.

She holds an M.S. in computer science from the State University of New York at Binghamton and an MBA from the Lake Forest Graduate School of Management.

Introduction: History and Overview of DFSS

A BRIEF HISTORICAL PERSPECTIVE ON SIX SIGMA AND DESIGN FOR SIX SIGMA (DFSS)

Six Sigma arose during a time when the executive leadership, engineering, manufacturing, and research and development at Motorola were particularly receptive. It came about following several events and efforts. The starting point for this incubation period was in 1979; during the annual Motorola officers meeting that year, Vice President Art Sundry said: "Motorola's quality stinks." This assessment had followed dramatic improvements in the quality of television sets produced after the sale of Motorola's Quasar brand of televisions to Matsushita. Matsushita's dramatic improvements in quality and reduction in defects, from essentially the same manufacturing process but with new leadership, provided a startling, eye-opening picture: that leadership directly influenced the quality of products.

Against this backdrop, statistical methods started "rising from the ashes," like the mythical phoenix, in the sites of Motorola located near Phoenix, Arizona. Engineers from the Government Electronics Group in Scottsdale, Arizona, the corporate research and development group (MICARL—Motorola Integrated Circuits Applications Research Laboratory), and the Semiconductor Group in Tempe, Arizona, attended classes on design of experiments (DOE) taught as a "yield enhancement seminar" by consultant Mike Johnson. This class was noteworthy for its cookbook approach for designing experiments, provided through a bound report from Dr. Hahn and Dr. Shapiro at

General Electric.[1] It is ironic that this document, along with a method to focus on variance reduction inspired by a chapter from Hahn and Shapiro's *Statistical Modeling in Engineering*,[2] helped in the development of Six Sigma in Motorola, and Six Sigma was later embraced by General Electric.

The first documented application of Fractional Factorial Design screening experiment in Motorola was directed toward simplifying the Radiation-Hardened (RadHard) CMOS process for integrated circuits (ICs) used in satellites, and thereby reducing the manufacturing cycle time for the set of integrated circuits. These ICs were to be used in communications satellites, and the excessively long cycle times had impacted or threatened to impact satellite launches! This first example of DOE led to the removal of a dozen non-value-added processing steps that had been thought to be required to ensure that the integrated circuits were robust against the radiation that they might encounter in the satellite's environment. It is perhaps ironic that the first documented DOE example provided the benefits associated with lean methods, about 20 years before the terms "lean" and "Six Sigma" were linked in common usage.

It is also interesting that some of the earliest success stories involved new process development (for new processes for manufacturing "computer chips"), using the set of methods that were evolving into the Six Sigma toolset. In the MICARL pilot lines, statistical methods such as DOE and sources of variability (SOV) studies were used, sometimes in combination with process simulators, such as SUPREM (Stanford University Process Engineering Model), to reduce the time that it took to develop new semiconductor processes from a historical average of about a year to about three months, to increase yields from about 25 percent to more than 80 percent. Some of these efforts were driven by Dr. J. Ronald Lawson (coauthor of an early book on Six Sigma methods) and Eric Maass (coauthor of this book), and were dramatic enough to be noticed by Motorola engineers and managers around 1980.

Based on these and other dramatic success stories, a network of statistical users grew, mainly in the section of Motorola located in the Phoenix area. Antonio (Tony) Alvarez and Eric Maass were among the nexus points for this network. This informal network provided a means for rapid sharing of success stories, problems, and solutions, and started brainstorming similarities among the successes toward developing a systematic approach. Growing interest from other engineers led to the development of internal courses on statistical methods and bringing in external consultants for training classes, largely taught by the highly respected Tony Alvarez

---

1. Dr. Gerald J. Hahn and Dr. Stanley S. Shapiro, "A Catalog and Computer Program for the Design and Analysis of Orthogonal Symmetric and Asymmetric Fractional Factorial Experiments," Information Sciences Laboratory (report number 66-C-165, May 1966).

2. G.J.Hahn and S.S. Shapiro, *Statistical Models in Engineering*, John Wiley and Sons, 1967.

and coordinated by Janet Fiero, who was responsible for training and development for Motorola's semiconductor product sector in the Phoenix area.

Among the systematic approaches that were developed was the recognition of the vital need for a focus on reducing variation, and a step-by-step approach for this. This culminated in a key article and presentation, "A Strategy for Reducing Variability in a Production Semiconductor Fabrication Area."[3,4] This was probably the first published article on the Six Sigma focus on variance reduction, and it was inspired by the book (mentioned earlier) by Gerald Hahn and Samuel Shapiro. The case studies from this earliest article on Six Sigma and variance reduction are used as examples in the section on variance reduction in Chapter 14 of this book.

By 1982, the success stories and statistical classes were being recognized throughout Motorola, and Janet Fiero was promoted to Corporate Director of MTEC (Motorola Training and Education Center) at Motorola's headquarters in Chicago. Fiero strongly promoted statistics training at Motorola. Among the courses she coordinated was a set of sessions by Dorian Shainin, which captured the imagination of a quality manager named Bill Smith—more on that in a moment.

From 1982 through 1985, executives and managers in this increasingly receptive environment started hiring experts on statistical methods, either as full-time employees or as consultants, including Dr. Mikel Harry (hired into the Government Electronics Group in Scottsdale, Arizona), Mario Perez-Wilson (hired into the Semiconductor Product Sector in Phoenix, Arizona), and Harrison "Skip" Weed (hired into the Semiconductor Product Sector Research and Development organization in Mesa, Arizona). Professors Dennis Young and Douglas Montgomery (newly arrived at Arizona State University) were actively engaged in this dynamic and exciting environment. A virtual Hall of Honor recognizes these early pioneers of Six Sigma at the Web site http://www.sigmaexperts.com/sixsigmahistory.

During this time frame, Janet Fiero led the MTEC team in rolling out a series of Statistics courses, including a course by the external consultant Dorian Shainin; this class captured the imagination of a senior quality engineer and manager named Bill Smith. Although Bill Smith passed away a few years later, his associate, John Forsberg, recalls a key sequence of events (see Figure 1.1 and Chapter 17):

> We were trying improve the overall reliability. Units would go through testing in repeated loops. Many failures matched what was going on in the field. Most were **early life failures** due to latent defects.

---

3. "System Moments Method for Reducing Fabrication Variability," *Solid State Technology*, August, 1987.

4. "A Strategy for Reducing Variability in a Production Semiconductor Fabrication Area Using the Generation of System Moments Method," *Emerging Semiconductor Technology*, ASTM, 1987.

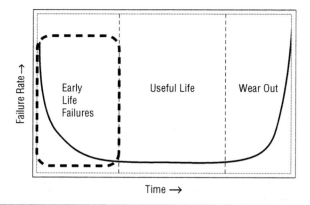

Figure 1.1 Bathtub curve, with the region of early life failures highlighted. This topic is discussed in detail in Chapter 17.

Bill Smith's key insight was to relate the early life failures he dealt with, as a senior quality engineer, to product characteristics such as length, voltage, and gain that were distributed such that a small percentage exceeded the product tolerance. Smith realized that the probability of having an early life failure was compounded if there were many components, as there were on the increasingly complex electronic products that Motorola was developing and manufacturing.

Based on what he had learned from the MTEC classes, including those led by Dorian Shainin, Bill reasoned that if typical product characteristics are normally distributed and mostly fell within ±3 standard deviations, and the product was designed so that it would still meet customer requirements at twice that range, or ±6 standard deviations (or **"Six Sigma"**), then many good things would happen:

Early life failures would disappear!
and
Manufacturing yields would increase!

To make his point, and because Motorola's management was still smarting from the success of Quasar in manufacturing higher-quality television sets from the former Motorola manufacturing line, Bill Smith gathered benchmarking data on rolled yields on Japanese TV sets. He determined that there could be several thousand components in a radio that could produce 99 percent yields if the critical characteristics had distributions that were so tight that they were six standard deviations away from the customer requirements. And so **Six Sigma** was born!

According to the recollections of Motorola's CEO, Bob Galvin:

Bill Smith called me asking for an appointment. He came to my office and explained the theory of latent defects. I called him back the next day to try to better understand what he was talking about. He soon became a sophisticated advisor in applying statistical methods to improve quality.

Bill Smith had completed perhaps the most critical step—achieving executive engagement! (This will be discussed in more detail in Chapter 2, on DFSS deployment.) In 1986, Bill Smith contributed material for a Motorola class, Eng 123, Design for Manufacturability (copyrighted in 1986). This topic, relevant to DFSS, is summarized in Chapter 19.

In 1986, with Bob Galvin's backing as the CEO, the Six Sigma concept was communicated top-down and accepted at virtually all levels of the organization. This communication focused on the goal of achieving a process capability (Cp) index of 2.0 or better (this is discussed in Chapters 10 and 16), and a defect rate of 3.4 ppm (parts per million) or better. It also introduced the idea of using "Sigma Level" as a metric, and provided benchmarking of various industries and products according to defect rates and associated opportunities for errors.

Figure 1.2 Photograph of Bill Smith, may he rest in peace

Dr. Mikel Harry documented some aspects of the Six Sigma concept in "The Nature of Six Sigma Quality" in 1987 and "Six Sigma Mechanical Design Tolerancing" in 1988; some aspects of the latter are discussed in the tolerance allocation section of Chapter 10. Dr. Harry and Dr. J. Ronald Lawson wrote "Six Sigma Producibility Analysis and Process Characterization," published in 1992.

On January 15, 1987, Motorola officially launched and announced Six Sigma and set the Six Sigma goal: achieving less than 3.4 defects per million opportunities. The original Six Sigma process involved these "Six Steps to Six Sigma":

SIX STEPS TO SIX SIGMA

Step 1: Identify the product you create or the service you provide.
Step 2: Identify the customer(s) for your product or service, and determine what they consider important.
Step 3: Identify your needs (to provide the product/service so that it satisfies the customer).
Step 4: Define the process for doing the work.
Step 5: Mistake-proof the process and eliminate wasted effort.
Step 6: Ensure continuous improvement by measuring, analyzing, and controlling the improved process.

Steps 1, 2, and 3 focus on gathering and understanding the voice of the customer (VOC), and align with the Define phase of DMAIC (Define, Measure, Analyze, Improve, Control; see detail on following page) and the VOC gathering step of DFSS discussed in Chapter 7. Step 4 involves process mapping, a key tool in the Define phase of DMAIC. Step 5 is compatible with lean concepts, and with the Improve phase of DMAIC, and step 6 is aligned with the Control phase of DMAIC (but also mentions the future Measure and Analyze phases).

Mario Perez-Wilson developed a five-step method called "M/PCpS" (Machine/Process Capability Study) for characterization in manufacturing,[5,6] which was described in the context of a comprehensive application of statistical process control in manufacturing. His approach was widely adopted at Motorola sites in Phoenix, Arizona, and in Asia:

- Process delineation
- Metrology characterization
- Capability determination
- Optimization
- Control

---

5. http://www.mpcps.com/

6. Mario Perez-Wilson, "A Case Study—Statistical Process Control," Motorola Inc., *The Round-Up*, Vol. 13(3), January 21, 1988.

These steps are very similar to the DMAIC steps adopted by a council composed of representatives from Motorola and other adopters of the Six Sigma philosophy several years later, which focused on improving an existing process or product:

- **Define** and identify the improvement opportunity, the business process or processes involved, the critical customer requirements, and the team that is chartered to focus on this opportunity.
- **Measure** the current level of performance in meeting the critical customer requirements and developing a method and plan to effectively collect relevant data.
- **Analyze** the data and information related to the opportunity and identify and validate the root cause(s), sources of variation, and potential failure modes that form the gap between performance and expectations.
- **Improve** the performance of the process or product by identifying, evaluating, selecting, and piloting an effective improvement approach, and to develop a change management approach to assist the organization in adapting to the changes.
- **Control** the improved process to ensure that the improvement is sustainable, standardized, institutionalized, and disseminated to potentially benefit similar processes and products.

The Malcolm Baldrige Award was inspired by and based on the Deming Prize awarded to companies in Japan for noteworthy accomplishments in quality. In 1988, Motorola won the first Malcolm Baldrige National Quality Award, given by U.S. Congress to recognize and inspire the pursuit of quality in American business (see Figure 1.3). Shortly thereafter,

Figure 1.3 Photographs of Bob Galvin, CEO of Motorola, receiving the very first Malcolm Baldrige National Quality Award from President Ronald Reagan

a very proud Bob Galvin announced his willingness to share what we had developed and learned:

> We will share Six Sigma with the world, and it will come back to us . . . with new ideas and new perspectives . . .

In 1989, Motorola established the Six Sigma Research Institute (SSRI) in Chicago, with Dr. Mikel Harry as its director. Other companies joined within the next year, including IBM, TI, Westinghouse, and Kodak. In 1993, Larry Bossidy, the CEO of Allied Signal, decided to use Six Sigma; in 1995, Jack Welch, CEO of GE, invited Larry Bossidy to discuss Allied Signal's Six Sigma experience at GE's corporate executive council meeting. After evaluating the potential financial benefits of successful Six Sigma implementation, Jack Welch led an ambitious and intensive focus on Six Sigma institutionalization. Jack Welch's leadership, and the heavy alignment with the management structure and leadership development at GE, along with the involvement of bright consultants who fleshed out the concepts for Six Sigma, led to an impressive, astonishingly successful Six Sigma program that may have fulfilled Bob Galvin's vision.

HISTORICAL PERSPECTIVE ON DESIGN FOR SIX SIGMA

For the first few years, Six Sigma efforts were directed toward improving existing processes, particularly manufacturing processes. However, there was an ongoing discussion regarding the possibility of moving from a reactive approach to a more proactive approach, wherein the products were designed to meet Six Sigma expectations. There were a few paper studies, including one study that seemed to indicate that developing an integrated circuit to Six Sigma levels for all key requirements would be cost-prohibitive because of a large die size. The study claimed that larger transistors and more circuitry was required in order to have every circuit element achieve Six Sigma capability, making the surface area of an integrated circuit larger. For integrated circuits, a larger die size drives the cost for each integrated circuit much higher. Nonetheless, the first successful DFSS project involved the development of a family of integrated circuits—and the high yielding integrated circuits did not require larger die sizes.

In 1990, Motorola established the total customer satisfaction (TCS) competition, which provided a company-wide forum to recognize teams. In the first year of the competition, over 5,000 teams from across Motorola competed, giving a 12-minute presentation on their projects. The best teams were selected at each level of the competition to progress to the next level, until the best-of-the best reached the corporate level competition in Chicago. The first gold medal was won by the very first

Design for Six Sigma (DFSS) team project, the FACT TOPS Team, led by Eric Maass and David Feldbaumer. This DFSS effort set a record for the introductions of new integrated circuits (ICs) (57 new ICs developed and qualified in 28 weeks, with all 57 devices being first pass successes, with an average yield of 92 percent), and resulted in more than $200 million profit over the next five years. To achieve this goal, the team leaders developed a novel approach to model and predict composite yields and composite sigma level with multiple responses. The story of this first DFSS project is told in Chapter 14, as part of the discussion of multiple response optimization.

The second DFSS effort was for a power amplifier module for a microwave local area network. The approach was referred to as Six Sigma Design Methodology (SSDM), and the successful team project, led by Craig Fullerton, won Motorola's TCS gold medal in 1992 with a team that referred to itself, tongue-in-cheek, as the "NERDS." Craig Fullerton received a grant from DARPA (Defense Advanced Research Projects Agency, part of the U.S. Department of Defense) to standardize the SSDM process (1993 DARPA MIMMIC PHASE III Contract Statistical Design Methodology), which provided a course within Motorola (ENGR290) and a course taught to SSRI partners including TI and Raytheon. In 1995, Craig Fullerton gave a presentation on Six Sigma Design Methodology at the DARPA MIMMIC Conference.

DFSS PROCESSES

After these first DFSS projects, several successful DFSS deployments and projects have occurred at companies including Kodak and 3M through the efforts of experts such as Dr. Clyde "Skip" Creveling. General Electric, building on strong management support for Six Sigma programs from the CEO, Jack Welch, and his staff, developed a very strong DFSS program using a DFSS process abbreviated DMADV, which they later renamed DMADOV: Define, Measure, Analyze, Design, (Optimize), and Verify. Other abbreviations for key phases of a DFSS project include CDOV and IDOV (Concept or Identify, Design, Optimize, Verify). Recently, there have been efforts made to define a software DFSS process, with notable efforts at GE, Raytheon, and Motorola, involving experts such as Patricia McNair and Dr. Neal Mackertich.

DFSS is a structured method for developing robust new products that are aligned with the VOC, using predictive engineering and anticipating and managing potential issues. Robustness refers to the relative insensitivity of the product and its subunits to variations in the manufacturing, environment, and usage of the product. A robust laptop PC will work in Alaska and in Arizona, in Switzerland and in Swaziland; it can handle the uses and abuses by knowledge experts and novices and the wear and tear from mishandling by a child. The VOC refers to information regarding expectations gathered from customers: what is important to them and what they would like to see and hear, feel, and be able to use in a product or with a service. Predictive engineering refers to developing and using a mathematical model of a requirement and being able to determine whether that requirement will be satisfactorily met before the product is actually built; if not, then predictive engineering provides a means to improve or optimize the performance to the requirement. Predictive engineering is intended to minimize the use of the primary alternative approach used in product development, often referred to as "build—test—fix." Predictive engineering has proven very beneficial for both hardware and software development; software development also involves efforts to manage defects, through prevention but also through detection and fixing the defects as early in the process as possible. A software DFSS black belt, Vivek Vasudeva, provided Table 1.1 to show the financial leverage of finding software defects as early as possible.

Figure 1.4 provides an overview of the structured method for Design for Six Sigma, shown in the form of a DFSS flowchart. Table 1.2 shows the alignment of various DFSS processes to some key tools, methods, and deliverables for each step from the flowchart.

The DMAIC process is a well-established abbreviation and mnemonic that serves as a standard for improving existing processes and products. By contrast, there are many

Table 1.1 Financial leverage of finding software defects in earlier phases

Cost Depending on Phase Injected and Phase Detected	Phase Injected		
Phase Detected	**Requirements**	**Design**	**Code**
Requirements	$240		
Design	$1,450	$240	
Code	$2,500	$1,150	$200
Development Test	$5,800	$4,450	$3,400
System Test	$8,000	$7,000	$6,000
Post Release Test	$30,000	$20,000	$16,000
Customer	$70,000	$68,000	$66,000

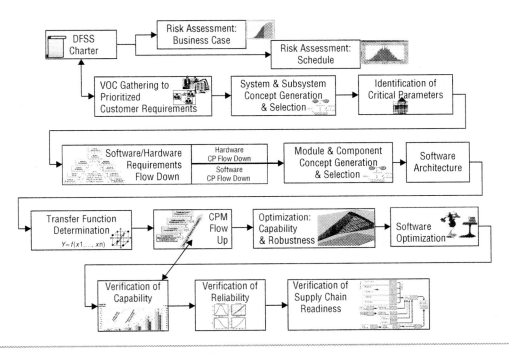

Figure 1.4 DFSS flowchart, as used in this book

DFSS processes as illustrated in Table 1.2, and none has emerged as the standard. This table also illustrates that each of these processes can be used as a structured process for DFSS for hardware solutions; however, most of them have some gaps associated with applying DFSS for either software products or systems comprised of both software and hardware, as discussed later in this book.

There are several possible alternatives for merging the nomenclature for hardware and software DFSS, which could lead to many tongue-twisters and tongue-contortions. However, the authors chose RADIOV—for the simplicity, ease of remembrance and pronunciation, and for the gentle recognition of the origins of Six Sigma, as described earlier in this chapter. RADIOV is an abbreviation, mnemonic, and acronym for Requirements, Architecture, Design, Integration, Optimization, and Verification. The DFSS process can best be illustrated with an example; the example will follow the flowchart in Figure 1.4.

Table 1.2 Some key tools and methods associated with steps for various DFSS processes

DFSS Step	DFSS Processes			Key Tools and Methods
	CDOV	**DMADV**	**RADIOV**	
DFSS Charter	Concept	Define	Requirements	DFSS Charter, Deployment Plan
Business Case: Risk Management				Monte Carlo Simulation—Business Case
Schedule: Risk Management				Monte Carlo Simulation—Critical Chain/TOC-PM
VOC Gathering				Concept Engineering, KJ Analysis, Interviews, Surveys, Conjoint Analysis, Customer Requirements Ranking
System Concept Generation & Selection				Brainstorming, TRIZ, System Architecting, Axiomatic Design, Unified Modeling Language (UML), Pugh Concept Selection
Identification of Critical Parameters		Measure		Quality Function Deployment (QFD), Design Failure Modes and Effects Analysis (DFMEA), Fault Tree Analysis (FTA)
Critical Parameter Flow Down	Design	Analyze	Architecture	Quality Function Deployment (QFD), Critical Parameter Management, Fault Tree Analysis (FTA), Reliability Model
Module or Component Concept Generation and Selection				Brainstorming, TRIZ, System Architecting, Axiomatic Design, Universal Modeling Language (UML), Pugh Concept Selection
Software Architecture				Quality Attribute Analysis, Universal Modeling Language (UML), Design Heuristics, Architecture Risk Analysis, FMEA, FTA, Simulation, Emulation, Prototyping, Architecture Tradeoff Analysis Method (ATAM)

Table 1.2 Some key tools and methods associated with steps for various DFSS processes (continued)

DFSS Step	DFSS Process			Key Tools and Methods
	CDOV	DMADV	RADIOV	
Transfer Function Determination	Optimize	Design	Design	Existing or Derived Equation, Logistic Regression, Simulation, Emulation, Regression Analysis, Design of Experiments (DOE), Response Surface Methodology (RSM)
Critical Parameter Flow Up and Software Integration			Integrate	Monte Carlo Simulation, Generation of System Moments Method, Software Regression, Stability and Sanity Tests
Capability and Robustness Optimization			Optimize	Multiple Response Optimization, Robust Design, Variance Reduction, RSM, Monte Carlo Simulation with Optimization
Software Optimization				DFMAE, FTA, Software Mistake Proofing, Performance Profiling, UML, Use Case Model, Rayleigh Model, Defect Discovery Rate
Verification of Capability	Verify	Verify	Verify	Measurement System Analysis (MSA), Process Capability Analysis, McCable Complexity Metrics
Verification of Reliability				Reliability Modeling, Accelerated Life Test (ALT), WeiBayes, Fault Injection Testing
Verification of Supply Chain Readiness				Design for Manufacturability and Assembly (DFMA), Lead Time and on Time Delivery Modeling, Product Launch plan, FMEA/FTA for Product Launch

DFSS EXAMPLE

A cellular base station is a complex system involving both software and hardware. When someone uses a CDMA cellular phone, the cellular phone maintains wireless communication through one or more CDMA base stations, which are also connected to a communication network using cables and fiber optics. Figure 1.5 shows some CDMA base stations and base station controllers. The base station on the left is similar to a base station that one might find around a city, suburb, or rural area. In Figure 1.5, the white bell-shaped object in the upper left of the leftmost CDMA base station is an antenna for a GPS subsystem (to ensure proper timings). The base station in the middle is called UBS (universal base station), and was associated with a DFSS project led by Rick Riemer, a Motorola master black belt. After the successful UBS project, Rick Riemer coached Subramanyam Ranganathan to DFSS black belt certification for applying DFSS in the development of the IP-based base station controller (IP-BSC), shown to the right of Figure 1.5. The IP-BSC had several competitive advantages, driven from the VOC, as summarized in excerpts from a press release:

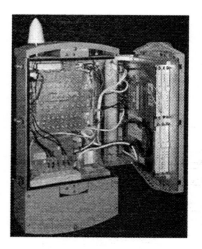  

Figure 1.5 CDMA base station and base station controllers. The base station in the middle, UBS, was developed using DFSS methods, as was the IP-based base station controller (IP-BSC) shown to the right, which represents the DFSS example in this section.

Motorola Introduces New CDMA EV-DO Base Site Controller

ARLINGTON HTS., Ill.—27 March 2008—Motorola (NYSE:MOT) today announced the availability of a new IP-BSC-DO, a highly scalable EV-DO base site controller (BSC) platform that enables CDMA operators to quickly and cost effectively upgrade the DO broadband data capacity. The new IP BSC provides 12 times the site capacity and IP data session capacity compared to the first generation DO BSC, while using significantly less floor space . . . . The new IP-BSC-DO product offers a higher bandwidth platform that not only supports existing operators' high speed data applications, but also is well suited to support advanced data applications like voice over IP, video telephony and multimedia push-to-X services [multimedia services in response to pushing a button]. . . . "The increased capacity and bandwidth of Motorola's IP-BSC-DO significantly reduces the number of EV-DO base site controllers an operator needs to operate their network," said Darren McQueen, vice president, Home & Networks Mobility, Motorola. "This means they can trim both capital and operational expenses while delivering an enhanced broadband experience to their customers."

Preliminary steps for DFSS include a charter and analysis of the risk associated with the business case and schedule (Figure 1.6). This project was initiated with the DFSS charter shown in Figure 1.7 (DFSS charters are discussed in Chapter 3).

The Requirements phase of the RADIOV DFSS process (Figure 1.8) corresponds to the Concept or Identify phase of CDOV or IDOV, respectively, and the Define phase of DMADV or DMADOV. The VOC is gathered and translated into measurable technical requirements, using the methods described in Chapter 7; the quality function deployment (QFD) deliverable for this example is shown in Figure 1.9. The customers for base stations and base station controllers are the cellular providers—companies such as Sprint and Verizon in the United States and SK Telecomm in South Korea. Interviews were conducted with customer representatives. The key customer requirements were found to be increased session capacity (understandable, as network capacity is directly linked to income for the cellular service providers), coverage, and availability.

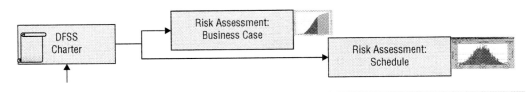

Figure 1.6 Preliminary steps for a DFSS project include a charter and assessments of project risks

Project: | DFSS for CDMA EVDO System

Business Case for entire Project (New Product Development) – Financial Objectives (Check all that apply to this project)

- [ ] Initiate Market Share by Opening New Markets with New Product (Top Line Growth)
- [ ] Obtain Market Share in New Markets with Existing Product (Top Line Growth)
- [x] Improve Margins & Revenue Growth in Existing Markets with New Product (Profit Growth)
- [x] Decrease Cost with Replacement for Existing Product (Hold Price)
- [x] Decrease Cost of Poor Quality with Replacement for Existing Product (COPQ Reduction)

Project Goal (Check all that apply to this project)	Scope
	System, Subsystem, Assembly, Component or Process focus areas:
[x] Reduce Performance Issues for customers (forecasted)	EVDO System
[ ] Reduce CRUD (forecast)	
[x] Improve Reliability (forecast)	
[x] Improve Availability (Forecast)	Critical Parameters for this DFSS Project Focus Area:
[x] Improve linkage to VOC	Data session capacity
[x] Improve stability of requirements/specifications	Data throughput
[x] Improve design capability for Critical Parameters (forecasted Cpk)	Base Station Controller (BSC) Coverage
[ ] Increase confidence in achieving or exceeding profitability goal	Handoff Signaling Delay
[ ] Increase confidence in achieving schedule goal (projected development time)	Call Processing Availability
[ ] Reduce number of iterations compared to historical average	
[ ] Increase confidence for Supply Chain Lead Time and On-Time Delivery goals	

DFSS Deployment Plan – Project Deliverables and Tools (note: please click the "Project Deliverables and Tools" tab and use the checkboxes. The left hand side will be automatically filled in)

Phase	Deliverable	Tool(s)	Status	Purpose/Goal	Expected Completion Date	Who
Concept	Identified Critical Parameters	KJ Analysis and for Kano Analysis				
Concept	Identified Critical Parameters	QFD/House of Quality				
Concept	System Requirements Document, DFSS Strates	QFD/House of Quality/CPM				
Design	Documented Flow Down	CPM Tree				
Design	Documented Risk Analysis	Design FMEA				
Design	Identified Critical Parameters, Initial Xfer fns	1st Principles Modeling (Mathematical Equation)				
Design	Identified Critical Parameters, Initial Xfer fns	Regression				
Design	Identified Critical Parameters, Initial Xfer fns	DOE (on simulator and /or prototypes)				
Design	Predicted Performance	Cp/Cpk with Monte Carlo				
Design	Predicted Reliability (or Availability)	System Reliability Modeling/Prediction/Allocation w/ Monte Carlo, Fault Tree Analysis				
Optimize	Optimized Critical Parameters/Xfer fns	DOE				
Optimize	Optimized Critical Parameters/Xfer fns	Comparative Methods, Regression				
Optimize	Optimized Critical Parameters/Xfer fns	Engineering Simulation Tools combined with Monte Carlo				
Optimize	Updated Capability	Cp/Cpk with Monte Carlo (CPM)				
Optimize	Optimized Reliability (Availability)	Updated Availability/Reliability after improvement (Fault Tree Analysis)				
Verify	Pilot Plan and Results	Comparative Methods, MSA, Conf. Intervals				
Verify	Updated Capability	MSA and Cpk (CPM) results from pilot or early production				
Verify	Updated Risk Analysis	Design FMEA defect resoultion				

DFSS Deployment Plan – Training Coaching and Support Plan

Participants	Training	Data for Training	Status	Coach/ Mentor	Project Start Data for Coaching	
Technical Architects	CPM		Y	Rick Riemer		
Technical Architects	DFMEA		Y	Rick Riemer		
Other Support Needs and Plan (possibly including Change Mangement, Management Support):						

Figure 1.7 DFSS Charter for IP-BSC-DO base station controller

Figure 1.8 The Requirements phase for a DFSS project include VOC gathering, concept generation and selection, and identification of the vital few critical parameters on which the DFSS project will focus

First House Of Quality
Rating Links Legend - Strength of the Relationship
H (9) = High Effect (Strong)
M (3)= Medium Effect (Medium)
L (1) = Low Effect (Weak)
0 = No Effect (None)

Roof Links Legend
Equal Effect: =
No Effect: 0 or Blank String
Relative Effect: +, ++, −, − −

Voice of Customer	Priority	Base Station Controller Coverage	Data Throughput	Data Session Capacity	Call Processing Availability	Cage Failover Time	Migration Time per BSC-DO	Handoff Signaling Delay	Number of Flows and Reservations	Calls per second per IP-BSC-DO
Direction of Goodness		+	+	+	+	−		+	+	+
Increased Session Capacity	15	3	3	9	0	0	0	0	0	3
Increased BSC Coverage	15	9	3	0	0	0	0	0	0	1
High Availability and Reliability	10	0	0	0	9	3	0	0	0	0
No Geographical RF Coverage Loss	10	0	0	0	3	9	0	0	0	0
Sufficient Support for Multi-flow Applications	10	0	3	0	0	0	0	0	9	1
Seamless Migration between BSC-DO and IP-BSC-DO	10	0	0	0	0	0	9	9	0	0
Increased Operability	10	0	0	0	0	0	0	0	3	0
Scoring Totals		180	120	135	120	120	90	120	90	70
Relative Scores		17.2%	11.5%	12.9%	11.5%	11.5%	8.6%	11.5%	8.6%	6.7%
Normalized Scores		10	7	8	7	7	5	7	5	4
Target Norminal Values		1008	2948	1.6		30		2		
Lower Spec Limit		108	316	0.8	Five 9's					
Upper Spec Limit						30		2		
Units		Number	Mbps	Million	Number	Sec		Sec		

Critical to Quality Parameters

Figure 1.9 QFD first House of Quality for the IP-BSC-DO base station controller

The Architecture phase of RADIOV (Figure 1.10) corresponds to the Design phase of IDOV and CDOV and the Measure and Analyze phases of DMADV and DMADOV. Figure 1.11 illustrates the alignment of the critical parameters (derived from the QFD in Figure 1.9) with the selected architecture. Table 1.3 summarizes the DFSS tools that Subramanyam Ranganathan used during the stages of this DFSS project.

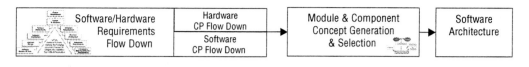

Figure 1.10 The Architecture phase of a DFSS project

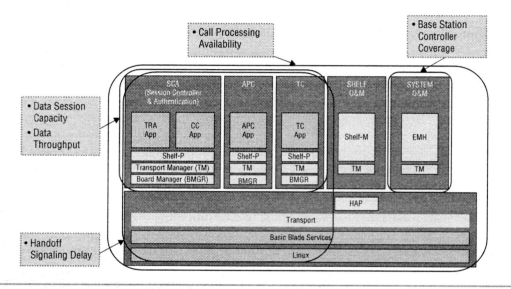

Figure 1.11 Alignment of the selected architecture for the IP-BSC-DO base station controller with the critical parameters

Integration and Optimization (Figure 1.12) for data session capacity involved the development of a model based on theory (sometimes referred to as "First Principles Modeling"), which was then used in conjunction with Monte Carlo simulation (Figure 1.13). Initially, the forecasted process capability index (Cpk) was 0.32, far short of the Six Sigma minimum expectation of 1.5. Sensitivity analysis showed that most of the variation was related to the memory needed per SCA (session controller and authentication) session. The optimization involved reallocating memory across functions, leading to forecasted Cpk of 1.54, consistent with Six Sigma expectations (Figure 1.14).

Integration and Optimization for system availability involved the use of design failure modes and effects analysis (DFEMA) (Figure 1.15) and prioritization using the associated risk priority numbers (RPNs). Software design changes in response to the prioritized risks included:

- Implementing a spanning tree algorithm to detect a transport layer overload and perform recovery actions.

Table 1.3 Alignment of critical parameters for the IP-BSC-DO base station controller with some DFSS tools that were used in the stages of the DFSS process

Critical Parameters	Requirements	Architecture and Design	Integration and Optimization	Verification
Data Session Capacity	Kano, NUD, QFD	CPM Flow Down, Flow Up, First Principles, Monte Carlo, Cp/Cpk	Capability Cp/Cpk	Capability Cp/Cpk
Data Throughput	Kano, NUD, QFD	CPM Flow Down, Flow Up, First Principles, Monte Carlo, Cp/Cpk	Prototype, Capability Cp/Cpk	Capability Cp/Cpk
Base Station Controller Coverage	Kano, NUD, QFD			Comparative Methods
Handoff Signaling Delay	Kano, NUD, QFD	CPM Flow Down, Flow Up, First Principles, Monte Carlo, Cp/Cpk		Capability Cp/Cpk
Call Processing Availability	Kano, NUD, QFD	DFMEA Fault Tree Analysis, DOE, Regression, Monte Carlo, Cp/Cpk	Fault Tree Analysis, DOE, Regression, Monte Carlo, Cp/Cpk	Defect resolution Cp/Cpk

Figure 1.12 The Integrate and Optimize phases of a DFSS project

- Including logic to arbitrate if duplicate IP addresses were received.
- Modifying initialization software procedures to set valid SNI attributes to avoid call processing errors during standby transitions.

Design FMEA was followed by fault tree analysis (FTA), as shown in Figure 1.16. Call processing errors were further analyzed using DOE (Figure 1.17), which provided an empirical model for call processing. This was integrated into a mathematical model for

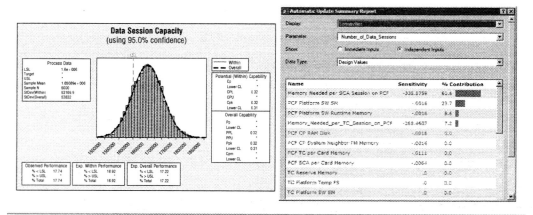

Figure 1.13 Initial Monte Carlo simulation and sensitivity analysis for data session capacity

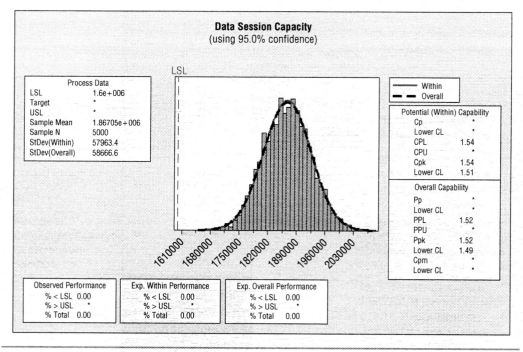

Figure 1.14 Monte Carlo simulation for data session capacity after memory reallocation

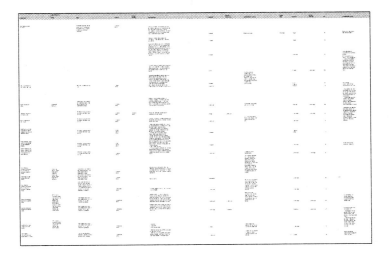

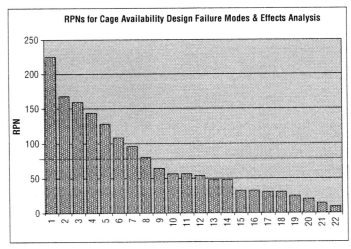

Figure 1.15 Example of DFMEA and Pareto Chart for RPNs obtained

availability measured as the percent uptime, described as the number of 9s—where "Five 9s" corresponds to the system being up and available 99.999 percent of the time, consistent with Six Sigma expectations. The Monte Carlo simulation for the predicted availability is shown in Figure 1.18.

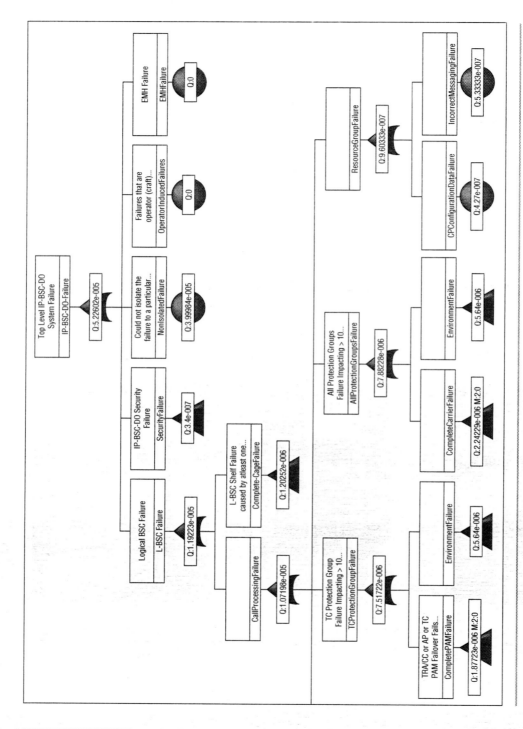

Figure 1.16 Fault tree analysis for system failures that might impact system availability

```
R-Sq = 98.73%   R-Sq(pred) = 96.82%   R-Sq(adj) = 98.16%

Estimated Coefficients for CallProcessingService
Failure using data in uncoded units

Term                               Coef
Constant                    5.52569E-0.6
Cagefailover                    1.00415
PAMFailoverFailure              1.13153
CarrierFailoverFailure         0.895048
SSCFailoverFailure           -0.0100666
SAMFailoverFailure           -0.0000097
```

Regression Equation

Call_Processing_Failure =
   5.53E−0.6
   + 1.132 × PAM_Failover_Failure
   + 0.895 × Carrier_Failover_Failure
   + 1.004 × Cage_Failover_Failure

Figure 1.17 Results from analysis of DOE (design of experiments) for call processing failures

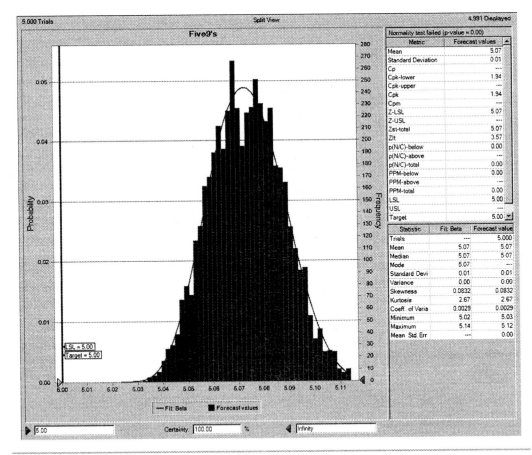

Figure 1.18 Monte Carlo simulation for system availability in terms of the number of 9s. The Monte Carlo simulation gives high confidence that five 9s (99.999 percent availability) will be achieved.

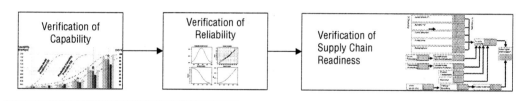

Figure 1.19 Summary of substeps involved in the Verify phase of a DFSS project

Verification of reliability (Figure 1.19) corresponded to verification of availability for this system. The test team performed extended life testing, and there were neither card failures nor failovers. Availability during the ELT was 100 percent under load test. The results of availability analysis projected the availability to be 99.99935 percent, meeting the Five 9s/DFSS criteria.

The substantial and significant benefits provided using DFSS for the IP-BSC-DO base station controller system are summarized in Table 1.4.

Table 1.4 Summary of benefits obtained from Subramanyam Ranganathan's DFSS project for the IP-BSC-DO base station controller

Critical Parameters	Initial Design without DFSS	Design Optimized without DFSS	Design Performance Improvement with DFSS	Business Benefits	
				Development Cycle Time Improvement	Cost Saving
Data Session Capacity	• No model for session capacity • Original design did not have adequate margin for minimum customer requirement of 1.6 million sessions	• Optimized memory design with existing hardware to support 1.6 m data sessions	• DFSS helped sizing memory to meet session capacity • Demonstrated that 2.0 m internal target was not attainable • Actual capability = 1.87 m sessions; 17% additional capacity	• Use of DFSS allowed earlier detection of defects • Requirements changes communicated to 3rd party Hitachi and fixed	• Cost avoidance due to earlier detection approx. $28k–$45k • Incremental benefits due to 17% additional session capacity with same h/w
Data Throughput	• No model for data throughput • Capability was estimated just based on 2 data points—2948 Mbps	• Performed prototype testing and used this data to create a model and generated more realistic prediction—6170 Mbps—100% more	• DFSS modeling and simulation helped realize that original (pre-DFSS) target was significantly lower than actual capability (80% more throughput)		
Base Station Controller Coverage	• Target 1008 modems at <=50% CPU	• 1008 modems at <= 20% CPU	• Identify root causes for CPU spikes and implement software changes	• DFSS allowed earlier detection of defects— Code change SRs implemented	• 32% additional CPU can be used for future expansion based on current analysis.

continued

Table 1.4 Summary of benefits obtained from Subramanyam Ranganathan's DFSS project for the IP-BSC-DO base station controller (continued)

Critical Parameters	Initial Design without DFSS	Design Optimized without DFSS	Design Performance Improvement with DFSS	Business Benefits	
				Development Cycle Time Improvement	Cost Saving
Handoff Signaling Delay	• No model for handoff signaling and hence we were unaware of the associated risks	• Model predicted Average Delay = 0.918 seconds	• Model set the stage for software behavior & expectations • The model then helped isolate the defect when actual behavior was different from predicted		
Call Processing Availability	• No model or formal measure for availability • Cage failover design was not derisked	• First availability model; Predicted Availability = 5.08 nines • Many software design changes implemented for cage failover	• Reduced high-RPN risk by 70% • 22 unique failure scenarios with root causes identified and SRs raised	• FMEAs enabled early detection of defects for R7 and were drivers of requirements for R8	• Cost avoidance due to earlier detection approx. $180K to $260K

Summary

This chapter provided a historical perspective of Six Sigma and its linkage to the development of Design for Six Sigma (DFSS). The various mnemonics for DFSS processes were compared, and the RADIOV acronym was selected to provide a context for reviewing and discussing the methods and steps used in a DFSS project for a complex system involving hardware and software aspects.

DFSS Deployment

Some would say that there are two separate types of experts and types of expertise involved in Design for Six Sigma (DFSS): experts in the DFSS tools and methods ("technical experts"), and experts in DFSS deployment ("deployment experts"). However, virtually anyone involved in DFSS will encounter some resistance to change, and preparation for the deployment and change management aspects are important to enable success. While other chapters will deal with some of the more technical aspects such as the DFSS tools and methods, this chapter and the next chapter will delve into the somewhat arcane and mysterious art of DFSS deployment.

DFSS deployment can involve just a single project or a portfolio of new product development efforts. Management engagement can range from upper management making the executive decision, to middle management driving the deployment, to middle management providing support (perhaps even lukewarm support), to management support being contingent on meeting some criteria, to minimal or nonexistent management support. In the spirit of Six Sigma, let's start by defining the ideal scenario; this will assist in describing deviations from ideality, and how these deviations can be addressed.

IDEAL SCENARIO FOR DFSS DEPLOYMENT

The ideal DFSS deployment scenario might include these elements:

- Strong management support, from the CEO through and including engineering, marketing, and quality executives.

- Adequate resourcing for the new product development effort, which is aligned with a prioritized product roadmap.
- Adequate, appropriate, and well-timed training for the DFSS leaders and the team.
- Strong support from the new product development team members, including marketing, engineering, and quality organizations.
- Methods and champions to deal with issues and roadblocks.
- Sharing of "lessons learned"—both best practices and mistakes to avoid repeating.
- A successful result, including a successful product launch and acceptance in the market.
- Celebration of the success, punctuated by recognition of the DFSS leaders and the team.
- Well-deserved recognition for the product, the corporation, and the management.

Note that the first four elements can be considered "inputs," whereas the last four elements might be considered "outputs" or results. Strong management support is clearly important, but is it necessary, and can it also have negative ramifications? As will be discussed later in this chapter, it is possible to deploy DFSS without strong management support—but this is much more difficult, and not highly recommended. Strong management support helps with all aspects of deployment: buy-in from organizations directly and indirectly involved in product development, the ability to provide adequate resourcing, and the ability to knock down barriers and roadblocks. By contrast, some negative ramifications that have been experienced with strong management support include a tendency toward overlaying bureaucracy and paperwork, excessive management reviews, and an overdependence on one driving executive . . . with a consequent palpable drop in support if the driving executive leaves.

STEPS INVOLVED IN A SUCCESSFUL DFSS DEPLOYMENT

The steps involved in a successful DFSS deployment are very similar to the business changes that are addressed in Dr. John Kotter's book *Leading Change*, which lists eight steps that have been expanded to eleven steps below (Steps 4, 5, and 8 were implicit in Dr. Kotter's work but are rendered explicit for our purposes).[1]

Step 1: "The Burning Platform": Confronting Reality

Campsites in the Okefenokee swamps of southern Georgia often consist of floating platforms separated by expanses of water with decaying vegetation and smiling alligators. Imagine that you are on one of the better-floating platforms, and you'd like to convince your friends and fellow campers to cross the watery expanse and join you. You shout, you cajole, you recite all of the advantages of your platform: "It's more stable, it's less risky, it's

---

1. John P. Kotter, *Leading Change*, Harvard Business School Press, 1996.

comfortable, it's alligator resistant, and it has a fire-proof grill"; but, alas, your fellow campers are not willing to take the plunge. You assure them safe passage . . . you even offer to paddle a boat and ferry them and their supplies to your better platform, but they are resistant to change.

But when their platform catches fire, they are motivated to make a change.

As highlighted in the book *Who Moved My Cheese?*,[2] people (and mice) tend to feel comfortable with the current situation, and resist change—unless there is a "burning platform."

What can be the "burning platform" that might drive the need for using DFSS for new product development? If the current new product development is quickly and efficiently chugging out superior and compelling new products that customers are hungry to grab up, then there may be no reason to change.

By contrast, most new product development efforts are not quite so stellar, and it is possible that some new approaches are worth considering. This is the step where the current reality is confronted, and the issues that impede success become the burning platform. Whether products are late, whether they have cost overruns, whether they have quality issues, or whether they are not embraced by the customers—DFSS can be tailored to specifically address those issues. The set of success metrics discussed in Chapter 3 could be a good starting point for confronting reality and setting the stage for the later steps.

The outcome of Step 1 is, in effect, a compelling business case for change. Chapter 1 mentioned how embarrassed recognition of the quality of Quasar televisions provided compelling motivation for change at Motorola. Years ago, Raytheon's financial challenges provided a burning base for implementing Six Sigma. GE medical received and responded to the voice of the customer (VOC) in reaction to scanning equipment that required patients to remain immobile for nearly an hour—an intolerable requirement for a patient suffering pulmonary disease.

A few notes of caution here:

DFSS does not come with a 100 percent guarantee. Although DFSS will increase the chances for success, there will nearly always be the risk that any given new product can fail.

Moreover, if a product succeeds after using DFSS, there is no guarantee that the product would not have succeeded anyway. Unless one develops the products twice, in parallel, with and without DFSS, the argument can be made that DFSS did not change the results . . . and, unlike the DMAIC problem-solving method, there usually is no before-and-after to show improvement.

There are exceptions where there have been before-and-after: the very first DFSS project at Motorola was in response to poor yields and major quality issues for the first of a family of integrated circuits that had been developed. The application of DFSS for

---

2. Spencer Johnson and Kenneth Blanchard, *Who Moved My Cheese?*, G. P. Putnam's Sons, 1998.

all subsequent products produced a very clear and well-recognized contrast. More recently, DFSS was applied to a CDMA base station software development effort involving session capacity as a critical parameter. The original design, not using DFSS, couldn't meet customer expectations, whereas the redesign using DFSS methods provided a fast improvement. Moreover, Monte Carlo simulations and emulations can help a development team visualize the impact of a design before implementation, providing a simulated before-and-after scenario for a product feature.

Step 2: The Guiding Coalition—Early Stakeholder Engagement

It's a funny thing about people—they generally don't like other people imposing something on them. This is more than a "Not Invented Here" (NIH) syndrome—it's the added resentment because an imposed new approach seems to imply that their previous efforts were insufficient, ineffective, and inferior.

Identifying the stakeholders and making them part of the guiding coalition changes that dynamic. The new approach is not being imposed on them—they are the champions for this effort. "A powerful guiding group has two characteristics: It is made up of the right people, and it demonstrates teamwork."[3] For DFSS deployment, the right people will likely include the people most involved in new product development—generally, the executives and key managers for engineering, marketing, quality, technology development, research, process or business improvement, supply chain and manufacturing, sales, and customer service: the people who are responsible and accountable for the success of the product or service. These people are also likely to meet the criteria for the guiding coalition team members of wielding position power, expertise relevant to the tasks at hand, credibility, and leadership in terms of ability to drive change.

Building teamwork among the stakeholders involves developing trust among them, and having a unifying purpose—the "burning platform" identified in Step 1. It also might help to involve one more stakeholder—a key customer. This personalizes the unifying purpose in a dramatic and highly effective way. Trust can be built through off-site events, planned opportunities for talk and team-building activities, and developing a shared goal—starting from the issues from Step 1, and having the guiding coalition team defining the vision that is both sensible and appealing. Because new products generally represent the future for the business, for these stakeholders the opportunity to define a vision for developing successful new products efficiently and effectively will appeal to their hearts and minds.

Step 3: Defining the Vision

As mentioned in Step 2, defining the vision for new product development is a team-building activity that leads to a vital deliverable: the compelling vision that the

---

3. John P. Kotter and Dan S. Cohen, *The Heart of Change,* Harvard Business School Press, 2002.

stakeholders, the guiding coalition, would share with the larger organization involved in new product development. Although simply handing the team a prewritten vision might be efficient and perhaps desirable for very busy people, the team-building for the guiding coalition is part of the desired outcome, and the vision development should involve the stakeholders such that they feel they own the vision they defined. The vision could include such concepts as doing things right the first time, dramatically reducing new product development time, or ensuring that the VOC is incorporated into the development process.

An effective set of steps for defining the vision could be:

a. The senior manager must clearly articulate the "burning platform." The stakeholders then brainstorm the issues, starting with the issues from Step 1 to initiate the brainstorming.

b. The stakeholders brainstorm key words, phrases, and terms that seem to capture the direction they would like to take new product development. It could be particularly effective to set the stage for this brainstorming to involve a lot of dreaming—perhaps including some fanciful props that invoke memories of childhood (many of us admit that we really are simply overgrown kids!), providing background music, or starting off with a video or a live speaker that evokes a spirit of envisioning the future.[4]

c. Either the team begins to construct a first-pass vision statement, or a stakeholder or a pair of stakeholders volunteer to work on a first-pass, "strawman" version of the vision statement for the team to review, amend, replace, or whatever.

d. The team reviews, edits, modifies, and finalizes the vision. The vision should be easily remembered, so brief, clear, and, hopefully, compelling—powerful!

The team can validate that the vision is effective by reviewing it against this set of criteria:

- Paints a picture of what the future product development will look like.
- Appeals to the longer-term interests of those involved in new product development, and to the customers.
- Includes realistic and attainable, although challenging, goals.
- Provides clear guidance for decision making.
- Is flexible enough to allow for alternative approaches and innovations to deal with challenging and dynamic situations.
- Is easy to communicate, and can be fully explained within five minutes.

---

4. An example of a video that might evoke the spirit of vision would include Joel Barker's "The Power of Vision," available at http://www.joelbarkertrainingvideos.com.

Step 4: Analyzing Potential Concerns, Issues, Roadblocks, and Impediments
Before communicating the vision (Step 6), the team or a subteam empowered by the guiding coalition team should begin preparations that would include anticipating the issues, counterarguments, and concerns of the broader range of stakeholders. Although some of these issues will be specific to the business, there are some general concerns and issues that are common:

"We already have a new product development process"

"DFSS will take too much time—we are already under too much time pressure"

"This is just another corporate initiative. Wait this out, and the next will come along"

"We don't have enough resources"

"Will I be forced to change my job responsibilities and title? Will I no longer be [a design engineer, a systems engineer, a project manager, a . . .]?

For each of the general concerns, it's effective to consider the value "What's In It For Me?" from the perspectives of various stakeholders, and prepare a counterargument. If the current new product development process involves excessive paperwork, bureaucracy, sign-offs, and unnecessary steps, then the value can include simplification: the new product development process can be mapped and simplified as part of integrating DFSS with the process (see the next subtopic). If the current processes involve considerable firefighting and significant personal sacrifices (such as 80-hour work weeks), then value can be provided through proactive aspects of DFSS: DFSS shifts resources earlier in new product development (see Figure 2.1). If the current new product development process allows living-dead projects to linger despite failing to meet customer requirements, then DFSS can add value by providing clear "Go/No Go" criteria for formal project reviews into the new product development process, as described in Chapter 3.

"We already have a new product development process."

If the company already has an existing new product development process, such as a staged gate process, then an individual or a subteam should be asked to integrate the DFSS process into the existing process. This generally takes between a few days and a few weeks, and is rather laborious. The key seems to be to consider the existing new product development process as the set of "Whats": what the expectations are for the product development process at each step, whereas DFSS supplements the process with a set of "Hows": how to provide confidence that those expectations will be met. As mentioned earlier, this activity also provides an opportunity to simplify the existing process, which can help reduce product development time.

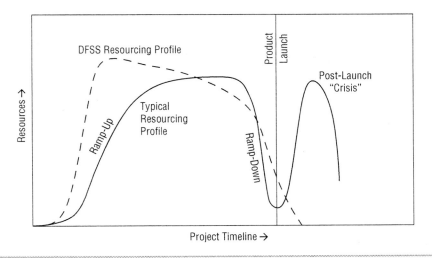

Figure 2.1 Resourcing profile for a project, illustrating a slow initial ramp-up, and the requirement to re-engage dispersed resources after product launch due to post-launch problems and crises

The alignment of DFSS to the existing new product development process can be done using a spreadsheet such as Excel, listing out the stages in one column, the deliverables and items required for completion of that stage in the next column, and the associated DFSS tools and methods in another column. Then gaps can be identified, including deliverables that have no associated tool, DFSS methods that imply a deliverable that is missing from the current process (should it be added?) and reviewing past and especially recent new product development projects and considering lessons learned and how they can be better addressed through an adjustment to the process or the addition of an effective approach, perhaps using DFSS tools and methods.

"DFSS will take too much time—we are already under too much time pressure."

This issue will inevitably arise, and there is perhaps some validity to it. It takes time to approach new product development methodically, and it takes time to train people in new methods. If it takes time to train people, and it takes time to use the tools, isn't it obvious that it will take longer to develop the new product using DFSS?

Interestingly, the data shows otherwise. After gathering new product development time for dozens of new products associated with DFSS black belt candidates, and hundreds of DFSS green belt projects that associated with subsystems for new products, development time was improved or significant schedule delays were avoided in 70 percent of the instances. There was neither negative nor positive impact of implementing DFSS in

30 percent of the instances . . . and no instances were found in which implementing DFSS led to schedule delays.

In one notable case, a new product development team was in the early stages of a project developing a new base station but had already completed several key gates through quickly checking off some deliverables for those early gates as being satisfactorily completed. After receiving some DFSS training, the product development team decided to hold a formal, rigorous review of those deliverables to determine whether they had high confidence in their deliverables, had sufficient confidence in the business case, and had gathered the VOC. After this formal review, the team unanimously decided to move the project backward, "uncheck" deliverables from those early stages of the stage-gate process (that had previously been approved), and effectively restarted the new product development project—intentionally injecting several months of delay. The team subsequently analyzed the business case using the approach described in Chapter 5 and gathered the VOC using the methods described in Chapter 7. Within a few months, the new product team was back on schedule, and the product was completed on time— a recovery through a team effort of diligent effort using DFSS methods.

There are at least three keys to dealing with the concern regarding the time requirements and the possible project schedule slippage:

- Involve the project and program managers as stakeholders, and request that they plan DFSS methods into the schedules.
- Analyze historical schedule issues to understand what really impacted performance to schedule. Chances are, the real problems were not that the team spent too much time using rigorous methods, nor that the team was too diligent in anticipating and preventing problems.
- Review Chapter 6, which discusses schedule risk in more detail, and work with the project and program managers, using the historical information to determine what approaches might be most appropriate to use DFSS to improve performance to schedule. DFSS is a strength in meeting commitments, not a weakness.

"This is just another corporate initiative. Wait this out, and the next will come along"

If the company has a history of centralized organizations driving "initiatives," then this comment is almost guaranteed to arise. Unfortunately, denial is almost counterproductive: if one denies that DFSS is just another initiative, the doubter will likely roll their eyes and disbelieve.

Perhaps a more promising rebuttal was offered by Ralph Quinsey, a vice president at Motorola before he moved on to become CEO at TriQuint. He suggested that any corporate initiative can be used to the advantage of the local organization and can provide an opportunity to address real issues. Then, as the doubter considers that thought, transition into a discussion of what the real issues are, and engage the doubter or challenge them into thinking of ways in which the initiative, DFSS, can be used to deal with the issues that have been frustrating the doubter.

"We don't have enough resources."

Resourcing seems to be a chronic problem in many companies and institutions and exists regardless of DFSS deployment. The real issues might involve several factors such as:

- Ineffective prioritization, such that each new product development project is treated as equally important
- Reluctance to pull the plug on projects that have little chance of succeeding

A "Current Reality Tree," such as that shown in Figure 2.2, can illustrate how anemic project selection, poor prioritization, and a reluctance to prune doomed projects can

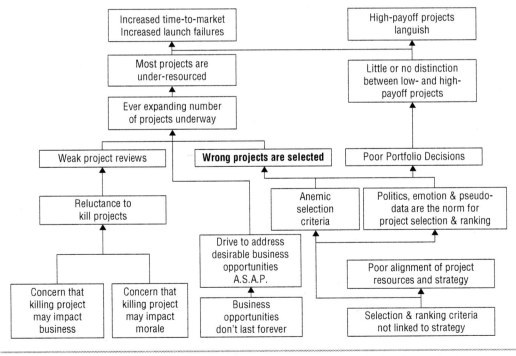

Figure 2.2 Current Reality Tree illustrating that poor project selection, prioritization, and reluctance to address and prune projects can cause low priority projects to consume resources with little chance of success, while underresourcing higher priority projects

lead to under-resourcing projects, which impacts the success of new product launches. Incorporating a strong project selection and prioritization approach (as discussed in Chapter 5), and disciplined, rigorous project reviews (as discussed in Chapter 3) can help the situation, as illustrated with the "Future Reality Tree" in Figure 2.3.

Another aspect of resourcing is the timing for deploying resources. If the project managers and program managers are engaged, they might be able to provide data for a graph similar to Figure 2.1. DFSS tends to promote a different resourcing profile, in which more resources are engaged earlier to capture and link requirements to the VOC, and to anticipate, manage, and make a serious effort to prevent problems. This can be considered something of a "pay me now, or pay me later" situation: the total resources required might be about the same for both approaches, but the area under the curve has been shifted earlier, and the resources that might be consumed in the post-launch crisis might be better used in anticipating and preventing problems earlier in the development process.

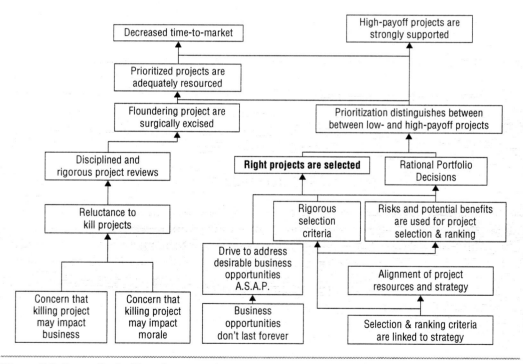

Figure 2.3 Future Reality Tree illustrating that rigorous project selection and prioritization, combined with disciplined and rigorous project reviews can help ensure that higher priority projects are adequately resourced

"Will I be forced to change my job responsibilities and title? Will I no longer be [a design engineer, a systems engineer, a project manager, a . . .]?

Six Sigma black belts engaged in DMAIC problem solving or business improvement are often requested to have full time jobs as black belts, so DFSS black belt candidates might be concerned that they might face a similar requirement. For some, giving up their job title as a Design Engineer or Systems Engineer is like losing their identity. There is prestige involved, and sometimes job security for technical expertise that is in high demand and relatively low supply. In other cases, managers worry about losing a very experienced and knowledgeable engineer in some very key focus area, and having them spend their time instead as statistical experts.

For DMAIC problem solving, the need to focus on improving processes often require a full-time focus; for DFSS product development, the design engineer or systems engineer is still developing their aspects of new products. Consequently, DFSS black belts do not generally give up their previous position and become full-time black belts; they can retain their title, continue their design or systems engineering or other responsibilities, but now adding some new capabilities and knowledge, growing and becoming a stronger engineer with a wider repertoire and more expertise to share with other engineers. There may also be a few full-time DFSS black belts or master black belts dedicated to coaching projects and developing methods that can benefit many new product development efforts or the product development process in general.

Step 5: Planning the Campaign—the DFSS Deployment Plan

The next several steps involve engagement of the wider product development community in a campaign for deploying the DFSS process and methods. Naturally, a campaign involves some planning; elements include a communication plan, alignment of DFSS to the existing development process (as described in Step 4), and developing a preliminary training plan, aligned to needs for new product projects. The training plans might include champion's training for managers, marketing training, project management training, DFSS black belt and green belt training, and training for engineers in key teamwork tools such as DFMEA, QFD, and CPM. An example of a campaign plan for DFSS deployment is shown in Figure 2.4; it is based on a DFSS deployment plan compiled by Tonda Macleod, a Six Sigma black belt responsible for deploying DFSS in one of Motorola's more entrepreneurial divisions.

Step 6: Communicating the Vision

Steps 1 through 5 have set the stage, and Step 6 begins the campaign (to mix metaphors). If the vision has been polished so that it is brief, clear, compelling, and easily remembered, then it has fulfilled the first part of the equation $Q \times A = E$—Quality times

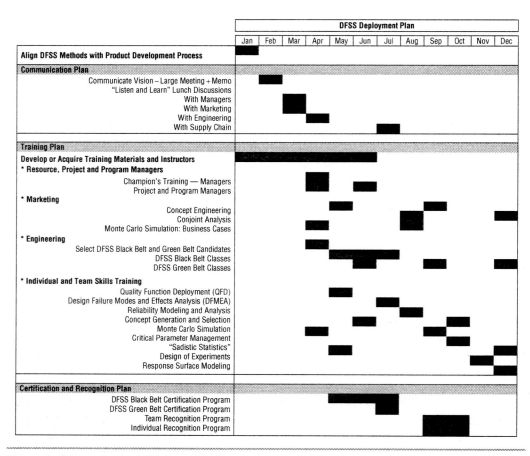

Figure 2.4 Example of a DFSS Deployment Plan

Acceptance equals Effectiveness.[5] This equation describes how the effectiveness of a proposed change within an organization depends not only on the quality of the change and the preparation for the change but also on the receptiveness of the organization to accepting and embracing the change. For example, this equation can describe how a training program can be ineffective if either the quality of the training materials and instruction falls short, or if the participants are resistant or dismissive to the training.

---

5. George Eckes, The Six Sigma Revolution, How General Electric and Others Turned Process Into Profits, Wiley, 2000.

Step 3 helped define the vision such that it should be of high quality, but for the vision to be effective, good communication enhances the likelihood of acceptance by the audience. Some key elements for good communication of the vision include:

- Simplicity of the communication.
- Use of analogies and examples to paint a picture in their minds—perhaps supplemented by images/actual pictures.
- Communication in a variety of forums and through a variety of media: large meetings, smaller meetings, memos, lunch sessions, e-mails, newsletters. Some of these forums should allow two-way communication, which is more powerful and provides a forum to address questions and concerns.
- Repetition of the message, as expressed through the vision.
- Leadership by example—nothing undercuts a vision like having leaders undercut the message by inconsistent behaviors or snide or countermessage remarks. Even expressions of lukewarm or contingent support undermine the credibility of the message, and subsequently its acceptance. The issue of lukewarm or contingent support among the leaders should be addressed prior to Step 6, particularly in Steps 2, 3, and 4; by Step 6, the direction should be that the leader had a chance to express his or her reservations and concerns, issues have been addressed as much as possible or plans have been made to address the issues, and—when the vision is being communicated is a time to show leadership, not hesitancy.
- Addressing seeming inconsistencies, which otherwise might also undermine the credibility.

Step 7: Executing the Campaign

After the vision has been communicated, beginning the campaign planned in Step 5 underlines the importance and seriousness of the leadership team in achieving the goals. The DFSS Deployment Plan example in Figure 2.4 can provide some guidance on how the execution of the campaign can progress.

Perhaps the best way to ensure that the deployment keeps on track and achieves the goals envisioned by the leaders is to define success metrics for new product development projects (as discussed in Chapter 3) and set up a system to first baseline and then measure these metrics on an ongoing basis. Figure 2.5 provides a flowchart for ensuring that the DFSS deployment meets the expectations set by the leadership team.

Step 8: Removing Roadblocks and Impediments

The term "frustration" is unusual in that it defines both the symptom and the cause. If, at some point in the DFSS deployment campaign or in an individual product development

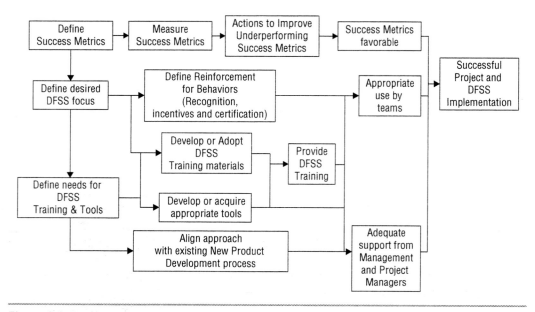

Figure 2.5 Steps involved in assuring that the DFSS deployment meets expectation, as defined by a set of success metrics

project, people begin to feel frustrated and discouraged, it probably means that something seemingly or actually beyond their control is preventing them from accomplishing the goals.

The role of the champion is to help individual project leaders and black belts remove roadblocks in their individual new product development projects. Although the DFSS black belt candidates or certified black belts might feel that they and their team are expected to handle everything themselves, the real expectation is to recognize roadblocks and impediments, handle what they can reasonably and confidently handle, but meet with the champion to request help with removing roadblocks beyond their control—and quickly, rather than potentially delaying the project schedule and allowing frustration and discouragement to fester within the team.

For roadblocks and impediments to the DFSS deployment campaign, the guiding coalition should keep its finger on the pulse of the organization and—when it senses frustration—it has the insight and the power to remove the roadblocks.

Sometimes, the roadblocks are organizational. DFSS is customer-focused, but the organizational structure can fragment resources and build walls between marketing and engineering. DFSS, like all Six Sigma processes, seeks to drive responsibility lower in the organization, to the teams and individuals responsible for product development, but the organizational structure can involve layers of middle management that might second-guess and criticize decisions, undercutting the willingness for employees to accept responsibility. DFSS seeks to make new product development more efficient, but the organizational structure can involve silos that impede communication—thereby impacting speed and efficiency. The leadership team can deal with these roadblocks by reviewing the organizational structure and either better align them with the vision for success, or make some changes as necessary—which can range from minor to major changes. The information and human resource systems also need to be aligned so as to be supportive of the vision; for example, the DFSS black belt and green belt certification approaches discussed in Chapter 3 must be aligned with the human resource systems.

Finally, and perhaps most unpleasantly, managers who undercut the changes must be confronted by the appropriate leaders. Not only do such managers undercut the needed actions directly, but when employees see behaviors and no response, no confrontation, they become discouraged, which further erodes the momentum.

Step 9: Generating Short-Term Wins
The advantages of having early, short-term wins are probably obvious. It provides evidence that supports and provides some justification for the campaign, provides a sense of accomplishment for the team or teams, provides helpful feedback for the leadership team, undermines cynics and critics, strengthens support from the managers, and generally helps build momentum.

Problem-solving, DMAIC Six Sigma projects lend themselves to "low-hanging fruit" in terms of finding opportunities for quick wins in the Define and Measure phases through process mapping, identification of non-value-added activities, and finding quick opportunities during team-building activities such as brainstorming and fishbone diagramming. The early phases of DFSS don't tend to lend themselves quite so readily to low-hanging fruit. Although activities that help gather the VOC, as discussed in Chapter 7, can provide insights and build rapport between marketing and engineering, the "proof of the pudding" may have to wait through the entire new product development cycle—which hardly qualifies such successes as "quick wins." There are a few approaches that can be used to generate short-term DFSS wins:

- Use DFSS with short-development projects (e.g., derivative products and specialty products).
- Use DFSS methods in the later stages for products nearing launch. Some of the DFSS methods discussed in Chapters 13 and 14 are particularly powerful to handle challenging situations where multiple requirements must be met with the same subsystem or component. Some of the methods discussed in Chapters 15 through 19 can enable the team to perform a "diving catch" on a project that is encountering troubles in later stages.

- Use DFSS methods on products from the beginning, but incorporate leading indicators that enable the team and stakeholders to see the progress before the product is released. Some suggestions for leading indicators are shown in a table for success metrics in Chapter 3.

Step 10: Consolidating Gains, Recognizing People and Teams and
Step 11: Anchoring the New Approach in the Culture

DFSS deployment is challenging, both as a change management process as described in this chapter, and in requiring people to learn and apply new tools, while working both as individuals and as members of teams to drive success. People rise to challenges—if they trust that:

- the leaders really care about the project
- the team will be supported
- the individual successes and the team's success will be recognized

The deployment plan should provide a recognition system for teams and individuals. The recognition system for teams does not need to be expensive nor complex. The DFSS black belt and green belt certification processes, as described in Chapter 3, are very effective.

People also feel recognized when they are encouraged to share their successes, through presentations to management and within their organization, and through presentations to other organizations. "Town halls" and management reviews are excellent forums for recognition, if available. Communication forums such as staff meetings, newsletters, bulletin boards, posters, and banners can be used to recognize people and teams and celebrate success. External forums may include working with the publicity department to publish articles. If the work is truly outstanding, significant recognition can be provided by selecting and recognizing their project as exemplary, having it videotaped as an outstanding success story, or having team members represent the company at a conference.

Monetary rewards could also be appropriate, and would reinforce the recognition—but recognition is not about "the money" so much as the recognition by management and peers that they and their project were successful.

One of the challenges of a sustainable DFSS deployment is to integrate within the existing product management methods and terminology of a product or technology development organization. DFSS terminology should embrace terminology with which the people are familiar: "project hopper" and "product pipeline" and "product portfolio" are synonymous. "Project selection" is the same activity as "portfolio optimization."

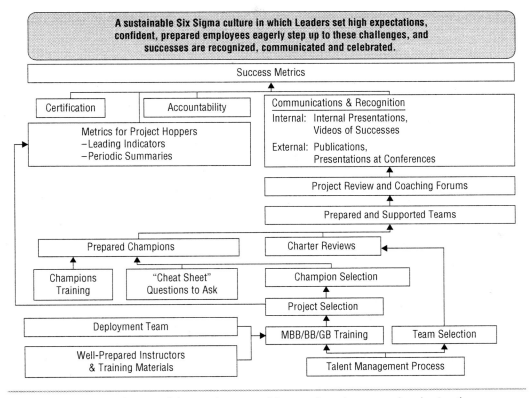

Figure 2.6 Flowchart for consolidating gains, recognizing people and teams, and anchoring the new approach in the culture (Steps 10 and 11)

Figure 2.6 integrates Steps 10 and 11 into a flowchart summarizing some of the key aspects for a sustainable culture that will support DFSS, including the certification and recognition aspects.

DFSS DEPLOYMENT: SINGLE PROJECT

The DFSS deployment approach discussed so far is basically a top-down process, in which the CEO or a senior executive decides to deploy DFSS within the organization.

However, there are instances in which DFSS deployment is initiated on a much smaller scale—starting with a single project, which may be the pilot project or may be aligned with a product development visionary such as an engineer, project manager, or engineering manager, or an individual with a goal for DFSS black belt or green belt certification. In these situations, the management support may be minimal, or conditional—contingent on the success of this first DFSS project. The key requirements are:

- Define success for the DFSS project, which can be aligned with the project charter, as discussed in Chapter 3, and made measurable (i.e., providing an operational definition of success for the project).
- Obtain agreement—perhaps conditional agreement—from stakeholders for the value to the product or the new product development program if success is achieved. Key stakeholders include your manager, some engineers involved in the new product development, and preferably at least one person from marketing.
- Obtain a champion, who can be a sounding board and help knock down barriers, and who can be a management advocate during the project.
- Obtain the tools and skills needed. Training and coaching can be obtained through resources listed at http://www.sigmaexperts.com/dfss/tools, which also lists sources and resources for tools such as statistical analyses, Monte Carlo simulation, critical parameter management, quality function deployment, TRIZ (Teoriya Resheniya Izobreatatelskikh Zadatch), UML (Unified Modeling Language), and reliability modeling
- Anticipate, develop, and provide approaches to address some of the concerns from stakeholders, such as those discussed for Step 4. Stakeholder analysis, as shown in Table 2.1, can be used to anticipate concerns and consider approaches for responding to those concerns. A forum approach can also be used: stakeholders are asked to come to a meeting, presented with a brief overview of the goals and DFSS methods

Table 2.1 Stakeholder analysis form

Stakeholder	Role	Expected (or Actual) Reaction	Concerns	Approach to Address Concern	Who	When, How

that were planned, and then the facilitator goes around the room asking each stakeholder to provide feedback, including their concerns. The concerns are listed on a large tablet, and after each stakeholder has a chance to speak, the facilitator and the DFSS candidate review and discuss each concern, in some cases addressing them immediately, in other cases committing to develop a response and communicate that response to the stakeholders within a short period of time. This forum provides the stakeholders with a chance to speak their minds, and know that they were heard. At the end of the meeting, the stakeholders can be asked if they would be amenable to a "willing suspension of disbelief," to be supportive for this pilot project, if the DFSS black belt candidate would take actions to deal with lingering concerns as the project progresses.

The individual DFSS project should involve a deployment plan, perhaps not of the scale discussed for Step 5 in this chapter but with the same elements as those shown in Figure 2.4. The deployment plan would be integrated into the DFSS project charter, as discussed in Chapter 3.

MINIMUM SET OF TOOLS, AND THE "ONE TOOL SYNDROME"

Some key stakeholders, concerned about the amount of work that will be added to the burden through using DFSS, may request that the team use the "minimum set of tools"—perhaps ranging from just one or two tools, to the top five tools or the top ten tools.

Similarly, some training consultants, enthusiastic about the tool they are teaching, may be prone to declaring that "DFSS is all about using DFMEA," or "Monte Carlo simulation is the heart of DFSS," or "TRIZ *is* DFSS." Although the enthusiasm from these instructors is laudable, and bodes well for the effectiveness of their training, the real issues here involve confusion about "What is DFSS?" so the need here might be for an operational definition for DFSS. Such an operational definition can be developed based on the goals for DFSS discussed in the next section of this chapter and the success metrics discussed in Chapter 3.

In terms of the minimum set of tools, the issue is whether they address the goals that, together, constitute higher likelihood of success for the new product. In some cases, current best practices may be sufficient to meet some or even most of these goals, and a subset of DFSS tools and approaches may be selected which address the goals that are not being effectively addressed through the current new product development process.

Consequently, the following section provides a set of goals for DFSS, in the context of considering whether the current product development process is effective in meeting each goal, or whether the addition of DFSS methods might be appropriate.

GOALS FOR DFSS

For the development of a new product or new service to be considered to be a DFSS project, the development project must incorporate efforts and rigor to ensure that the new product or service will meet or exceed expectations from the customer or customers and from the business in terms of the performance, availability, competitive value, manufacturability, and business case.

The operational definition can be further defined in terms of specific expectations and recommended tools or methods to achieve those expectations. These recommended tools may not be required for each DFSS project, but the management/leadership should demand credible evidence that the expectation has been met. In most cases, the recommended tools and methods are the most straightforward and proven way to provide this credible evidence and meet these expectations.

- **Risks are anticipated and managed through rigorous gate reviews, management engagement, and risk management approaches.**
  Management reviews and demands evidence from the development team regarding the expectations and risks. In particular, risks that could affect the business case and the project schedule are anticipated and managed in addition to risks that could impact performance, customer satisfaction, and reliability/availability.

  The recommended approach is a rigorous gate review process that reviews risks and includes Monte Carlo simulation and FMEA for both the business case and the project schedule. Alternative approaches for the project schedule risk could include theory of constraints project management.

  The business case is updated at later stages. or as assumptions, competitive environment, and product performance and features change. The updated business case is reviewed by management to enable data-driven decisions regarding the viability and priority of the project.
- **The new product or new service is expected to provide competitive value based on the VOC.** This will generally involve an identifiable competitive edge.

  The recommended approach is concept engineering, which includes customer selection matrix, customer interview guide, gathering the VOC, and Jiro Kawakita (KJ) analysis. However, the intent can occasionally be met by other means: for example, if there is only one customer, and that customer provides a clear specification that reflects the VOC.

- **Several alternative concepts or architectures were considered, and the team agrees that the selected concept is one of the best approaches in terms of its capability to meet or exceed customer expectations.**
  The recommended approach is the Pugh concept selection process; however, the intent can be met by other means: for example, if the customer expectations can be met quickly and with low risk by using a slightly modified version of a product or service that has already been developed and proven—high reuse.
- **Critical parameters have been identified and "flowed down" to requirements for subsystems, modules, assemblies and/or components such that the design sub-teams and vendors can receive clear requirements.**
  The recommended approach is critical parameter management; however, approaches such as QFD and DOORS can meet this expectation.
- **The performance of components, assemblies, modules, subsystems, and systems have been forecasted as robust:** the product or services is predicted to meet customer requirements in the presence of variations or "noises" including manufacturing variability and variations in the environment and usage, and the team is confident that the product or service will consistently meet customer requirements.
  The recommended approaches involve predictive engineering/robust design and critical parameter management, including or in addition to Monte Carlo simulation, system moments method, multiple response optimization or yield surface modeling. Robust design includes p-diagrams combined with designed experimentation for average and variance of key parameters. Design FMEA (DFMEA) and fault tree analysis (FTA) can supplement this expectation.
- **The product is shown to be reliable and consistently available to the customers**, with predictions as to the likelihood that the product will fail to meet customer expectations after time or events that could degrade performance.
  The recommended approach is reliability modeling combined with FMEA or FTA.
- **The product is reviewed for manufacturability prior to release to production.**
  The recommended approach is inclusion of appropriate representatives from manufacturing or supply chain in at least the later reviews before release to production, and providing access for them to the robust design, critical parameter management, and probabilistic design results and databases.
  An alternative approach is to build the product in pilot runs and review the data to verify manufacturability.

"THE DFSS PROJECT WAS A SUCCESS, BUT . . ."

There is a saying, particularly among surgeons: "The operation was a success, but the patient died." Unfortunately, this saying reflects a sad and hard reality: it is possible for a surgical procedure to be performed successfully, but for the patient to die as a result of other factors. A human being is a very complex organism, consisting of many vital organ systems; the surgery focuses on one aspect, but other aspects can also go terribly wrong.

With new products, there is also a high level of complexity, involving many vital subsystems and engineering quality and market dynamics aspects. DFSS necessarily focuses the team on some key aspects of the product and the product development effort . . . but it entirely possible that some aspect that is not addressed will go terribly wrong, and will cause the product to fail, either technically or in the market. One DFSS project involved a fantastic effort to develop and optimize a new type of display for a mobile device—the project was successful in developing a very robust display, but the product itself suffered from other issues including cost overruns and market acceptance issues. The optimized, robust display was subsequently incorporated into a different product that has been an extraordinary success, but in terms of the original product—the DFSS project was a success, but . . . .

The best hope is to make the criteria for success for the DFSS project clear; however, any situation in which the new product is unsuccessful, for whatever reason, will inevitably cast a dark cloud over the declaration of success for the DFSS project.

DFSS can be considered to be analogous to insurance: you pay up front, with real resources and real efforts, against something that may or may not happen—and something that you didn't cover (with insurance or with DFSS efforts) may still go wrong. Insurance is worthwhile—it is better to realize that there really are risks out there, and take action to deal with these risks, than to do nothing and passively accept the risks. Similarly, it is better to realize that there are risks involved in developing new products, and take action to anticipate and deal with those risks through approaches such as predictive engineering and risk management, than to simply leave the situation to fate.

SUMMARY

DFSS deployment deals with the "people side of things," particularly the change management aspects. DFSS introduces new methods and approaches, and therefore stimulates resistance to change. Both types of DFSS experts—experts in DFSS tools and methods ("technical experts"), and experts in DFSS deployment ("deployment experts")—can benefit from a step-by-step strategy for dealing with resistance to change associated with DFSS. This chapter suggests a step-by-step approach for a top down implementation, and also discusses the DFSS deployment for a single DFSS project. In each case, a key aspect is considering the perspectives and anticipating the concerns of those who will be affected, then developing responses that can refine the DFSS approach to clearly address those concerns. Goals for DFSS can provide criteria for an operational definition of DFSS, to handle tendencies toward the "one tool syndrome," and can lead to a set of success metrics that can make successes visible and enhance recognition for the people and teams involved in the successes.

Governance, Success Metrics, Risks, and Certification

Some of the goals for Design for Six Sigma (DFSS) were discussed in Chapter 2; this chapter will refine those goals into a set of success metrics for individual DFSS projects, and define the elements of a governance process to support meeting those goals. The success metrics allow the teams and the leadership to gauge the progress towards achieving success; a governance process provides a mechanism for the team and the leaders to make adjustments or, in some cases, more major decisions regarding the project. The governance process is also aligned with recognition and certification processes that support the goals of the individuals who are leading and contributing to the success of the product development efforts. Figure 2.5 illustrated these aspects through success metrics, governance through project review and coaching forums, and a certification process.

Deployment leaders may need to clarify, perhaps repeatedly, that a DFSS project follows the schedule and aligns with the expectations of the associated new product development. This alignment includes not only alignment of schedules but also alignment of the DFSS governance process with the oversight and reviews for development of the new product, technology, or service.

DFSS GOVERNANCE

DFSS governance is a mechanism to sustain the deployment and institutionalization of DFSS, but it can also be considered a feedback mechanism. The progress and current situation is discussed, and adjustments are made to help keep the DFSS projects and the overall program on track. If the project is encountering difficulties, then the champion can try to deal with the issues, and, if not, then either the broader leadership team can try to help, or if the issues pose a high risk of not meeting customer expectations for the product or service, then the leadership team can make some hard decisions.

The nature of this process and these decisions lends itself to two types of reviews—supportive project reviews with the champion, which can be formal or informal, and formal gate reviews with the decision makers (Figure 3.1). The champion is a leader in the organization who is respected for driving positive change and influential in the new product development process and resources. He or she is an advocate for DFSS and for the DFSS projects who is ready and willing to support and possibly mentor project team leaders and green and black belts, who actively participates in project reviews, and who is ready and willing to remove barriers.

SUPPORTIVE PROJECT REVIEWS

The supportive project reviews with the champion can be periodic—perhaps weekly, for about an hour. Additional project reviews can be event-driven: the team might expect

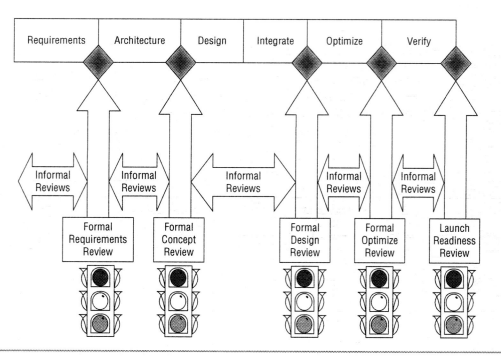

Figure 3.1 Arrangement of formal and informal reviews

somewhat more intense project reviews with the champion before (and as preparation for) the formal gate reviews; also, special project reviews should be held to consider changes in circumstances that could impact the success or viability of the project—such as a change in the competitive environment (a competitor announces a product almost exactly like, or even better than, your product; or, alternatively, a competitor drops out) or any dramatic change in the technical risk, market risk, financial risk, or schedule risk. These and other risks will be discussed later in this chapter and in upcoming chapters (particularly Chapters 5 and 6).

The supportive project reviews discuss progress toward the key deliverables for the current phase of the development project, and perhaps review any presentations being put together for upcoming meetings, such as formal gate reviews. Then the team discusses the issues and barriers, and any other things that are causing frustration. A feeling of frustration is a clear signal for champion involvement, but the champion can either work with the DFSS black belt or black belt candidate or take responsibility directly for resolving the issues and removing barriers. Next, the team reviews the project plans and associated risks. The meeting is then summarized with a set of action items that constitute a plan, a recovery plan, or assistance.

Champions, either by training or by nature, are especially skilled in **resolving conflicts** and **removing barriers** to assure the success of the projects. DFSS black belt and green belt candidates should also learn effective barrier removal skills as part of their leadership development. Champions should define what the protocol will be in barrier removal as part of the project kick-off or early meetings; the first supportive project review should generate an initial **list of barriers**. Over time, this list will be whittled down as issues, including new ones that arise along the way, are dealt with. The champion needs to develop open communication and relationships built on trust and honesty—so that the champion hears the truth, team members "telling it like it is." There should be no reluctance or fear among team members about telling the truth. The first step in managing issues is to know what the issues are. The informal gate reviews can and should be open, nonthreatening, supportive environments in which issues can be raised and addressed.

FORMAL GATE REVIEWS

Formal gate reviews address the overall new product development effort, of which the DFSS methods play a key part. The timing of formal gate reviews relate to key transitions in a phase gate product development process or to key decision points for the product, technology, or service, which could include the decision to begin resourcing the project, the decision to approve the requirements and architecture, the decision to approve the design and begin the integration and optimization of the design, and the decision to approve the readiness for product launch.

The formal gate review starts by reviewing the phase that is being completed: the deliverables and the level of confidence (or, contrarily, the level of risk) associated with the deliverable. For example, deliverables in the Requirement phase might include the business case, the project schedule, the selected product concept, and the set of critical parameters. The business case and project schedule deliverables might include Monte Carlo simulations that estimate the confidence that the commitments will be met. The product concept deliverable might be accompanied by the Pugh Matrix showing why it was selected and also an FMEA summary for the selected concept. The set of critical parameters deliverable might include a first pass of the critical parameter scorecard (as discussed in Chapters 10 and 14), and perhaps some initial benchmark and baseline information that indicates the initial, not-yet-optimized confidence level for each critical parameter.

The team should review the critical parameter scorecard in formal gate reviews for later stages as well. The team discusses issues and barriers, along with plans and any appropriate requests for help. Next, the team reviews the project plans and risks, possibly through a summary of the risk priority numbers (RPN) from the initial or updated FMEA. Finally, the team reviews the success metrics (discussed in the next section of this chapter) as leading indicators for the forecasted success of the product. The development team provides evidence that the goals and success metrics will be met, while honestly sharing issues and concerns. The primary outcome from a formal gate review is a decision, which can be:

1. "Full Steam Ahead": The key deliverables and success metrics are acceptable.
2. Conditional Acceptance: The concerns warrant an action plan; return with results and updated risk assessment after key action items are addressed.
3. The Project Is in Jeopardy. Proceed through a period of evaluation (perhaps a few weeks). If the situation improves considerably, develop an action plan for conditional acceptance; if not, the decision makers will reluctantly cancel the project and, after a transition period, the resources will be assigned to a more promising project.

If the primary outcome is favorable or conditional, the concluding part of the formal gate review is a look forward, to preview what is coming in the next phase and to anticipate and prevent problems in the next phase.

SUCCESS METRICS

The intent of success metrics is to provide visibility of how the project is doing and how likely it is that the project and product will be successful, and then to highlight areas of substantial risk to the success of the project. The ideal set of DFSS success metrics would meet these criteria:

1. Orthogonal: there would be no overlap between metrics, and so little or no confusion as to which metric is affected by each activity, risk, or situation.
2. Comprehensive: if the set of success metrics are all favorable, the team and leadership should feel confident that the project is likely to experience success.
3. Predictive: if the success metrics are applied to past projects, those projects considered more successful should have better values than products considered less successful.
4. Diagnostic: the set of success metrics should provide valuable guidance on which issues and aspects are risky and need to be addressed.
5. Readily accepted: the success metrics should make sense to the team and managers, and preferably align with hardware and software metrics collected in the past.
6. Leading indicators: the success metrics would be measured during the course of product development, as opposed to lagging indicators that might not provide values until the product has been launched and in production for a period of time.

The "requirement flow down" concept, associated with critical parameter management (see Chapter 10), provides a way of achieving these criteria of orthogonality and providing diagnostic and predictive value. With this concept, there would be a primary index of success for the new product development effort (the "big Y"), which would flow down into an orthogonal set of success metrics (the "little ys"). In some cases, these "little ys" could align with individual efforts, or perhaps with individual DFSS black belt or green belt projects. The success of each "little y" effort would be clearly linked to the success for the new product development effort (the "big Y"), but there would be the realistic risk that the product's success could be impacted by other little ys that were not addressed.

This flow down/flow up concept could use an additive or a multiplicative model: an additive model could tie to the business case (discussed in more detail in Chapter 5) through a set of success metrics such as those shown in Table 3.1.

Table 3.1 Success metrics for a model tied to the business case

$ Say − $ Do Impact =		
	$/Day Late × Projected Days Late	[Schedule Slippage Metric]
+	$/Day Late × # Iterations × Delay/Iteration	[Iteration Metric]
+	$/Day Late × # releases to customer	[Software Release Readiness Metric]
+	$/Feature Delivered × # Features Delivered	[Feature Metric]
+	$COPQ/Bad Part × (1-Projected Composite Yield) × Projected Volume	[Cpk Metric]
+	$/Stop Ship × Pr(Stop Ship)	[Stop Ship Metric]
+	$/Customer Line Down × Pr(Line Down)	[Supply Chain Readiness Metric]

Table 3.2 Success metrics for a model tied to the probability of success

Pr(Project Success) =		
	Pr(Project Success|On Time) × Pr[No Schedule Slippage]	[Slippage Metric]
×	Pr[No Excess Iterations]	[Iteration Metric]
×	Pr[No Excess SW releases to customer]	[Software Release Readiness Metric]
×	Pr(Project Success|Sufficient Features delivered) × Pr(Sufficient Features delivered)	[Feature Metric]
×	Pr(Project Success|Acceptable Cpk's for Critical Parameters)	[Cpk Metric]
×	Pr(Project Success|No Stop Ships) × Pr(No Stop Ships)	[Stop Ship Metric]
×	Pr(No Line Down situations)	[Supply Chain Readiness Metric]

A multiplicative model could tie to the probability of project success through a set of success metrics such as that shown in Table 3.2.

Alternatively, the success metrics could be linked into the recognized risks for the project, as summarized in Table 3.3.

PRODUCT DEVELOPMENT RISKS

The risks associated with metrics in the previous section are risks that can impact any new product development project. The point is that DFSS is an integral part of the new product introduction (NPI) process and can provide tools and methods to help the team manage these product development risks. The NPI process guides the overall development and

Table 3.3 Risks and associated metrics for a project

Risks	Metrics
Business Case Risk	Forecasted Pr(Profits)
	Forecasted Costs
	Forecasted Profit or Gross Margin
	Real/Win/Worth
	Forecasted Market Share
Technical Risk	Composite Cpk
	Forecasted Composite Yield
	FMEA RPN Improvement
Schedule Risk	Pr(Schedule Slippage)
	Forecasted Completion Data
	Forecasted Schedule Slippage
	Number of Iterations
Reliability Risk	Forecasted Reliability
	Forecasted Availability
Product Delivery Risk	Forecasted Pr(On-Time Delivery)
	Forecasted Lead Time
Instability of Requirements	Number of Spec Changes during Product Development
	Percent Alignment to Expectations

DFSS provides the disciplined methodology to make sure the product, technology, or service meets customer expectations. NPI guides the overall development by:

1. providing appropriate checkpoints or gates
2. ensuring that resources are sufficient for the job
3. coordinating cross-discipline teams

DFSS provides a disciplined methodology to predict and verify that a product will meet customer quality requirements. These are two highly compatible processes. DFSS really enhances the NPI process. Typical NPI processes (e.g., marketing studies, risk assessments, FMEAs, gate reviews) are aligned with DFSS. Managing the risk for NPI is well served by tailoring the risk management process for DFSS.

Risk management is the planned control of risk. It involves monitoring the success of a project, analyzing potential risks, and making decisions about what to do about potential risks.[1] It is a formal documented methodology to:

- **Identify**
- **Analyze (assess/rate)**
- **Plan and communicate abatement activities**
- **Implement and track resolution of risk issues throughout the life of a program**

Risk identification and abatement activities are planned and occur throughout the DFSS project timeline, increasing the probability of success.

Integrating formal risk management with DFSS project management is a new phenomenon in the software engineering and product management community. It requires that project managers be involved in a DFSS project from the Concept phase to the product's retirement.

RISK MANAGEMENT ROLES

Projects support a business's goals and objectives, have defined, unique scopes of work, agreed-upon starting and ending dates, and committed resources. Project managers should develop new roles and responsibilities in risk management. These roles ensure responsibility for overall risk management and should be defined in a project management plan document. The following describes three key players in project risk management.

Project Manager

The project manager is the person who is directly accountable to the sponsor and program manager for successful execution of a project. The term "project manager" refers to both project leader and project manager throughout the rest of this section. The DFSS project manager (DPM) determines the responsibilities for risk management within in a given project. The DPM has overall responsibility for ensuring risk management occurs on the project, and directs or approves risk management planning efforts and all substantive changes to the risk control plans. The DPM assumes overall responsibility for all risk management actions associated with risks assigned to the project.

---

1. Controlling Risk, Patricia Duhart McNair, ACM, Ubiquity, 2001.

Development Manager

The systems development manager (SDM) coordinates risk management planning to ensure that risks are identified, risk assessment data is prepared, and control plans are complete. The SDM supports the DPM in reporting risk status to senior management and the customer. The SDM should prepare and maintain the risk plans and coordinate all subsequent risk status, reporting actions as well as evaluating risk control action effectiveness and recommending control actions. The SDM should coordinate the continuous review of the risk plan to keep the set of risk issues and associated risk assessment data and control plans current with project conditions.

Risk Owner

The risk owner or green belt or black belt is the individual identified by the SDM or Master Black Belt (MBB) as the responsible individual for overseeing the risk. The risk owner identifies and assesses "own" risk to produce probability and impact information. He or she should develop risk mitigation and contingency plans, provide status data for respective risk issues, and assist in evaluating risk control action effectiveness. The risk owner also documents threshold criteria of high and medium risk and supports identification of new risks.

Oftentimes in projects, a plan is created for risk every time a system is developed, and an assessment is done for that risk. Both managers and technical personnel develop risk plans for current projects, based on historical data. However, quite often the risks are identified and then forgotten throughout the product's life cycle. They are listed and put on the risk watch list but not addressed until they become problems. To prevent this from happening, at each "project review" the risk owner should give the status of the mitigation or contingency plan, and the project team should decide to (1) take action on the risk, (2) eliminate the risk, or (3) retire the risk. Monitoring the risk should only be done when the risk does not have a high impact. As the risk management planning activities begin addressing risk identification, and assessment and control plans near maturity, use a spreadsheet to prepare a risk management plan that captures the risk planning data and provides a means to collect and report risk status.

In order to effectively evaluate risks for a project the basic risk attribution of probability and impact is not enough. A third element called time frame should be given in order to provide a basis for comparison to other risks. The time frame allows the project to readily assess prioritization. Prioritizing the risks assists the team in deciding how to allocate resources for mitigation, particularly if the project team has identified a large number of risks.

The DFSS formula for risks is Probability $\times$ Impact $\times$ Time frame ($P \times I \times T$). These three elements determine each risk's risk priority number (RPN) by multiplying attribute ratings for probability, impact, and time frame. Time frame is directly related to when a DFSS solution will be needed for the critical success of the product.

During risk tracking and control, the project leaders must ensure that risks stay visible to the project and business management teams. They should recognize that those risks and their attributes (probability, impact, and time frame) will change over time. This helps to eliminate or retire risks when applicable. The project manager should track new identification of risk and create or modify mitigation and contingency plans on a regular basis. The identification of risks that become a problem should be deleted from the list of risks. The risk control actions defined by the mitigation and contingency plans should be implemented in accordance with the details of those plans. The risk owner is commonly responsible for the implementation of these actions.

Passive acceptance of risk is not an option. The likelihood of a product defect is reduced when risk management applies to all product development and spans the life cycle of a product.

DFSS CERTIFICATION

Design for Six Sigma green belts, black belts, and master black belts become leaders, coaches, experts, and change agents for applying DFSS methods. The certification process provides a means to recognize the people who have learned, mastered, and applied these skills. Successful completion of the certification requirements verifies that the candidate is ready, willing, and able to meet future challenges, lead product development efforts, and coach others in these methods. On a personal basis, DFSS certification can be career-enhancing in that the successful candidate has shown that he or she has mastered skills that go beyond the usual, static palette of skills for a development engineer, and has the tools and skills—and motivation—to deal with a dynamic world full of challenges and uncertainties.

The training requirements will vary depending on the certification level and the certification approach/methodology. The candidate should be able to pass a test or tests to verify understanding, and then apply these methods by completing a DFSS project. A project report(s) summarizing the DFSS process used, key findings, demonstrated benefits, lessons learned, and identification of other project opportunities would be reviewed with the mentor and appropriate master black belt and senior management.

The project has success metrics that provide expectations and visibility that will demonstrate significant favorable impact for the product development. A DFSS black belt candidate will develop a DFSS charter than incorporates a DFSS deployment plan to meet these objectives, which includes the deliverables and tools expected to be used as well as those who will be involved in using those tools. The plan should include anticipated needs for training, coaching, data collection, and support. An example of a DFSS black belt charter was shown in Figure 1.7 in Chapter 1. Excel templates for DFSS black belt and DFSS green belt candidates can be downloaded from http://www .sigmaexperts.com/dfss/certification.

A black belt candidate will manage the deployment of the DFSS methods and tools on product development programs through all phases of the DFSS methodology, identify the critical parameters (high-risk technical performance and functional parameters), mitigate associated risks, and optimize and monitor those critical parameters.

The DFSS black belt project might involve several smaller-scale DFSS projects that could be assigned to DFSS green belt candidates. The black belt candidate would coach the assigned green belt candidates, and would ensure that the smaller-scale projects are coordinated and the results integrated to meet the success criteria of the larger-scale product development program.

A DFSS green belt project can be a stand-alone project or part of a DFSS black belt project (as just described). The DFSS green belt candidate will develop a charter summarizing the tools that will be used that is similar to the charter shown in Figure 1.7.

Certification also requires that the DFSS project provides value that is recognized by key stakeholders. For a DFSS green belt project, this can be verified through having the technical lead write or agree with a "Statement of Value" that summarizes the key benefit or benefits that the DFSS project provided. For a DFSS black belt project, the technical lead and the business lead should each provide statements of value reflecting both the technical value and the business or marketing/customer value provided by the project. Examples for the content of statements of value are provided with the certification template at the Web site mentioned earlier.

"Technical lead" refers to an engineering leader or manager who is held responsible for the technical success of the system, subsystem, or module for which the candidate is applying a DFSS approach. The technical lead might have a job title such as Mechanical Engineering Lead, Software Engineering Lead, Software Engineer Manager, Electronics Engineering Manager, Program Director, Director of Engineering, or perhaps Team Leader for a Subsystem. He or she will generally be involved in partitioning the engineering tasks to the individual design and development engineers and will generally be held accountable for successfully meeting expectations to the technical specifications for the critical parameters.

"Business lead" refers to a leader or manager who is accountable for the business results and/or the marketing, and customer acceptance or receptiveness for the product. The business lead might have a job title like Business Manager, Operations Manager, Marketing Manager, Program Manager, Project Manager, Director of Marketing, or Vice President.

These roles, responsibilities, and potential job titles are summarized in Table 3.4.

Table 3.4 Roles and responsibilities for champions, technical leads, and business leads

	Role	Responsibilities and Functions	Job Titles
Champion	A leader in the organization, respected for driving positive change and influential in the new product development process and resources. An advocate for DFSS and for the projects.	Define and scope DFSS projects; resource projects for success Support and mentor Project Team Leaders, Green and Black Belts Actively participate in Project Reviews Remove barriers; Communicate results to executives	Director, Vice President, Program Manager, Senior Manager, Product Development Manager.
Technical Lead	An engineering leader or manager who is accountable for the technical success of the system, subsystem, or module for which the candidate is applying a DFSS approach.	Partition the engineering tasks to the individual design and development engineers and resource for success. Ensure that technical specifications for the critical parameters are achieved. Verify that the project provided value for the new product, service, or technology.	Team Leader, Mechanical Engineering Lead, Software Engineering Lead, Software Manager, Electronics Engineering Manager, Program Manager, Director, or Vice President.
Business Lead	A leader or manager who is accountable for business results or marketing, and customer receptiveness for the product.	Verify that the project provided value for the new product, service, or technology either financially or in terms of customer satisfaction and market acceptance.	Business Manager, Operations Manager, Marketing Manager, Program Manager, Project Manager, Director of Marketing, or Vice President.

SUMMARY

This chapter discussed the governance and certification aspects of Design for Six Sigma projects, along with the associated success metrics and risks. Governance provides a support mechanism and a feedback loop that helps ensure that the project will be successful, leading to benefits for the company and certification and recognition for the individuals and teams involved.

Overview of DFSS Phases

DFSS FOR PROJECTS, INCLUDING SOFTWARE AND HARDWARE

DFSS is applicable to just about any development project involving a variety of engineering disciplines. An individual mechanical engineer can use DFSS while developing a relatively simple mechanical device like a pair of scissors or a mouse trap, or a team can use DFSS developing a complex mechanical system like a lawnmower or an electromechanical system like a motor. A single civil engineer can use DFSS for a relatively simple civil engineering project like designing a well or a team can use DFSS to develop a complex civil engineering venture like building a bridge or a large facility or a city in the desert. A team of biomedical engineers, electrical engineers, and mechanical engineers can use DFSS to handle the complex requirements for a prosthetic hand.

DFSS can be applied to technology development as well as product development. This is sometimes referred to as TDFSS—Technology Development for Six Sigma—and uses similar tools and methods as DFSS for a product, but with additional emphasis and tools for innovation (concept generation) and focus on transferring technology rather than verification of supply chain readiness in later stages.

There have been several cases in which DFSS has been applied to services; within Motorola, team leaders learned DFSS as applied to products, reviewed the DFSS flowchart followed in this book and selected and used a subset of the DFSS tools and methods. The tools they selected are consistent with published recommendations for applying DFSS to services.[1]

The examples in this book primarily relate to products and systems that involve electronics or software. Some electronic products, such as buzzers or music amplifiers or smoke alarms, might involve only hardware aspects. Hardware-only electronic systems might include input and output transducers, like a microphone or guitar for input and a loudspeaker for output, and electronic circuitry to modify the electrical signal—for example, amplifying the signal, as shown in Figure 4.1. Although "simpler" electronic systems like some music amplifiers may lack the complexity of software and hardware interactions, they can involve complexity in other respects; a debate between knowledgeable audiophiles on the performance of vacuum tube compared to solid state amplifiers can be a remarkable experience.

Other electronic products, such as computers and communication equipment, tend to incorporate both hardware and software. A complex electronic product that involves memory and computation, controlled by algorithms in software or firmware, is

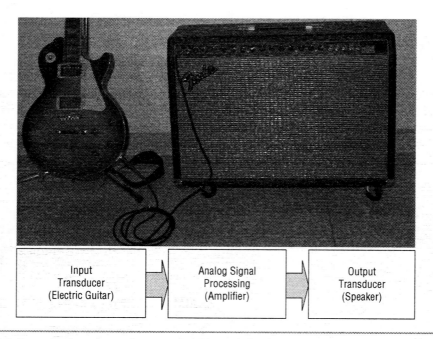

Figure 4.1 Example of an electronic system involving only hardware aspects

1. See, for example: Kai Yang, *Design for Six Sigma for Service,* McGraw-Hill, 2005.

described in Figure 4.2. The melding of software and hardware DFSS for a systems approach will work for classical electrical engineering systems such as cellular phones, base stations, and computers, but also for electromechanical and complex mechanical products with embedded electronic control systems such as jet engines.

Design for Six Sigma can also be applied to software alone; this application primarily arises if new software will be developed with an eye toward using the software on an existing hardware platform. This provides a supportive environment for iterative development, using Agile methods (discussed in Chapter 11), which benefits from the combination of rapid feedback from testing the software on the hardware or an emulator with the predictive engineering approaches on software technical parameters and quality attributes, as described in some examples and case studies later in this book.

In many situations, the new product development involves both software and hardware development. Although the DFSS flowchart reflects the common reality that requirements are usually flowed down and addressed by separate software and hardware

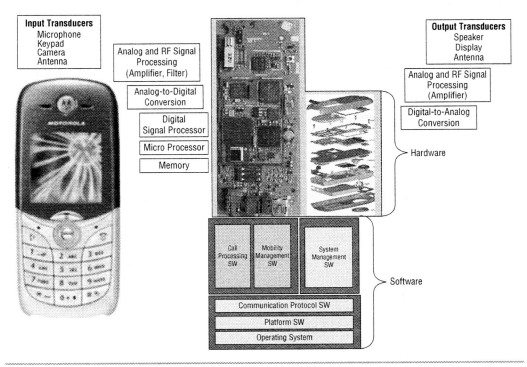

Figure 4.2 Cell phone as an example of a complex electronic system involving both hardware and software aspects

development teams, the fact remains that software—by its very nature—depends on the hardware, and the performance of the hardware tends to depend on the software. The hardware/software interactions and interfaces pose challenges and risks that can best be addressed in a straightforward manner—whether by integrating software and hardware teams and/or by using hardware emulation effectively, or by empowering special teams to handle these hardware/software interfaces and interactions.

The Design for Six Sigma approach for a simple electronic system is illustrated by Figure 4.3. The approach for a complex electronic system must allow the high-level requirements to be flowed down to hardware and software aspects that can be addressed by those with the appropriate expertise, while supporting the ability of the team to address the interactions and interfaces involved. The DFSS flow for a complex electronic system is illustrated in Figure 4.4, and will be discussed further in this chapter. Figure 4.5 shows the DFSS flow for a software development project using existing

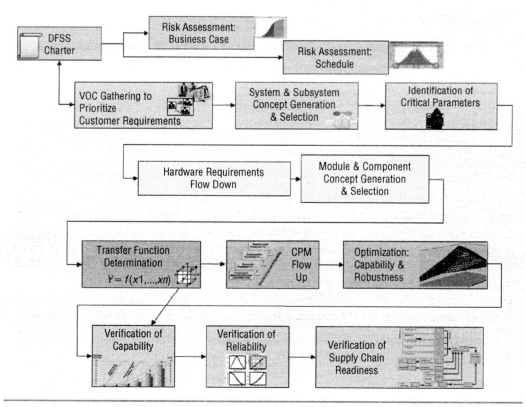

Figure 4.3 DFSS flowchart for a hardware-only electronic system

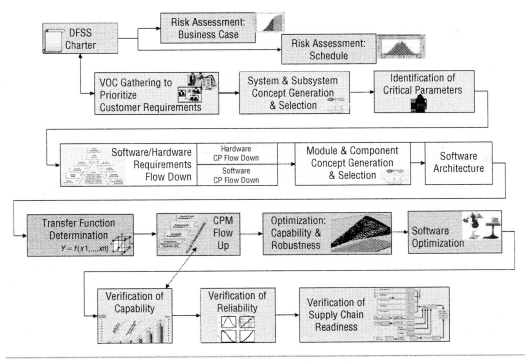

Figure 4.4 DFSS flowchart for a complex (hardware and software) electronic system

hardware. Comparison of Figures 4.3, 4.4, and 4.5 shows that the DFSS approach for a simpler electronic system can be treated as a subset of the approach for a more complex system, as can the DFSS approach for a software-focused development.

DFSS PROCESS NOMENCLATURES

For the process or business improvement and problem-solving aspects of Six Sigma, DMAIC has become recognized as the standard sequence of steps. By contrast, no clear standard process has emerged for DFSS. Alternative DFSS processes that have been proposed and used include: IDOV and CDOV, which involve the Identify or Concept phase, the Design phase, the Optimize phase and the Verify phase; DMADV and DMADOV, which share the Define, Measure, and Analyze phases of DMAIC, followed by Design, (Optimize), and Verify phases; and RADIOV, which was discussed with the case study in Chapter 1.

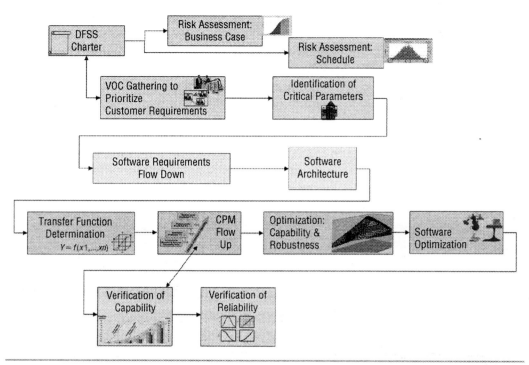

Figure 4.5 DFSS flowchart for a software development project using existing hardware

DFSS practitioners who feel comfortable with CDOV or DMADV can follow their preferred process with this book. The alignment of key tools and methods, and associated deliverables, to each of these processes is shown in Table 4.1, which is reproduced from Chapter 1. Figure 4.6 shows alignment of IDOV/CDOV with the DFSS flow, whereas Figure 4.7 shows a similar alignment for DMADV/DMADOV. The chapters of this book follow the DFSS flowchart, which can be viewed at http://www.sigmaexperts.com/dfss/.

For this book, the authors have chosen terminology that we hope is clearer for software and hardware people who might be unfamiliar with Six Sigma jargon, and might prefer a sequence of steps using terms with which they are more familiar. Systems, software, and hardware engineers commonly employ the terms requirements, architecture, design, integrate, optimize, and verify; using these terms reduce the need for memorizing a new set of terms, and also provide an easy mnemonic, RADIOV. The alignment of RADIOV with the DFSS flow, and with the sequence of chapters in this book, is shown in Figure 4.8.

Table 4.1 Key DFSS tools and methods associated with steps for CDOV, DMAD(O)V, and RADIOV

DFSS Step	DFSS Processes			Key Tools and Methods
	CDOV	**DMADV**	**RADIOV**	
DFSS Charter	Concept	Define	Requirements	DFSS Charter, Deployment Plan
Business Case: Risk Management				Monte Carlo Simulation—Business Case
Schedule: Risk Management				Monte Carlo Simulation—Critical Chain/TOC-PM
VOC Gathering				Concept Engineering, KJ Analysis, Kano Analysis, Interviews, Surveys, Conjoint Analysis, Customer Requirements Ranking
System Concept Generation & Selection				Brainstorming, TRIZ, System Architecting, Axiomatic Design, Unified Modeling Language (UML), Pugh Concept Selection
Identification of Critical Parameters		Measure		Quality Function Deployment (QFD), Design Failure Modes and Effects Analysis (DFMEA), Fault Tree Analysis (FTA)
Critical Parameter Flow Down	Design	Analyze	Architecture	Quality Function Deployment (QFD), Critical Parameter Management, Fault Tree Analysis (FTA), Reliability Model
Module or Component Concept Generation and Selection				Brainstorming, TRIZ, System Architecting, Axiomatic Design, Universal Modeling Language (UML), Pugh Concept Selection
Software Architecture				Quality Attribute Analysis, Universal Modeling Language (UML), Design Heuristics, Architecture Risk Analysis, FMEA, FTA, Simulation, Emulation, Prototyping, Architecture Tradeoff Analysis Method (ATAM)

continued

Table 4.1 Key DFSS tools and methods associated with steps for CDOV, DMAD(O)V, and RADIOV (continued)

DFSS Step	DFSS Processes			Key Tools and Methods
	CDOV	**DMADV**	**RADIOV**	
Transfer Function Determination	Optimize	Design	Design	Existing or Derived Equation, Logistic Regression, Simulation, Emulation, Regression Analysis, Design of Experiments (DOE), Response Surface Methodology (RSM)
Critical Parameter Flow Up and Software Integration			Integrate	Monte Carlo Simulation, Generation of System Moments Method, Software Regression, Stability and Sanity Tests
Capability and Robustness Optimization			Optimize	Multiple Response Optimization, Robust Design, Variance Reduction, RSM, Monte Carlo Simulation with Optimization
Software Optimization				DFMEA, FTA, Software Mistake Proofing, Performance Profiling, UML, Use Case Model, Rayleigh Model, Defect Discovery Rate
Software Verification				Software testing
Verification of Capability	Verify	Verify	Verify	Measurement System Analysis (MSA), Process Capability Analysis, McCabe Complexity Metrics
Verification of Reliability				Reliability Modeling, Accelerated Life Testing (ALT), WeiBayes, Fault Injection Testing
Verification of Supply Chain Readiness				Design for Manufacturability and Assembly (DFMA), Lead Time and On Time Delivery Modeling, Product Launch plan, FMEA/FTA for Product Launch

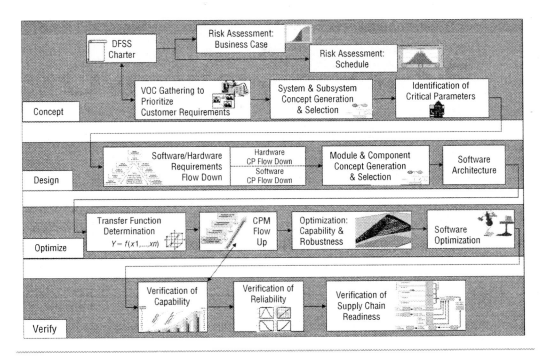

Figure 4.6 Alignment of IDOV or CDOV processes with the DFSS flow

REQUIREMENTS PHASE

During this early and key step, the requirements and critical competitive differentiators are identified and prioritized, leading to the selection of a superior product concept that is based on the voice of the customer (VOC). Implicit customer expectations and business expectations (VOB) are also incorporated as appropriate.

This sequence of steps includes gathering, understanding, and prioritizing the VOC, generating alternative concepts and selecting among these alternatives, and identifying measurable critical parameters requiring intense focus. Some key tools and methods for the Requirements phase are summarized in Table 4.2.

Some requirements may be stable, particularly those involving the "must-be" requirements, the basic functions of the product, and requirements that handle historical problems with similar products. Some other requirements—particularly those involving optional features—might change over the course of the project; as mentioned in Chapter 6, studies have indicated that software requirements typically

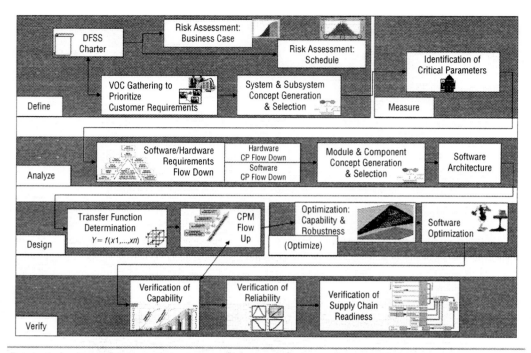

Figure 4.7 Alignment of DMADV or DMADOV process with the DFSS flow

change at least 25 percent.[2,3] Benchmarking of semiconductor companies found that products with a stable set of requirements required less than half as many iterations and were developed in approximately two-thirds of the time.

The general goal, then, should be to develop as close to a finalized set of requirements as possible during the Requirements phase at the beginning of the project, and then have an approach that can effectively respond to any changes, such as the agile development discussed in Chapter 11. At a minimum, the requirements developed during the Requirements phase need to be sufficient to provide guidance for the team when they select the architecture.

---

2. C. Jones, *Applied Software Measurement,* McGraw-Hill, 1997.

3. B. Boehm and P. Papaccio, Understanding and Controlling Software Costs, *IEEE Transactions on Software Engineering,* October 1988.

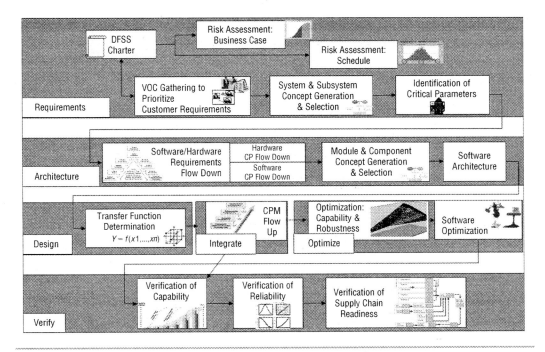

Figure 4.8 Alignment of RADIOV with the DFSS flow, as followed in this book

ARCHITECTURE PHASE

During the Architecture phase, the vital few system level requirements, called critical parameters, are flowed down to measurable requirements for hardware and software subsystems. Other key stakeholders are also engaged, including test engineering resources, to ensure that the requirements can be measured, tested, and evaluated in a production environment. Supply chain resources are also engaged in order to begin working toward assuring supply chain readiness, and the capability of the manufacturing processes to achieve the flowed-down requirements is analyzed. Concepts and architectures for the software and hardware subsystems are selected that support the required functions, features, and customer expectations flowed down from the system level requirements. Some key tools and methods for the Architecture phase are summarized in Table 4.3.

For the hardware aspects of the electronic system, the system level critical requirements are flowed down to measurable requirements at the subsystem, module or subassembly, and component level. Measurement system analysis (MSA) is performed on these floweddown, measurable requirements, and baseline or historical performance is assessed. If a

Table 4.2 Steps and DFSS tools and methods associated with the Requirements phase

DFSS Step	Key Tools and Methods	Purpose
DFSS Charter	DFSS Charter	Define the business case and expectations for the project
	Deployment Plan	Plan the tools and methods to be used, and support needed
Business Case: Risk Management	Monte Carlo Simulation—Business Case	Estimate confidence and prioritize risks to business goals
Schedule: Risk Management	Monte Carlo Simulation—Schedule	Estimate confidence and priortize risks to meet schedule
	Critical Chain/TOC-PM	Handle multiplexing, distractions, Parkinson's Law—schedule
VOC Gathering	Concept Engineering	Method to obtain & translate Voice of the Customer (VOC) to prioritized requirements
	Interviews	Gather VOC, rich stories and insights
	KJ Analysis	Filter many voices and images from interviews to a vital few
	Kano Analysis	Assess impact of customer requirements on purchase decisions
	Conjoint Analysis	Prioritize customer requirements and predict impact on sales
	Customer Requirements Ranking	Prioritize customer requirements
	System Level House of Quality	Translate prioritized customer requirements to system requirements
System Concept Generation & Selection	Brainstorming	Generate set of concepts that can help fulfill the requirements
	TRIZ	Generate concepts and overcome requirements trade-offs
	System Architecting Axiomatic Design	Generate concepts by partitioning the system using heuristics
	Functional Modeling, Unified Modeling Language (UML)	Create architecture-independent model with flows & functions
	Pugh Concept Selection	Select an optimal concept that helps fulfill the requirements
Identification of Critical Parameters	System Level House of Quality	Translate customer requirements to measurable requirements
	Design Failure Modes and Effects Analysis (DFMEA)	Assess and prioritize technical risks
	Critical Parameter Selection—Risk and Importance	Select key System Requirements for predictive engineering

Table 4.3 Steps and DFSS tools and methods associated with the Architecture phase

DFSS Step	Key Tools and Methods	Purpose
Critical Parameter Flow Down	Quality Function Deployment (QFD)	Find measurable requirements to fulfill customer requirements
	Critical Parameter Flow Down	Identify which subordinate parameters (y's), control factors (x's) and noises (n's) affect the Critical Parameters
	Fault Tree Analysis (FTA)	Flow down failures: determine how events can cause failures
	Reliability Modeling	Use system reliability model to focus on key reliability risks
Module or Component Concept Generation and Selection	Brainstorming, TRIZ, System Architecting, Axiomatic Design, Pugh Concept Selection	Generate concepts that fulfill module or component requirements, and select optimal concept
Software Architecture Selection	Quality Attribute Analysis	Measure quality aspects: availability, usability, security, performance
	Design Heuristics	Develop alternative software architectures using rules of thumb
	Simulation, Emulation, Prototyping	Evaluate functioning version of candidate architecture
	Architecture Tradeoff Analysis Method	Consider requirement trade-offs as selecting software architecture

prototype is developed during this phase, the goal for the prototype evaluation is to ensure feasibility—that functionality and performance can be achieved, without considering variations due to noises such as manufacturing variability, environmental conditions, or use cases. The selected concept is evaluated in terms of potential issues, with efforts to anticipate and prevent problems, using tools such as Design FMEA and FTA.

If the product involves only hardware, then the architecture may be defined in the system concept generation and selection step of the Requirements phase. Otherwise, the hardware architecture involves modules and components that can be defined during the module/component concept generation and selection step for the Architecture phase.

ARCHITECTURE PHASE FOR THE SOFTWARE ASPECTS

The goals of the Architecture phase for the software aspects are to flow down system-level requirements that the software must satisfy, and to select an excellent high-level software architecture that supports satisfying those requirements. Technical risks inherent in the product architecture are identified using techniques such as DFMEA and FTA. Critical parameters are passed to the Architecture phase, where quality attributes (non-functional requirements) are identified. Concept selection techniques aid in design trade-off analysis and in selecting the most cost-effective architecture for the product/platform. The selected architecture is evaluated using modeling, simulation, emulation, and prototyping. The architecture is documented, which is crucial for the subsequent detailed design. Architectural patterns, styles, and functional modeling facilitate the employment of standard solutions.

DESIGN PHASE

The term "design" in DFSS terminology might involve some confusion. In the CDOV and IDOV processes, the "Design phase" refers to the steps after concept selection and involving the flow down of critical parameters and functional performance for a beginning baseline. The term "Design phase" in the DMADV and DMADOV processes refers to a later stage involving the determination of transfer functions to be used in predictive engineering. For DMADV, the Design phase encompasses the entire predictive engineering effort, while predictive engineering is split among the Design and Optimize phases for DMADOV.

For the RADIOV process (as for the DMADOV process), the Design phase refers to the transfer function determination aspect of predictive engineering. Some key tools and methods for the Design phase are summarized in Table 4.4.

INTEGRATE PHASE

The integrate phase begins the process of combining and integrating the components. Integration is particularly important for the software aspects, as summarized in Table 4.5.

OPTIMIZE PHASE

During the Optimize phase, "flow up" is performed to assess the capabilities of the design allowing for variability in manufacturing, environment, and use cases. This flow up involves the latter aspect of predictive engineering, using the transfer function from the Design phase in conjunction with statistical modeling (such as Monte Carlo simulation). Transfer functions for the critical parameters and/or flowed down measurable

Table 4.4 Steps and DFSS tools and methods associated with the Design phase, for determination of the transfer function

DFSS Step	Key Tools and Methods	Purpose
Transfer Function Determination	Existing or Derived Equation	Use equation or mathematical model for predictive engineering
	Logistic Regression	Use equation for probability of event from history, experiment
	Simulation	Use a computer simulation program for predictive engineering
	Emulation	Try software functions with substitute hardware and/or software
	Regression Analysis	Develop equation for continuous parameter from history
	Design of Experiments (DOE)	Analyze planned experiment, identify significant factors,-> equation
	Response Surface Methodology (RSM)	Analyze planned experiment, develop polynomial for model

Table 4.5 Steps and DFSS tools and methods associated with the Integrate phase

DFSS Step	Key Tools and Methods	Purpose
Critical Parameter Flow Up	Monte Carlo Simulation	Use repeated runs with variations of x's to predict distribution of y
	Generation of System Moments Method	Use calculus, statistical variations of x's to predict distribution of y
Software Integration	Software Regression	Find software defects (after code changes) that impact functionality
	Stability and Sanity Tests	Check that software functions as expected and doesn't crash

technical requirements are determined using appropriate methods such as regression, DOE (design of experiments), and RSM (response surface methodology) for continuous parameters, and logistic regression for discrete parameters. Robust design and stochastic optimization approaches are used to build high confidence that critical parameters will meet expectations and have adequate capability, allowing for variations in manufacturing, use cases (and perhaps abuse cases), and environmental noises. Some key tools and methods for the Optimize phase are summarized in Table 4.6.

Verify Phase

At least three expectations should be quantified and assured during the Verify phase for electronic systems: that the product will meet performance expectations with high confidence, that the product will be reliable, and that the supply chain is ready and able to deliver the product dependably in the quantities and time frames required. Verification often includes both testing and evaluation to assess the capability and reliability of the product with pilot and early production samples. For software, testing is particularly important in assessing that the software functions as intended, over a range of conditions and scenarios. Key tools and methods for the verify phase are summarized in Table 4.7.

Table 4.6 Steps and DFSS tools and methods associated with the Optimize phase

DFSS Step	Key Tools and Methods	Purpose
Capability and Robustness Optimization	Multiple Response Optimization	Find set points for x's to co-optimize several y's
	Robust Design	Find set points for x's that render y much less sensitive to noises
	Variance Reduction	Choose and use a method to reduce variation of the y
	RSM	Analyze experiment, find set points for x's with polynomial model
	Monte Carlo Simulation and Optimization	Find set points for x's that reduce variation of one or several y's
Software Optimization	DFMEA, FTA	Assess, prioritize, and manage risks to software functionality
	Software Mistake Proofing	Anticipate and prevent errors on inputs, interfaces, processes
	Performance Profiling	Measure, analyze, benchmark function calls' frequency & duration
	UML, Use Case Model	Model functions by/for the user and other systems, and dependencies
	Rayleigh Model	Model and predict failures that will be observed after software release
	Defect Discovery Rate	Model and graph number of defects discovered over time

For hardware, the performance capability on an appropriate sample of early production can be used to verify that the optimized parameters perform consistently within acceptable limits, and approaches such as ALT (accelerated life testing) can provide confidence that the hardware aspects will meet reliability expectations. Approaches such as FIT (fault injection testing) enhance confidence in meeting software reliability expectations.

Supply chain resources anticipate and take preventative action for potential supply chain issues, and verify that the supply chain, including vendors and internal manufacturing and testing facilities, is ready with appropriate and dependable lead times. Some key tools and methods for the Verify phase are summarized in Table 4.7.

Table 4.7 Steps and DFSS tools and methods associated with the Verify phase

DFSS Step	Key Tools and Methods	Purpose
Software Verification	Software Testing	Provide customers confidence that the software works as intended
Verification of Capability	Measurement System Analysis (MSA)	Estimate variation introduced by measurement system
	Process Capability Analysis	Assess confidence y's will meet specifications despite variability
	McCabe Complexity Metrics	Measure software complexity, determine test cases needed for coverage
Verification of Reliability	Reliability Modeling	Predict and assess reliability
	Accelerated Life Testing (ALT)	Accelerate failure mechanisms to assess reliability in normal use
	WeiBayes	Provide lower confidence limit if reliability testing has no failures
	Fault Injection Testing	Inject most likely fault conditions into software code to detect issues
Verification of Supply Chain Readiness	Design for Manufacturability and Assembly (DFMA)	Design to simplify and immunize manufacturing and assembly
	Lead Time and On Time Delivery Modeling	Apply predictive engineering to a model for the supply chain to provide confidence in lead time and on time delivery
	Product Launch Plan	Develop supply chain and marketing plans for a successful launch
	FMEA/FTA for Product Launch	Assess, prioritize, and manage risks to the product launch

SUMMARY

DFSS has a wide range of applicability and can be relevant for a variety of engineering disciplines and interdisciplinary efforts. The RADIOV process provides a comprehensive process that is flexible enough for simple or complex projects for developing systems, products, technologies, and services. A DFSS flowchart can provide guidance for practitioners as they plan and follow through on developing a robust, reliable product. This chapter provided some perspectives on how the DFSS flowchart would align to various DFSS processes, including RADIOV.

Tables were provided to serve as a glossary of terms and to provide descriptions of the purposes of key tools used for each step of the DFSS process.

Portfolio Decision Making and Business Case Risk

POSITION WITHIN DFSS FLOW

Chapters 5 and 6 deal with two of the more common and pernicious types of risk: the risk that the product will be released late, perhaps missing the market window. . . and the risk that the product will not even have a market window to miss. This chapter will focus on the latter.

Product portfolio decision making is somewhat analogous to portfolio decision making for investments by a business or an individual. On an individual basis, investing all of one's hard-earned savings in one risk investment is considered very risky. Similarly, counting on one "home run," one miraculous product success, is very risky for a business. Portfolio decisions can propel the business forward—or can set the business back for years.

Whereas Figure 5.1 positions the evaluation of the associated business case risk early in the new product development process, this evaluation should be repeated periodically as circumstances, events, progress on the product, progress by competitors, and other risks evolve. A proposed new product that looked great six months earlier may have become a wasted effort, whereas a long-shot, "skunk works" project may become a home run. It can happen. . . and it has happened, many times. Chapter 3 discusses governance, which encompasses processes to reassess the project at various decision points during the product development effort.

Monte Carlo simulation provides a way to evaluate the business case and financial results for a new product or service, allowing for uncertainties. Monte Carlo simulation virtually "rolls the dice" thousands of times, and provides a histogram summarizing the probabilities associated with achieving a range of financial results. Monte Carlo

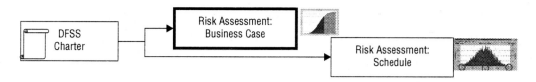

Figure 5.1 Assessment of the risk and opportunity for the business case begins early in the new product development process; however, it should be reviewed and updated as the project progresses and as circumstances change (favorably or unfavorably)

simulation allows someone to simulate experiencing versions of the future business results over and over again, with variations in business conditions and manufacturing costs for each version of the experience. Monte Carlo simulation is discussed in more detail in Chapter 14. Crystal Ball, from Oracle, is an Excel add-on that performs Monte Carlo simulation using formulas within spreadsheets.

Many executives are reluctant to kill a project once it is under way. Although this may seem considerate in a sense, there is a reasonable rationale supporting the need to "surgically remove" a doomed project from the product portfolio. Although the development team may initially go through the stages of mourning for a project ended prematurely, the alternative is investing a year or more of one's life working on a project that resembles a zombie: a "living dead" project that everyone intuitively knows is doomed, and yet keeps going, consuming resources. Working on a "living dead" project is demoralizing in the longer term. When a project is "surgically removed" from the portfolio, and after "mourning has broken," the development team will be ready to work on a more promising project—and, chances are, the team working on that more promising project could use a healthy injection of experienced resources.

PORTFOLIO DECISION MAKING AS AN OPTIMIZATION PROCESS

Portfolio decision making is an optimization process. Optimization involves maximizing or minimizing some measurable criteria, subject to certain constraints. For portfolio optimization, the measurable criteria will likely be a financial metric, and the constraint will likely be resources. Portfolio optimization fits into a class referred to as the "knapsack problem"; in a more interesting variation of this problem, a thief is looking to maximize the value of his "take" but is constrained by the limited capacity of his knapsack. Should he try to take the huge big-screen TV? Or should he consider the small but valuable jewels? The obvious answer ties to a key criteria: the thief would do best if he focused on items with the highest ratio of the measurable criteria he wants to optimize (dollar value) divided by the constraint involved (volume or weight). The ratio of value/volume or value/weight is high for the small, valuable jewelry. By analogy, projects with a high ratio of the financial metric to resource requirements would be preferred through portfolio optimization.

With new products, there are also strategic considerations involved. Some new product efforts may be tied to contractual obligations; others may be important for penetrating a new market or building the value of the brand name. These strategic considerations can be incorporated into the decision-making process. The trick lies in distinguishing between constraints, goals, and distractions.

FINANCIAL METRIC

Most enterprises are interested in optimizing some financial metric or metrics. Simultaneous optimization of several financial indices requires a complex approach such as defining a new optimization parameter that is a mathematical combination of several financial parameters. This increased complexity can be a source of confusion and error. For simplicity, it seems reasonable to select only one financial metric for optimization, such as:

- Revenue = Total Available Market (TAM) × Market Share × Unit Price   (5.1)
- Profit = TAM × Market Share × (Unit Price − Unit Cost/Yield) − Development Costs   (5.2)
- Percent Profit = Profit/Revenue   (5.3)
- Payback Time (in months) = Development Costs/(Profit/Month)   (5.4)
- Percent ROI (return on investment) = TAM × Market Share × (Unit Price − Unit Cost/Yield)/Development Costs   (5.5)

Optimizing the selected financial metric should generally lead to favorable business results, such as growth, viability, and increased shareholder value. By contrast, it should be difficult to find reasonable and realistic situations in which optimizing that financial metric would lead to unfavorable business results. Let's consider decisions involving choosing between projects, and decisions about pricing.

Imagine there are only enough resources available to develop one of two products. Both have similar prices, but one is expected to have twice the unit volume—with a cost that is about equal to the price. Some people might dub this a "loss leader." The other product, with half the unit volume, has a very high margin—a large difference between the price and the cost. The "loss leader" would maximize revenue, but would the enterprise be well-served to fill its portfolio—and its supply chain—with loss leaders, and downplay highly profitable products?

Some companies set goals for sales and marketing organizations, to maximize revenue—the presumption being they can be held responsible for revenue, over which they have some control or influence. How are profits impacted if pricing decisions are made to maximize revenue? Figure 5.2 shows the results of Monte Carlo simulations for revenue and profits when the price was set to optimize revenue.

If we assume that higher prices generally result in reduced market share, then there is a price that will maximize revenue. A selling price of zero could be expected to lead to high unit volume but zero revenue. Increasing the price would increase revenue, but an excessively high price would result in zero volume and thus take revenue back down to zero. Assuming a linear relationship between price and market share, the revenue can be optimized by setting the price midway between these two extremes (Figure 5.3). In this example, based on an actual product, the price set for optimizing revenue results in a loss rather than a profit. The optimal price for revenue is different than the optimal price for profits, and Figure 5.4 compares Monte Carlo simulations for revenue and profits when the price was optimized for revenue versus for profits.

Clearly, portfolio decisions based on optimizing revenue could lead in the wrong direction. By contrast, portfolio decisions and price decisions based on optimizing profits seem to consistently lead to favorable business results under reasonable conditions.

Similar analyses would show that decisions based only on the payback time, percent profit (the ratio of profit to revenue), or percent ROI (essentially the ratio of profit to investment) may drive the business toward smaller scale projects, each of which has a high ratio of profit to revenue or investment or a short payback time, but that could provide meager profits in total for the enterprise while tying up resources among a smorgasbord of small projects. If you had a choice between starting a project with an expected 15 percent return on $100,000 for one year, or a project with an expected 13 percent return on $100M for five years, which would you choose?

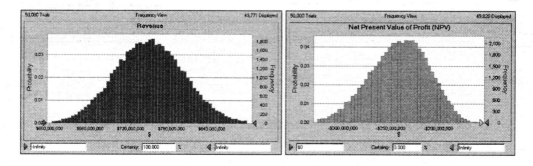

Figure 5.2 Monte Carlo simulation results for an actual business case where the selling price would be set to maximize revenue (left) but would lead to losses rather than profits (right)

In our personal lives, the best payback time, percent ROI, and percent profit items might be rebate coupons—two minutes of a person's time can lead to a net gain of several dollars. Yet, redemption rates are much lower than one would expect if people were making decisions based on these financial metrics. Decisions based on payback time encourage short term projects, and discourage strategic projects (such as research and development). Portfolio decisions based on optimizing payback time, percent profit or

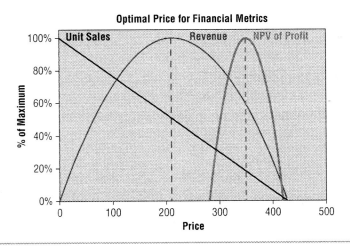

Figure 5.3 The assumption that market share varies linearly with pricing leads to two different relationships and two different optima for revenue and profit as functions of pricing, as illustrated here for an actual business case

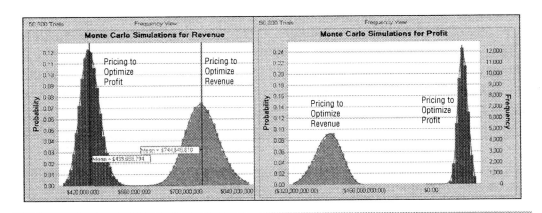

Figure 5.4 Monte Carlo results for a business case, comparing forecasted revenue (left) and forecasted profit (right) depending on whether the selling price would be set to maximize revenue or profit

percent ROI as the sole metric can lead you astray. These cases are summarized in Table 5.1, and collectively suggest that the best single financial metric for portfolio (and pricing) decisions is profit; the bottom line is the bottom line.

A variation on the use of profit is to use the net present value (NPV) of the profit, which allows for the time value of money. The return on investment of a million dollars tomorrow has more value than a return of a million dollars ten years from now. With NPV, the delayed return is discounted:

$$\text{Present Value of Cash Flow} = \frac{\text{Future Value of Cash Flow}}{(1 + \text{Discount Rate})^{\text{Number of years}}} \tag{5.6}$$

Setting the discount rate very high can result in a negative NPV for virtually any long-term project or research and development project. The discount rate should reflect a reasonable estimate of the discounted value for each year of delay in receiving a return. Rather than try to adjust for riskiness by using exorbitant discount rates in NPV calculations, the riskiness can be directly addressed using the economic commercial value (ECV):

$$\text{ECV} = \text{NPV}\{[\text{TAM} \times \%\text{MS} (\text{UP} - \text{UC/Y}) \times P_{CS} - C] \times P_{TS} - D\} \tag{5.7}$$

Where:
- TAM = Total Available Market (units)   %MS = Market Share (%)
- UP = Unit Price ($)   UC = Unit Cost ($)
- Y = Yield for the product (%) P_{CS} = Probability of commercial success (%)
- P_{TS} = Probability of technical success (%)
- C = Commercialization (launch, marketing) costs ($)
- D = Development costs remaining ($)
- NPV function adjusts the values to the present values using equation 5.6

The probabilities of technical success and commercial success are estimates, or perhaps more appropriately, guesstimates. The marketing team can come up with their best estimate of the likelihood that a new product or service will be successful in terms of market acceptance—for example, a new product in a new, unfamiliar market may be estimated to have a 25 percent probability of commercial success, whereas a new product in an established market where the company is dominant may have a 75 percent probability of commercial success. The engineering team can similarly estimate the likelihood that the product will be successfully technically. Guidelines for both of these probabilities can be found at http://www.sigmaexperts.com/dfss/.

Table 5.1 Comparison of projects and activities that might be preferred if each of several financial metrics were used as the sole criteria for optimization

Optimize	Project or Activity	Revenue	Profit	ECV	% Profit	Payback	% ROI
Revenue	Communication Device, Priced to Optimize Revenue	$750M	Loss	Loss	Loss	NA	NA
Profit	Communication Device, Priced to Optimize Profit	$425M	$75M	$37M	18%	5 months	455%
ECV	Communication Device, Priced to Optimize Profit	$425M	$75M	$37M	18%	5 months	455%
% ROI	Send in $25 Rebate Coupon	$25	$24	$21	96%	.25 month	2400%
% Profit	Loan coworker $10 Tuesday, get repaid $12 Friday	$12	$2	$1.80	17%	.10 month	17%
Payback Time	Loan coworker $10 Tuesday, get repaid $12 Friday	$12	$2	$1.80	17%	.10 month	17%

PORTFOLIO DECISIONS AND RESOURCE CONSTRAINTS

Assuming crisp decisions wherein either a project is fully committed (with adequate resources committed), or it is not committed, portfolio decision making can use a deterministic optimization approach called integer programming. Alternatively, portfolio decision making can involve an optimization approach, as illustrated in Figure 5.5. This approach considers the uncertainty of the terms in the model (TAM, market share, price, cost, and yield). In Figure 5.5, there are nineteen projects under consideration, and OptQuest was used with Crystal Ball to find a subset of those projects that optimize the expected (mean) NPV of the profits, allowing for variations in the profit based on the uncertainties.

	A	B	C	D	E	F	G
1	Project Nickname	Project Description	Profits - Likeliest	Profit Std Dev	Resources Required	Resources Applied	Selected?
2	Lancelot	Wireless LAN	67	5	50	50	1
3	Roadster 2	Telematics System with GPS (derivative)	14	1	15	15	1
4	Lancelot 2	Wireless LAN (derivative)	12	0.5	15	15	1
5	Untethered	ISM Band Cordless Phone	15	3	40	40	1
6	Chirper	CDMA Cellular Phone - Medium Tier	22	7	100	100	1
7	Europea	GSM Base Station	140	5	700	0	0
8	Ladybug	GSM Cellular Phone - Low Tier	12	0.6	70	70	1
9	Sybil	Quad Mode Cellular Phone - Medium Tier	23	0.2	170	170	1
10	G.I.F.T.	Global Infrastructure for Telecommunication	74	20	650	0	0
11	Cadmium	CDMA Base Station	45	4	900	0	0
12	Boldringer	Multi Mode Cellular Web-phone - High Tier	100	25	1000	0	0
13	Telepath	Future Cell Phone	30	20	100	100	1
14	Gipper	GPRS Base Station	55	4	70	70	1
15	C BSCuit	GSM/GPRS/EDGE/WCDMA BSC	45	3	170	170	1
16	HUB	Home UltraWideband Network	20	15	40	40	1
17	*T Rex*	*Customizable Telecommunication Unit*	*75*	*14*	*50*	*50*	*1*
18	Websurfboard	Circuit Board for VOIP and Data Comm	150	25	650	0	0
19	CARgonaut	Automotive Electronics System	15	4	5	5	1
20	Lifeline	Medical Telemetry System	20	2	15	15	1
21		Total for Selected Projects	425		910		

Figure 5.5 Results of optimization using OptQuest with Crystal Ball software to find the subset of 19 projects under consideration that can be selected to optimize the expected value for the total NPV of profits for the selected projects, subject to a resource constraint of 1,000 people. Columns C and D describe the distribution for profit for each project, if selected, in millions of dollars. Projects selected for resourcing in the optimal portfolio are designated with a "1" (and shaded) in column G (the column titled "Selected?").

For portfolio decisions, the decisions are "Go/No Go": either do a project, or don't do a project. In the portfolio optimization example shown in Figure 5.5, the decisions are carried out in terms of allowing one or the other of two levels of resourcing. Either all of the resources required (column E) are applied (column F), or zero resources are applied. Applying all of the required resources is equivalent to selecting the project (column G), and applying no resources is equivalent to choosing not to do that project.

OptQuest tries a potential portfolio (a combination of projects selected and resourced), checks that the total resources required for all of the projects in that potential portfolio is feasible (that is, doesn't exceed the constraint), and if feasible, runs Monte Carlo simulation for that combination to determine the mean of the distribution of results. Then, OptQuest assesses the mean compared to prior results with other potential portfolios, and uses an algorithm to systematically explore other possible portfolios. At some point, OptQuest detects that the results are converging towards one optimal portfolio, and recommends that portfolio of projects (shaded cells in column G of Figure 5.5). Alternatively, OptQuest can be used to optimize another parameter, such as ECV or the probability that the NPV will exceed a lower limit (such as the profit committed to upper management).

GOALS, CONSTRAINTS, CONSIDERATIONS, AND DISTRACTIONS

The discussion so far indicates that there should only be one financial goal, which should be related to the goal of profitability—such as the NPV of the forecasted profit (equation 5.2, incorporating the time value of money represented by equation 5.7). Other risks, such as schedule slippage, could similarly be incorporated into NPV or ECV as the financial goal (albeit with some additional complexity).

The resources or the investment requirements can be treated as constraint(s). This is straightforward, but the assumption that a project can only be either fully committed and fully resourced, or not committed at all also assumes considerable self-discipline and maturity among the decision makers and the leadership team.

Other considerations, often referred to as strategic considerations, are often brought up in this project selection process. The first step is to filter these between important considerations and distractions (or counterproductive detractors). Examples of important strategic considerations might include a proposed new product that could be vital for penetrating a promising new market; although this first new product in this promising new market might not have stellar financial forecasts by itself, opening the new market might be a brilliant long term strategy. Another strategic investment might involve developing and building something like a "concept car"—a working, impressive prototype that may never lead to production, but that would provide highly valuable publicity, build the value of the brand name, perhaps lead to increased sales of other products in the short term, and perhaps lead to variations on the "concept car" prototype that can provide sales in the longer term.

Some strategic considerations have often proven to be more of a distraction than a smart investment. One author has observed several efforts on developing new products in order to "fill a factory"; a manufacturing facility or "factory" is projected to remain underutilized, posing a drain on the business enterprise, inflating manufacturing costs for the few products that use the underutilized factory, and scaring off other potential products that might similarly face discouraging projected manufacturing costs.

In one effort to "fill a factory," a business unit looked at a high-volume, high-margin power management chip for GSM cell phones that was being sole sourced by a competitor. The business unit decided to try to become a second source for the chip, and allocated a large and talented team of design engineers to design this chip. The competitor and prime source naturally wanted to protect its 100 percent market share for this chip, and decided on a strategy. As the business unit worked on designing the chip, matching the specifications for the competitor's power management chip, the competitor developed the next evolution of the chip, changing the specifications that the second source would need to chase in order to be acceptable as a second source for the illusive sale. While the second source was still chasing, the prime source would be on the next stage of evolution.

This illustrates a key insight: It is far better to be the matador than to be the bull.

Other strategic efforts that proved to be distractions have included designing commoditized products in hopes of becoming an additional supplier for a high-volume commodity product. This is a situation in which neither the management nor the development team were engaged, energized, nor focused on developing a (yawn) "me-too" product that did not align with the vision and future for the business.

Some strategic considerations involve contractual obligations; others may be important for penetrating a new market or building the value of the brand name. These strategic considerations can be incorporated into the decision-making process. Once again, the trick is in distinguishing among constraints, goals, and distractions.

ADJUSTING PORTFOLIO DECISIONS BASED ON EXISTING COMMITMENTS AND THE ORGANIZATION'S STRATEGIC DIRECTION

On reflection, and often with considerable discussion and argument, some of the strategic considerations may be recognized as being distractions—as illustrated with the "factory filling" stories. In general, if the strategic consideration does not align with the current product thrust, nor with major future thrusts, and if the strategic consideration puts the company's efforts at the tender mercies of one or more competitors (see "the matador and the bull" analogy earlier), then the strategic consideration is almost certainly a distraction.

Once the distractions have been recognized and filtered from the set of considerations, there generally remain **three** major complications to prioritization and selection:

1. Some new product efforts may be tied to contractual obligations, and there is no option to terminate the effort. This situation is readily incorporated into the optimization process by simply agreeing that this project must be completed, and deducting the associated resources from the available pool.

2. Projects under way may involve customer expectations rather than commitments; terminating the project may have a major negative impact on relations with that customer—and customer retention is vitally important. Studies have generally found that it is about five times more difficult to gain a new customer than to keep an existing customer. This situation is discussed in a sequence of steps to be described shortly.

3. The final situation involves strategic projects to penetrate new markets, develop and integrate high-risk and potentially disruptive technologies in the product, or to build the value of a brand name. The first case may be the lifeblood for growth of the business—penetration of new markets (that is, new to the enterprise—not necessarily totally new markets from a global sense). Such efforts may require resources and, perhaps, some degree of secrecy. Development and integration of high-risk and potentially disruptive technologies may be similar; a primary caveat is that new products driven by customer needs and desires tend to be more successful in the market than new products driven by "technology push." If the technology aligns with such needs and desires, they may provide a substantial competitive edge. Strategic efforts to build the value of a brand name includes efforts that, in the automotive industry, are referred to as "concept cars"—developing fascinating cars that may never reach production as-is, but which elicit favorable publicity and recognition that may benefit the value of the brand name, and potentially assist with sales of related products that are in production or are planned for production.

The level of risk associated with these projects may not lend themselves to comparison with and selection from among projects associated with established markets. Consequently, the recommended approach would be to segregate such strategic projects, and prioritize among them. The highest priority project(s) should be selected, resourced, and perhaps draped with appropriate levels of secrecy. In terms of resourcing, determine the number and types of resources needed to make the highest priority projects successful; underresourcing high-priority projects leads to frustration, resources working excessive hours, and a disquieting atmosphere in which some key people wonder whether management really cares about the success of the project, if they are unwilling to provide appropriate resourcing.

Here is the suggested sequence of steps for a project portfolio that includes a set of projects already underway:

1. Review the current portfolio in terms of customer commitments and flexibility
2. Prioritize products currently in development in declining order of the ratio:

$$\frac{\text{(Current projected profit–remaining expenses)}}{\text{Remaining resource or investment requirements}} \qquad (5.8)$$

3. For highest priority projects, determine "resource for success" levels; assess resource gaps; assess probabilities of commercial and technical success
4. Develop Monte Carlo simulation models for the projects and the portfolio
5. Identify "floundering projects" through formal reviews, and provide criteria and conditions for a final decision whether to continue or end the project
6. Use Monte Carlo simulation and optimization to find an optimal portfolio; prudently prune the portfolio as necessary; prioritize the selected projects
7. Support and resource the committed and highest priority projects for success

SUMMARY: ADDRESSING BUSINESS CASE RISK

This chapter addressed a key challenge in most enterprises: which projects to pursue or stop pursuing.

Several key aspects were considered and several recommendations provided:

1. Distinguish among constraints, goals, and distractions.
2. The goal for optimization will generally be a financial metric; NPV of profit and ECV, a risk-adjusted version of NPV of profit, would be good choices for financial metric goals for purposes of optimization.
3. Whereas pricing decisions involve several considerations, including customer expectations and expected and anticipated actions by competitors, the decision makers should be aware that the price for optimizing revenue is quite different from the price for optimizing profit.
4. Optimization programs, such as OptQuest, can be used in conjunction with Monte Carlo simulation to assess risk and find the optimal portfolio of projects to support.
5. If a project is selected using these decision criteria, it is critical to resource it for success.

Project Schedule Risk

POSITION WITHIN DFSS FLOW

Among the risks associated with new product development that were discussed in Chapter 3, perhaps the risk that has the largest ongoing influence and impact on the new product development is the schedule risk. Like an uninvited ghost at a dinner party, it can haunt the entire new product effort, doing its level best to deflate the enthusiasm, drain the vigor, and inject fear and trepidation among the team.

Project schedule risk is also a major barrier to buy-in for Design for Six Sigma, as discussed in Chapter 2. Managers, project managers, and team members may be concerned that the rigorous methods involved with Design for Six Sigma might require too much time and lead to schedule slippage. Consequently, this chapter seeks to confront this issue, provide a different, hopefully enlightening perspective to project schedule risk, and propose ways to address schedule risk so that it is handled more as a manageable key project deliverable and less as a haunting experience.

Project schedule risk should be addressed in the initial stages of DFSS, before or in parallel with the Concept phase of RADIOV or CDOV, the Define phase of DMADV or DMADOV, or the Requirements phase of a software product development effort.

PROJECT SCHEDULE MODEL

To some extent, the challenge of meeting the committed project schedule can be treated like a critical parameter: a continuous, measurable parameter that is important to the customer (or the business, or both) and that involves challenges or risks, as discussed in Chapter 9. Just as critical parameters can be flowed down and then flowed up with a mathematical model or transfer function to predict the probability distribution of results, the overall project duration can be flowed down to the individual tasks and associated resources. At first glance, the transfer function might be expected to be a simple additive function for all tasks that reside in the critical path. However, there are a few subtleties, each of which can have a big impact on the schedule. These will be explored in conjunction with a slightly more complex additive model (see equation 6.1).

Total New Product Development time = (6.1)
  Time to Make Decision ("fuzzy front end")
+ Time for First Pass (Design to Prototypes or Evaluation samples)
  which includes:
  Σ(Committed Durations for each task on Critical Path)
  + Delay caused by skewed duration probability distributions
  + Σ(Delays caused by changing requirements)
  + Σ(Delays caused by Parkinson's Law)
  + Σ(Delays caused by multitasking)
  + Σ(Delays caused by wandering critical path and resource dependencies)
+ Number of Iterations $\times$ Duration for each Iteration
+ Time for Qualification and Release to Production

Each of these terms will be discussed in the following sections. "Parkinson's Law" refers to the tendency for activities to fill all the time made available to complete the activity. Multitasking refers to key resources—people—trying to do several activities at once, or switching back and forth between tasks. Wandering critical path refers to the situation in which two or more sets of deliverables from parallel chains or paths of activities are required to be completed before the project can continue. If two or more of the parallel paths take almost the same time, the critical path—the longest chain of activities that are required for project completion—might change as one path gets delayed and becomes the bottleneck.

Let us begin this discussion with an insightful, yet somewhat sardonic quote:

"Planning is an unnatural process. It is much more fun to do something.
The nicest thing about not planning is that failure comes as a complete surprise,
rather than being preceded by a period of worry and depression."
—Sir John Harvey-Jones

The "Fuzzy Front End" and Delays Caused by Changing Requirements

The term "fuzzy front end" has been used to describe the time it takes to review and assess an opportunity, and then make a decision to begin the new product development effort in order to pursue the opportunity. The fuzziness refers to both the uncertainty as to the duration of the phase (uncertainty as to when an opportunity is recognized), and also to the exploratory process that ends in a decision, but that often seems to be unfocused. Interestingly, a study of semiconductor suppliers found that the fuzzy front end took an average of about six months—perhaps particularly surprising as the semiconductors are moderately far down the food chain, where the semiconductor products are supplied to companies that incorporate the semiconductors in products for the ultimate customers. Consequently, the semiconductor suppliers are at least one stage removed from the market risk of the electronics companies, and often receive orders for the product wherein the electronics companies assume some portion of the market risk and provide estimates of the demand by time frame.

The time consumed by the fuzzy front end becomes more predictable, more manageable, and—very possibly—considerably shorter if a product/project activation or decision-making process is defined and followed. This could be a decision-making process for a single project, or a decision-making process for the portfolio of future products—the product roadmap—as discussed in Chapter 5.

During the fuzzy front end, a serious effort should be made to develop a set of requirements for the new product. A study of several new product development efforts showed that projects with early and sharp product definitions succeeded 85 percent of the time, whereas projects with relatively poor product definition succeeded only 26 percent of the time.[1] Indeed, one of the coauthors experienced an illustration of this when he was involved in two parallel product development efforts for virtually identical products by two customers, each of whom requested a custom product—such that the two products were developed in parallel by two virtually equivalent product development teams that were not allowed to share information or insights. One customer had a clearly defined specification that was handed to the corresponding team early in development, and never changed. The other customer changed the specifications about a dozen times, each time requiring reinitiation of detailed design activities. Can you guess which project resulted in a first-pass success, delivered on schedule?

---

1. R.G. Cooper and E.J. Kleinschmidt, *New Products: The Key Factors in Success,* Chicago: American Marketing Association, 1990.

Requirement changes may simply be a fact of life for complex products. Studies have indicated that software requirements typically change at least 25 percent.[2,3] Two approaches can be considered: using rigorous up-front requirements capture and management to develop clear, consistent, and hopefully unchanging requirements . . . or using an iterative, flexible process that can effectively respond to changing requirements. The first approach can use the concept engineering method described in Chapter 7, whereas the second approach is generally referred to as agile development in the software world, although the approach could be equally relevant for both hardware and software development. Set-based concurrent engineering, as applied by Toyota product development, similarly deals responsively and flexibly to changing requirements in hardware and software aspects. Agile development will be discussed in Chapter 11.

TIME FOR FIRST PASS: CRITICAL PATH VERSUS CRITICAL CHAIN

Without a statistical approach to modeling the new product development time, a project schedule is often based on the sum of most likely durations for the sequence of tasks in the critical path. There are several issues with this approach, each of which will be described and illustrated with the example shown in Figure 6.1.

DFSS CDOV Project		Durations-Triangular Distributions		
Phase	**Key Deliverables**	**Minimum**	**Most Likely**	**Maximum**
Concept	Customer Selection	1	2	3
	VOC Gathering	30	40	50
	KJ Analysis	1	4	20
	Customer Prioritization	5	15	20
	Pugh Concept Selection	1	1	10
Design	QFD and CPM Flow Down	5	20	40
	FMEA	1	4	10
Optimize	DOE Designed Expt	5	10	20
	RSM	5	10	20
	Robust Design	5	10	20
Verify	Reliability Evaluation	5	30	45
	Capability-CPM Flow Up	2	5	10
	Total Duration Expected	**66**	**151**	**268**

Figure 6.1 Project schedule example: critical path for a simple product, using proven components. Task durations for each key deliverable are estimated with three values: minimum, most likely, and maximum duration.

2. C. Jones, *Applied Software Measurement*, McGraw-Hill, 1997.

3. B. Boehm and P. Papaccio, Understanding and controlling software costs, *IEEE Transactions on Software Engineering*, October 1988.

There are several things to note from the start: if the requirement is to complete the product in 180 days, then the sum of most likely estimates for the critical path tasks is a favorable 151 days, whereas the "worst case" sum of maximum durations for tasks would extend considerably beyond the requirement for project completion.

As shown in Figure 6.2, the Monte Carlo simulation indicates a 90 percent probability of the product being completed within 180 days. Interestingly, the average duration from the Monte Carlo simulation is 162 days—considerably longer than the 151-day estimate that would be expected from the sum of most likely durations of critical chain tasks. This difference in the expected means is a result of the skewed distributions for task durations: some tasks shown in Figure 6.1 are not symmetrical, in that there is a longer tail to the right, indicating more time between most likely and potential late completion than between most likely and potential early completion. Because this lack of symmetry is common for tasks in a project, it is very common that the sum of most likely estimates will underestimate the most likely project duration that would be modeled using Monte Carlo simulation—and this underestimate is commonly problematic, putting the team in an unanticipated schedule performance deficit from the get-go!

Monte Carlo simulation of the project schedule also provides insight into the major sources of schedule uncertainty, through sensitivity analysis as shown in Figure 6.3. The sensitivity analysis can suggest where the project team might need some additional

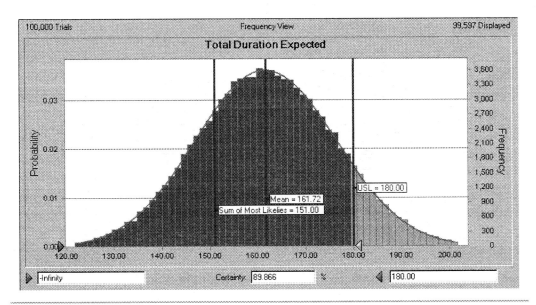

Figure 6.2 Monte Carlo simulation for the project schedule example. The more lightly shaded area is beyond the 180-day limit. The mean is more than the sum of most likely task durations.

DFSS CDOV Project		Durations - Triangular Distributions		
Phase	**Key Deliverables**	**Minimum**	**Most Likely**	**Maximum**
Concept	Customer Selection	1	2	3
	VOC Gathering	30	40	50
	KJ Analysis	1	4	20
	Customer Prioritization	5	15	20
	Pugh Concept Selection	1	1	10
Design	QFD and CPM Flow Down	5	20	40
	FMEA	1	4	10
Optimize	DOE Designed Expt	5	10	20
	RSM	5	10	20
	Robust Design	5	10	20
Verify	Reliability Evaluation	5	30	45
	Capability - CPM Flow Up	2	5	10
	Total Duration Expected	**66**	**151**	**268**

Figure 6.3 Sensitivity analysis from Monte Carlo simulation for the project schedule example. For this example, reliability evaluation contributes more than a third of the schedule variation.

resources or help in order to reduce the uncertainty in the duration for the key tasks highlighted as the main sources of schedule variability.

Delays as a result of Parkinson's Law ("Work expands to fill the time available for its completion")[4] can perhaps be best understood in a personal sense. Imagine that you have just been informed that you need to give a presentation to your boss, and his boss, and his boss' staff in two days. Being honest with yourself, when will your presentation be ready? If you are like most people, your honest answer will be that you'll still tweak, modify, and try to perfect the presentation until you run out of time.

Similarly, if they consider their tasks to be very important, the people working on their tasks will likely also want to keep fine-tuning, reevaluating, resimulating, and improving their work up until they run out of time. If their tasks are not considered to be quite as important, they may work on it for a while, put it on the shelf while they work on other critically important things, then get back to it just before they run out of time—inter-

---

4. C.N. Parkinson, *Parkinson's Law: The Pursuit of Progress,* John Murray, London, 1958.

ruptions and multitasking. It may also be similar to how many students deal with an assignment due for class on Monday morning—they may start it shortly after it's assigned but won't really finish it until the night before, or perhaps just before it is due—procrastination. Also, there is a perceived, and perhaps real, downside to completing a task early: it sets the expectation for the next time—so if a resource is going to complete a task early, there is an incentive to keep it quiet or put the project on the shelf until the committed date arrives. These behavior patterns are perhaps the primary reasons behind the phenomena referred to as Parkinson's Law. These phenomena mean that, if you plan for a task to be completed between a best case and a worst case timing, but give the resource involved in the task a commitment to a target completion date or duration, then the task is unlikely to be completed earlier than the committed date or duration. In a statistical sense, the left side of the probability duration for task durations is lost.[5]

Even if a task is occasionally completed early, it is unlikely that the project schedule will receive full benefit. The impact of Parkinson's Law on project schedules combines with other phenomena that tend to limit the benefit of early completions of tasks. The resource for the next task is probably busy, and won't interrupt to work on something that arrives earlier than expected. Furthermore, if the task that's completed early is near the beginning of the project, the schedule would show no issues, so there might be no strong drive toward starting the next task early. These effects can be simulated with Monte Carlo simulation: if the random value for the task completion is earlier than the committed duration or date, then replace the value with that committed duration or date; otherwise, use the random value. The impact of Parkinson's Law for the project schedule example is shown in Figure 6.4; whereas it tightens the distribution of project duration, it also shifts the mean about two weeks later for this example.

At least two other phenomena are mentioned in critical chain[6] (also known as theory of constraints) project management. Multitasking can impact the schedule by both distracting critical resources who should focus on completing their tasks on the critical path or critical chain, and also by those resources jumping from project to project or task to task (analogous to interleaving in communication protocols) in such a way that most projects are delayed compared to the completion that could be achieved if each resource remained focused on their task until completion, as illustrated in Figure 6.5.

The term "critical chain" refers to the longest chain of task *and* resource dependencies in a project, contrasted with the term critical path, which refers to the longest chain of task dependencies. If a task can only be completed by a certain person or

5. E. Goldratt, *Critical Chain,* North River Press, 1997.

6. C.N. Parkinson, *Parkinson's Law: The Pursuit of Progress,* John Murray, London, 1958.

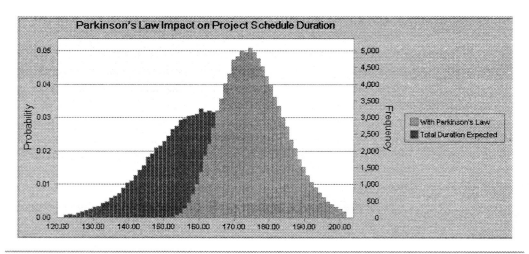

Figure 6.4 Monte Carlo simulation for the project schedule example, including a model for the impact of Parkinson's Law

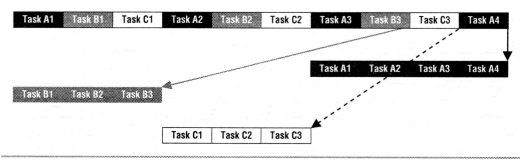

Figure 6.5 Multitasking among N projects can delay N – 1 projects compared to staying focused

with a certain resource, then the project will be delayed until that resource becomes available. Parallel to the critical chain might be several feeder paths; classical critical path methods assume that the critical path dominates the duration of the project, regardless of the feeder path durations. However, if one or more feeder paths have durations that are nearly as long as the critical path duration, then the project will encounter a phenomenon referred to as "wandering critical path," where the critical path might change or not change in response to variations in task durations and other events. The critical chain or theory of constraints project management (TOC-PM) approach uses buffers for the feeding paths to prevent this phenomenon.

CRITICAL CHAIN/THEORY OF CONSTRAINTS PROJECT MANAGEMENT BEHAVIORS

Some of the key behaviors involved in the critical chain/theory of constraints project management approach are summarized in Figure 6.6, and discussed in outline form here. More information is provided at http://www.sigmaexperts.com/dfss.

a. **Plan backward:** The project manager builds the project plan starting from the completion goal and listing the tasks and associated deliverables needed to achieve the completion goal. He or she checks whether each of the deliverables is necessary and whether the set of deliverables will be sufficient to proceed to the goal. Then, the project manager takes each of those tasks and works backward, listing tasks and associated deliverables required and checking whether that set is necessary and sufficient. This process continues backward in time until the present-day situation is reached.

Conventional Approach	Critical Chain/TOC-PM Behaviors
Plan forward	Start with completion goal, plan backward w/ tasks Necessary and Sufficient for completion
Identify the Critical Path (longest chain of task dependencies)	Identify the Critical Chain (longest chain of task and resource dependencies)
Multitask between tasks in several projects 	Synchronize projects so resources can focus on one task at a time 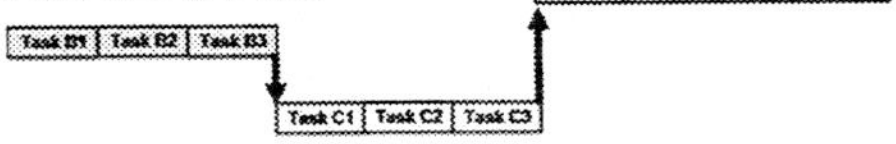
Personal Buffers 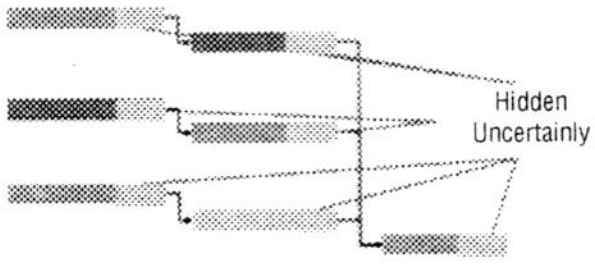	Aggregate personal buffers → Project Buffer 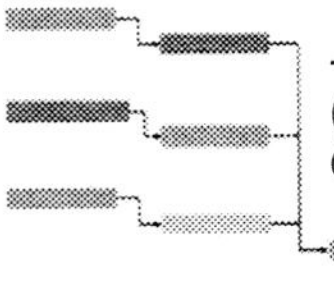 -Use it as control mechanism (react only as buffer nears depletion) 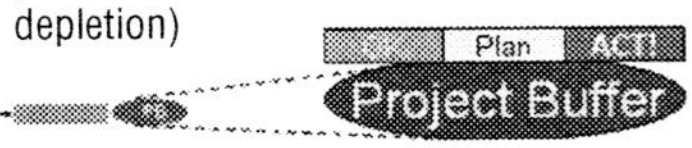
"Train Schedule" Behavior 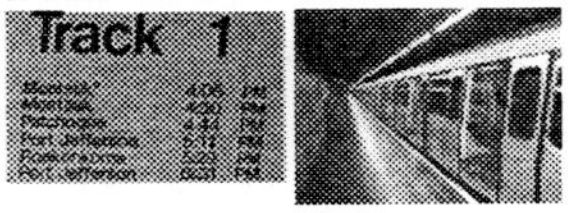	"Relay Runner" Behavior 

Figure 6.6 Summary of critical chain or theory of constraints project management approaches

b. The **critical chain can be identified** by project management software such as Microsoft Project if the project manager identifies the set of resources that can perform each task. Some tasks can only be performed by a limited set of resources and, in some cases, only one expert can perform certain tasks successfully. If that one expert or that limited set of resources will be tied up on other tasks, the project will wait for key resource availability.

c. Take actions necessary to **minimize multitasking;** free resources engaged in tasks on the critical chain from distractions, meetings, or monthly status presentations. Synchronize the projects to minimize and avoid resource contentions between projects through these steps:

1. Start with the **critical chain schedules for the highest priority projects.**
2. Select the most heavily loaded resource as the **synchronizer.**
3. **Stagger the contending projects based on the availability of the synchronizer:** Priority 1 synchronizer task goes as far back in time as feasible.
    Layer each individual resource of each resource type, add sequential task times, and subtract the gap/buffer between. A buffer will need to be added to a project as a synchronizer buffer, shifting that project later in time, to avoid resource contentions.
4. If necessary, **add a synchronizer buffer** to the entire project.
5. Review the projects and **set project commitments.**

d. **Aggregating personal buffers** involves a human aspect: building trust with the resources such that they may give up their personal buffer, and a technical aspect: **developing an appropriate shared buffer:**

1. Communicate with the development team resources the plan and rationale for using this critical chain/TOC-PM approach, and the need for moving from personal buffers to an aggregated project buffer that the team will share.
2. Using a script as shown at http://www.sigmaexperts.com/dfss/ get two duration estimates for each task: aggressive but possible (ABP) and highly probable (HP).
3. Let the program management software determine the appropriate project buffer, or use Monte Carlo simulation to estimate the aggregation of project buffers.
4. Have resource managers and project managers fulfill the trust they are building with the development team and resources, and ensure that the project benefits from this approach:
    - **Resources involved with critical chain tasks act as "relay runner" resources:** Begin work as soon as assigned, work without interruption until done, and announce the finish immediately when task completion criteria have been met.
    - **Managers review and update plan with the development team, and resources.**

- **Managers must know if a task is finishing early, running late or on schedule.**
  - **Regular if not "daily" updates to** critical chain tasks
  - **Regular if not "daily" updates** to the "next" resource in the critical chain
- If activities slip (from estimate), **managers must:**
  - **Check what went wrong**
  - **See what can be done to pull in the schedule**
  - **Managers MUST NOT "shoot the messenger" or punish!**
    - Consider the possibility that delays are attributable to normal variability.

e. **Use the project buffer as a control mechanism.** If the projected completion date (provided by the project management software and confirmed through Monte Carlo simulation of the schedule) encroaches on the first third of the buffer, consider that to be normal variability—that is the reason there is a shared project buffer. Once the projected completion date encroaches on the second third of the buffer, develop contingency plans and become very aware of issues and distractions that may be impeding the efficiency of the resources. The moment the projected completion date encroaches into the final third of the buffer, act based on the contingency plans.

More details and some success stories regarding critical chain/theory of constraints project management can be found at http://www.sigmaexperts.com/dfss/.

ITERATIONS, QUALIFICATION, AND RELEASE TO PRODUCTION

People who have been involved in new product development projects are generally surprised when they see a distribution of the number of iterations it has taken, historically, to complete projects. Iterations may be complete redesigns, or redesigns of subsystems or components, or board spins for printed circuit boards. For components such as integrated circuits, the average number of iterations (defined as a new set of semiconductor wafers going through semiconductor processing with a new set of photo-masks) is three. For portable devices, it is common to see between ten and twenty board spins from first prototype to production release.

If the development process is intended to be once-through (a "waterfall" model in software terminology), then equation 6.1 might be a fair mathematical model in which the first-pass prototype is modeled as a sum of task durations, supplemented by "build-test-fix" iterations, where the uncertainty is represented by both a distribution of the time required for each iteration and the discrete number of iterations that would be required, plus the time required for qualification and release to production.

By contrast, if the development process involves iterative development, a variation might be required in the model for development time (as described by equation 6.1). Iterative development for software is often referred to as agile development; agile methods include scrum, extreme programming, and unified process.[7] An analogous approach for hardware development might be referred to as set-based concurrent engineering.[8] With iterative development, the model for development time might involve the development time per iteration, which might be fixed, as some agile methods involve setting a fixed time per iteration. In this case, the uncertainty might be represented with a distribution for the discrete number of iterations required, plus the time for qualification and release to production. As mentioned earlier, the term "agile" refers to the responsiveness to changing requirements. Moreover, agile development typically involves a colocated team with a defined schedule for deliverables; hence, relay race behavior is readily accepted and encouraged, reducing the impact of Parkinson's Law and multitasking. Agile and iterative development will be discussed further in Chapter 11.

SUMMARY: ADDRESSING SCHEDULE RISK

Schedule risk can be modeled to help understand the impact of uncertainty and variation, prioritize the risks associated with individual tasks, and evaluate alternatives. Some other phenomena involving human behaviors and changing requirements can also impact the schedule, providing a deleterious tendency toward lateness. Approaches such as critical chain/theory of constraints project management and agile development provide some ideas on how to model and manage risks associated with these and other phenomena. The predictive engineering perspective associated with Design for Six Sigma suggests that tools such as Monte Carlo simulation can help with modeling, analyzing, and improving confidence for the project schedule estimates.

---

7. Craig Larman, *Agile and Iterative Development—A Manager's Guide,* Addison-Wesley, 2004.

8. James M. Morgan and Jeffrey K. Liker, *The Toyota Product Development System,* Productivity Press, 2006.

Gathering Voice of the Customer to Prioritize Technical Requirements

IMPORTANCE AND POSITION WITHIN DFSS FLOW

The success or failure of a new product often hinges on the degree of market acceptance. There are many examples of new products that were technical successes but failed to take off in the market. For Motorola, the Iridium project was a notable example of a brilliant series of technological, geopolitical, and project management successes that culminated in abject failure in the marketplace, and then was sold and rose again as a successful business—but at a fraction of the original business plan. Oftentimes, there are competing approaches, such as competition between Betamax and VHS videotaping, and more recently between Blu-ray disc and HD DVD based products for high definition video and data storage on optical disc media. According to survey results from the Product Development and Management Association (PDMA), about half of product development and commercialization resources are wasted on products that are canceled or fail to yield adequate return. Two of the top three reasons relate to inadequate or ineffective marketing.[1]

Unfortunately, there are no guaranteed approaches for ensuring a product will achieve market success; people's buying behaviors are not completely predictable. On the other hand, products that are differentiated to have moderate to strong advantages compared to their competition have three to five times the rate of success and market share compared to "me-too" products. Furthermore, products that have early, sharp,

---

1. Robert G. Gooper, *Winning at new Products: Accelerating the Process from Idea to Launch,* Perseus, 2001.

voice of the customer (VOC)-based definition in product development have two to three times the success rate compared to products with poor definition.[2]

Gathering the VOC might be the step that is most critical to the marketing success, and this is the first step in the DFSS process (Figure 7.1). Although there might not be solid proof that gathering the VOC helps minimize the "wrong product" risk, there is considerable anecdotal evidence for using the VOC rather than depending on the perceptions of marketing or engineering, without listening to potential customers. One story involves an engineering team that was preparing to develop a new sports car, and naturally started from what they themselves would want in a new sports car. They'd want it to have a powerful engine, like a race car, to get from 0 to 60 mph (0 to 100 kpm) very quickly—the better to win races and impress friends.

A new product manager arrived and suggested that the team talk with some potential customers, and listen to them before starting to design the new sports car. As they listened to their potential customers—high school students, college students, recent college graduates, and others—it became apparent that the customers really didn't care about a big, powerful engine. These customers didn't care all that much about how quickly they could get from 0 to 60 mph.

> These customers wanted a sports car that looked COOL . . . sounded COOL . . . and was relatively inexpensive.

The engineering team discarded their initial idea; then, led by their product manager, Lee Iacocca, the team developed the Ford Mustang—which went on to become the world's most successful car for the next 20 years.

A rigorous approach has been developed to formalize this approach of listening to the customers—"bringing the customers into your workshop," in the words of David

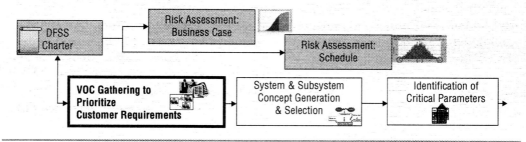

Figure 7.1 DFSS flowchart, highlighting VOC gathering step

2. R.G. Cooper, and E.J. Kleinschmidt, *New Products: The Key Factors in Success*, American Marketing Association, Chicago, 1990.

Garvin, professor of marketing at Harvard Business School. Concept engineering was developed at the Sloan Business School of MIT (Massachusetts Institute of Technology), working with Bose and Polaroid, and has been adopted by other major universities and disseminated through the Center for Quality of Management.[3]

Concept engineering has been used in a range of product types, from relatively low technology products like boots for hunters (L.L. Bean) and custom golf clubs (Calloway Golf), to complex products like truck engines (Cummins), plasma and LCD displays (Samsung), hard drives (Seagate, Western Digital), low noise speaker systems (Bose), and cellular base stations (Motorola). It has also been used for services like online trading services (Credit Suisse), on-site service experiences (Bank of America), and on-board experiences (Disney Cruise Lines).

The concept engineering process is illustrated in Figure 7.2. Gathering and using the VOC to drive product definition is a key to developing attractive new products

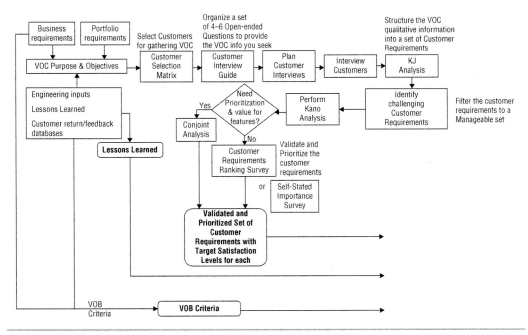

Figure 7.2 Detailed flowchart describing the process for gathering and prioritizing the customer requirements

3. Gary Burchill and Christina Hepner Brodier, *Voices into Choices—Acting on the Voice of the Customer,* Center for Quality of Management, 2005.

that reflect what the customers want and that meet their expectations. Sometimes this process goes beyond defining attractive new products and helps the team discover opportunities to exceed expectations, delight the customer, and to develop new products that are both attractive and compelling.

VOC PURPOSE AND OBJECTIVES

Before gathering the VOC for a new product, some initial effort may be required to review relevant information including business requirements, portfolio requirements, engineering inputs, lessons learned, and information that can be derived from customer returns and customer feedback for products that may be similar to the proposed new product (Figure 7.2). One key deliverable from this effort in reviewing and understanding the business requirements can be a set of voice of the business (VOB) criteria, which can be used later in the quality function deployment (QFD) method and as criteria for the concept selection process. The VOB is often rather predictable, including expectations that the project should be completed on time, the development and marketing costs should stay within the budget, and there should be no unpleasant surprises—including major recalls after the product is launched. The latter requirement might flow down into reliability, availability, and defect requirements that might be important to customers but might not be specifically mentioned when the VOC is gathered.

THE VOC GATHERING (INTERVIEWING) TEAM

Once the relevant business requirements, portfolio requirements, engineering inputs, and historical information have been reviewed, a team is empowered to gather the VOC. This provides a fantastic opportunity for marketing and engineering to work together in preparing to define the new product. Many organizations seem to have some level of tension between the marketing and engineering organizations. While the new product is being developed, that tension may devolve into doubts about the product definition, questions as to whether certain properties or features of the product are worthwhile, and even finger-pointing when problems arise or when managers question whether the product will be viable and successful in the marketplace.

Does it matter whether engineering and marketing are on the same page in terms of new product development? A study of 289 new product development projects from 53 firms found that the success rate for the new products was more than four times better and the failure rate was more than five times lower if there was organizational harmony

Table 7.1 Results from study of the impact of engineering–marketing relationships on new product success or failure (see footnote 4).

Result	Success	Partial Success	Failure
Harmony (41%)	52%	35%	13%
Mild disharmony (21%)	32%	45%	23%
Severe disharmony (39%)	11%	21%	68%

between marketing and engineering (Table 7.1)[4]; these results were significant with 99.9 percent confidence per chi-square analysis.

Gathering the VOC provides an ideal opportunity for marketing and engineering to work together in defining the product and to reach a common understanding of what is important to the customer, and why. For example, a few teams from marketing and systems engineering visited customers to gather the VOC for a new cellular base station. During these visits, the systems engineers came to learn that some of these base stations might be placed inside high rise buildings in Asia—and that, in some cases, people might live next to, above, and beneath the rooms that might house the base stations that were handling portions of cities with high population densities. Through these VOC interviews, the engineering community came to understand the importance of making these base stations quiet enough not to wake people who might be sleeping nearby— and light enough lest people worry about base stations dropping unexpectedly through ceilings onto their beds.

CUSTOMER SELECTION

Concept engineering involves interviewing customers in their own surroundings, so the team would first select which customers to visit and interview using a customer profile matrix. A customer profile matrix lists traditional market segments across the left side and nontraditional segments along the top. Traditional market segments along

---

4. W.E. Souder, "Managing Relations between R&D and Marketing in New Product Development." *Journal of Product Innovation Management,* Vol. 5, No.1, March 1988. As quoted by Ronald B. Campbell, Jr.,"Architecting and Innovating," Center for Innovation in Product Development, Massachusetts Institute of Technology, 2004 Symposium.

the left side might be listed by industry, by size, by income, by education level, or by geographic region. Nontraditional market segments along the top might be listed as lead users (these are users who are generally at the forefront of using products similar to the product you are developing; they would be referred to as innovators and early adopters, the people on one side of the "chasm" mentioned in the book *Crossing the Chasm*[5]), satisfied customers, dissatisfied and disgruntled customers, customers who have been lost, and customers never had. In business-to-business situations, the customer profile matrix can be enhanced to include specific job functions that might influence the purchase decisions; for example, the supply chain manager, engineering manager, and business manager of a manufacturing enterprise, or the medical director, physicians, and nurses in a hospital. A customer profile matrix template is available at http://www.sigmaexperts.com/dfss/chapter7voc. Because of time and expense constraints, the team will likely interview a subset of the potential customers enumerated in the matrix; according to marketing research, about 95 percent of the useful information is obtained from the first 15 to 20 interviews.[6]

VOICES AND IMAGES

A major advantage of interviewing customers in their own surroundings is that it provides and enhances opportunities for the team to gather "images." What the customers say, directly, constitutes "voices"—these need to be captured in the VOC gathering process. However, nonverbal communication provides additional information that often provides clues on ways that the product can be differentiated from its competitors. When the customer shows how they use current, similar products, do they seem to struggle with using certain features or doing certain steps? Do they (perhaps unconsciously) use some workarounds because something is missing or set up to be inconvenient? At what points do they grimace? At what points do they seem to smile?

In concept engineering, "images" are customer needs that are not articulated by the customer but show up in nonverbal communication, needs and opportunities that are observed as described in the previous paragraph, and needs and opportunities that pop up in the interviewers minds, that the interviewing team pick up on as they listen to the customer answer questions and describe how they might use the product, and as they observe the customer using a similar product or showing how they might use the

5. Geoffrey A. Moore, *Crossing the Chasm*, Harper Business, 1991.

6. Abbie Griffin and John R. Hauser, "The Voice of the Customer", *Marketing Science,* Vol. 12, No. 1, Winter 1993, pp. 1–27.

product. Some of the best opportunities to delight the customer come not from the articulated statements from customers—voices—but from images.

In one instance, a team—including some managers who were considered somewhat resistant to change—visited a customer site to gather the VOC, and to observe the customer installing an upgrade. After watching the trials and tribulations of the installation process, these managers returned as the strongest champions for the need to simplify, clarify, and mistake-proof the upgrade process.

In another instance, a team developing equipment used in stores and shops interviewed people working in shops in Europe, and found that their equipment would need to be smaller and more compact than originally planned.

CUSTOMER INTERVIEW GUIDE

Once customers have been identified for purposes of interviewing and VOC gathering, the team can prepare for the interviews by developing a customer interview guide. The first step is to clarify the purpose and objectives of the VOC gathering and the type of information the team is looking for. An example of a purpose statement might be, "To discover what needs and expectations teenage customers have for a mobile device," or "To hear what types of needs, features, and capabilities might be needed by undercover agents whose only communication lifeline might be their 'bug.'" This latter example will be explored later in this and some other chapters—primarily because class participants have particularly enjoyed, even relished the role playing involved in gathering the VOC for this case. The communication device for undercover agents is actually based on a real case study—but some of the details are necessarily hidden, lest the reader discover things too dangerous to know.

The objectives would involve what specific things the team would like to learn. However, the customer interview guide will consist of questions that are open-ended; that is, they are designed to elicit a flow of relevant stories, examples, and information from the customers. The intent is to allow the customer to tell their story, from their perspective. By contrast, close-ended questions, such as might be found on a survey, tend to provide yes/no or multiple choice answers, or Likert scale responses for the degree to which the customer agrees or disagrees with statements that originated from the interviewers' rather than from the customers' perspective.

The customer interview guide is primarily a set of evocative, open-ended questions that would be likely to stimulate the customer to share his or her perspective and thoughts that might relate to the desired insight for the team. The first rule of the VOC gathering process is not to have your own mind made up about what the customer's business is. Avoid preconceived notions; allow the customer to tell his or her story and share his or her feelings, frustrations, issues, requirements, and nice-to-haves. **Don't walk in with a solution**! Examples of open-ended questions are provided here:

- What things come to mind as you imagine and visualize this new product?
- What problems have you had with prior, similar products?
- What features would you like to see that you would find attractive in a new product?

If a team wants to determine what applications the customer might want to use when using a proposed new laptop computer, an open-ended question might be, "Tell us about a very complicated day this week, when you had too much to do—and what you might have found frustrating on that day." If a team wants to hear about needs for a new software application for travel arrangements, an open-ended question might be, "Show us your current preferred or favorite way of setting up a trip, and please walk us through it and tell us what you like and don't like along the way."

A good way to start developing the customer interview guide is to first post the purpose in a visible location, brainstorm the sort of information the team members would like to obtain, and then ask team members to each write down their own suggestions for questions on Post-It notes.

The team can then use affinity diagramming, placing their notes with questions on a board or wall, combining or eliminating duplicates, and then organizing the questions along some themes. Some of the questions may be eliminated, others may become main questions, and still others may be selected as follow-up questions. Affinity diagramming is related to the KJ analysis approach discussed in an upcoming section of this chapter.

The intent is for the team to agree on perhaps four to seven main questions, along with a few follow-up questions, that seem to cover the full range of information that the team wants to gather from its customers. The team should check and refine each question—ensuring each question is open-ended and likely to evoke stories and discussions rather than quick "Yes/No" answers. It is worth noting that questions that begin with "Are," "Do," "Can," and "Should" seem to have an unfortunate tendency to elicit those short, "Yes/No" responses, while questions that begin with "What," "How," "Could," and "Please tell us about. . ." tend to elicit longer answers and stories as desired. Also, questions that start with "Why" may tend to come across like an interrogation rather than an interview.

After the team has a first-pass set of questions that they believe will meet the objectives, it is worthwhile to try them out—first on each other. The team members can rotate roles: interviewer, recorder, nonverbal observer, and the customer to be interviewed. This role-playing exercise can be a fun and team-building activity; if the customers include kids or teens, some team members can use this as an opportunity to get in touch with "the kid inside." If the customers include people who have interesting vocations, the role-playing might involve jogging through the park with a secret agent, or driving with a policeman through a rough neighborhood. Beyond these enjoyable aspects, the role-playing allows the team to judge the effectiveness of each question and further refine them.

PLANNING CUSTOMER VISITS AND INTERVIEWS

After the team has developed and is comfortable with the questions comprising the customer interview guide, the team needs to negotiate the time frame, budget the travel costs with the decision makers, and plan the customer visits. The team or team leaders should meet with the budget decision maker to summarize the concept engineering process, describe how the customers were selected, and provide an initial estimate for the costs associated with travel, and so on, and the time frame.

Oftentimes, concept engineering is replacing an unstructured fuzzy front end process that may take months, but with little visibility for this unstructured time investment. The interview process associated with concept engineering is more structured, and the costs are much more visible and clear—but will generally take the same or less time than the fuzzy front end it replaces, while providing more substantial information that the team can use to effectively develop new products that the customers want.

In addition to sharing this insight with the decision makers, it might be advantageous to describe the significant peripheral advantages: the team-building and the alignment of the perspectives between marketing and engineering, and the subsequent buy-in to the product definition and requirements by the broader new product development team. Moreover, the interview process often leaves favorable impressions on the customers, as they were contacted and consulted to help define a new product. Customers appreciate the opportunity and the respect they are given.

The decision maker may negotiate with the team regarding the number of people who will travel, how much traveling can be afforded, and the timing involved. After the team receives the "Go ahead," along with constraints on the resources, timing, and travel budget, they can finalize the list of customers to visit and interview, work out the travel arrangements, and schedule appointments with the interviewees.

The field salespeople might be helpful in selecting the interviewees and arranging the visits, but they will need to clearly understand the purposes of the visits and interviews, and that—while the customers will likely appreciate the chance to share their views and influence new product development—these visits are not to be confused with or mingled with sales calls. Each interviewed customer is being asked to share his insights, opinions, and stories—and the mood would be unfavorably impacted and perhaps fail if the customer suddenly feels that the visit had a hidden agenda—that the purpose was not to listen to him or her, but to try to sell something.

Once the salespeople understand the purpose of the visits and interviews, they might be able to set up the appointments and provide other resources to help the team.

CUSTOMER INTERVIEWS

The VOC gathering team will break into subteams of two or three people for interviewing. Before each interview, the team may select who will handle each role. For example, a marketing person might serve as the interviewer, a systems engineer might take notes, and another engineer or marketing person might observe the nonverbal communication (or the team might use two-person teams—interviewer and recorder). Sometimes, however, an interviewee might unexpectedly establish rapport and gravitate toward a different member of the team; perhaps the systems engineer/recorder builds rapport and credibility because the interviewee is a fellow engineer, or an alumni of the same college. In any case, the team members should be aware and open, and deftly rearrange roles to accommodate the interviewee, "go with the flow," and use the rapport to advantage in having a very effective interview session. Here are some guidelines that might prove helpful during the interview:

- Notice how the customer is using your product, free from any preconceived solutions.
- If the customer offers a solution, gently guide him or her back to discussing the issues; for example, "If we used that solution, would what that allow you to do?"
- Document observations, not how you would fix their problems (you are drawing conclusions prematurely if you are thinking of fixes already).
- Document exactly what the customer has stated.
- Bring your domain knowledge to understand the context of what you are observing. From there you need an *open mind*. Check your titles at the door.

After each interview, the schedule should allow enough time for the team to briefly debrief, catch their collective breaths, and arrive on time to their next interview appointment.

At the end of the interview process, the sets of interviewing subteams will generally have hundreds or even thousands of voices, plus quite a few images. The team is ready to pull it all together, during a KJ analysis session.

KJ ANALYSIS: GROUPING, STRUCTURING, AND FILTERING THE VOC

KJ analysis was developed by Jiro Kawakita, who encountered situations in the field of anthropology that are somewhat analogous to the situation the VOC gathering team finds itself in: having lots of information in the form of notes.

> In college, I learned how to do quantitative analysis; but, when I got out into the field, most of the data was qualitative. I had to develop this [KJ analysis] method to deal with the large amount of qualitative data.
> —Jiro Kawakita (very loosely translated from Japanese)

The intent of KJ analysis in concept engineering is to process this qualitative data in the form of hundreds or thousands of voices and images into a structured and filtered set of customer requirements. Each customer requirement that emerges from KJ analysis should be written as a positive statement ("the product should provide the approximate location," rather than "The product must not leave people wondering where they are"), should not involve solution-bias ("the product's output file must be compatible with common word processing programs," rather than "the product must use Microsoft Word 2003"), should use the active rather than passive voice ("the product must provide graphs," rather than "graphs should be output from the product"), and should be clear and easy to understand ("the product must be thin and have dimensions less than those of a wallet," rather than "it would be helpful if the product would have three-dimensional aspect ratios compatible with Euclidean ideals of beauty").

KJ analysis applies an affinity diagramming approach in a highly structured sequence:

- Select a purpose for the KJ analysis, like the purpose from the customer interview guide.
- Have each subteam recount their experiences while interviewing, and review the images with other subteams in that context. This will help the other subteams come up to speed and understand the relevance of the images.
- Write descriptions of the images on specially marked Post-It notes. If there are just a few images, incorporate the images with the recorded voices (Figure 7.3). If there are more than 20 images, follow the sequence twice, first with descriptions of images (image KJ analysis) to translate summaries of the images into customer requirements, then combining the summaries from the images with the voices (requirements KJ analysis).
- Translate the voices and images into customer requirement statements and write each on a Post-It note, using black pens with wide points. Customer requirement statements generally take the form "The product must do this" or "The customer expects the product to behave like this," as described in an earlier paragraph.

Customer Voice (as recorded): "I want the song I'm playing to automatically pause when a phone call comes in, and then I want to be able to continue after I hang up."

Image (as observed during interview):
Interviewer observed the customer growing frustrated as she tried to take a call, tried to pause the song, struggled to get back to the song after hanging up, and grew more frustrated when she had to start the song from the beginning.

Customer Requirement
(derived from the Voice and Image):

When the phone rings,
the cellular phone must save
the spot in the song,
switch from music to phone mode
(without user intervention), then
switch back to music mode and
resume from the spot in the song
within 3 seconds of the call ending.

Figure 7.3 Combination of recorded customer voice with KJ image, then translated into a customer requirement from a KJ analysis session for a new cellular phone

- Combine or eliminate redundant voices and images.
- Team members then peruse the voices and images silently, and group Post-It notes by what each perceives as common themes. Some individuals will have different perceptions and group somewhat differently, but over time the team will tend to converge on sets of similar and related customer requirements. There will also likely be some "lone wolves," singleton customer requirements that don't fit into any groups.
- Ending the silence, have team members write a title customer requirement statement that summarizes the expectation associated with each group with a wide-point red pen.
- If there are more than 10 to 15 groups, continue and combine groups by common themes, grouping similar and related red customer requirement statements.
- Write a title customer requirement statement that represents each group of groups, using a wide-point blue pen.
- Optionally, draw arrows between groups where one grouping tends to drive another grouping of customer requirements, and draw elongated Xs to show conflicting sets of requirements, as shown in Figure 7.4 for a cellular base station.

The goal for the KJ analysis session is to end up with between 4 and 10 important customer requirements—perhaps referred to as critical customer requirements (CCRs). A requirement is a specific function, feature, quality, or principle that the system must or should provide to either meet basic expectations or to satisfy and potentially delight its customers and its customers' customers or end users. Figure 7.4 depicts the results of a KJ analysis session for a base station, and Figure 7.5 shows the results for a cellular phone.

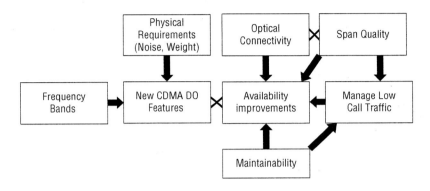

Figure 7.4 Results from KJ analysis session for a new cellular base station

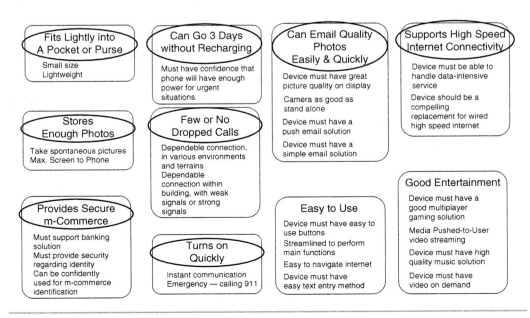

Figure 7.5 Results from KJ analysis session for a new cellular phone

If there are more than 10 groupings of summarized customer requirements, it is probably appropriate and useful to differentiate among the customer requirement sets using some combination of team member voting, new-unique-difficult (NUD) differentiation, or Kano analysis to select a smaller set of customer requirements for deeper focus. Team member voting involves giving each team member three to five "dots" to place on the higher level customer requirements he or she thinks are most important to the customers they interviewed. A team member can place one dot on each important item, or place more than one dot on an item to indicate higher importance. Requirement groupings with no or very few dots can be removed. NUD and Kano analysis filtering approaches will be discussed in upcoming sections of this chapter.

IDENTIFYING CHALLENGING CUSTOMER REQUIREMENTS (NUDs)

The team can separate the "vital few" from the "trivial many" requirements using criteria that helps to identify the most challenging requirements; these are the requirements that are more likely to benefit from the increased focus and application of predictive engineering approaches that are applied to critical parameters in the Design for Six Sigma process. The term "NUD"[7] represents a new, unique, or difficult customer requirement; "new" refers to a customer requirement that is both new to the market and new to the development team; it therefore involves both market risk and technical risk, since it is unproven in both respects. "Unique" refers to a customer requirement that is represented in the market, either through a competitor's product or a product that is used in addition to the current generation of the product. For example, at one time many business people were using a cell phone and a PDA (personal digital assistant), and at another time many were using both a cell phone and a GPS system, or both a cell phone and a camera. Adding the functionality of a PDA or a GPS or a camera to a cell phone would involve "unique" customer requirements, in that the PDA, GPS, or camera functionality and requirements were established in the market, but involved technical risk for the cellular phone development team because of a lack of direct experience with these added functions. A unique requirement thus involves technical risk more than it does market risk. A difficult requirement may or may not be new or unique, but regardless, represents technical challenges and consequently some level of technical risk.

By contrast, if a customer requirement meets none of the NUD definitions, has been done before, and isn't expected to pose difficulties this time, then it probably fits among

---

7. C.M. Creveling, J.L. Slutsky, and D. Antis Jr., *Design for Six Sigma,* Prentice Hall, 2003.

the "trivial many." The team believes the requirement may need to be incorporated in the product, but does not involve much market or technical risk because it has been represented in the market and does not raise concerns for technical challenges. These requirements don't require special, intense engineering efforts, but should be recorded in the requirements documentation such that the team will ensure and later verify that these requirements were met. In Figure 7.6, NUD criteria are applied to filter some customer requirements for a cellular phone into the NUD or ECO (easy, common, and old) categories. Figure 7.6 shows customer requirements for a cellular phone categorized according to the Kano model as well as through the NUD criteria.

Customer Requirement	NUD Category	Kano Analysis
Fits Lightly Into a Pocket or Purse	(N)ew, (U)nique, **(D)ifficult,** (ECO/SR)	(M)ust.Be (I)ndifferent **(L)inear Satisfier** (D)elighter
Can go 3 days without recharging	(N)ew, (U)nique, **(D)ifficult,** (ECO/SR)	(M)ust.Be (I)ndifferent **(L)inear Satisfier** (D)elighter
Can email Quality photos easily & quickly	**(N)ew,** (U)nique, (D)ifficult, (ECO/SR)	(M)ust.Be (I)ndifferent **(L)inear Satisfier** (D)elighter
Supports high speed internet connectivity	**(N)ew,** (U)nique, (D)ifficult, (ECO/SR)	(M)ust.Be (I)ndifferent (L)inear Satisfier **(D)elighter**
Stores reasonable number of photos	(N)ew, **(U)nique,** (D)ifficult, (ECO/SR)	(M)ust.Be (I)ndifferent **(L)inear Satisfier** (D)elighter
Few or no dropped calls	(N)ew, (U)nique, **(D)ifficult,** (ECO/SR)	**(M)ust.Be** (I)ndifferent (L)inear Satisfier (D)elighter
Provides secure m-Commerce	(N)ew, (U)nique, (D)ifficult, **(ECO/SR)**	(M)ust.Be (I)ndifferent (L)inear Satisfier **(D)elighter**
Turns on quickly	(N)ew, (U)nique, **(D)ifficult,** (ECO/SR)	(M)ust.Be (I)ndifferent **(L)inear Satisfier** (D)elighter

Figure 7.6 Classification of customer requirements for a cellular phone (from Figure 7.4) using both NUD (new, unique, or difficult) and Kano analysis categories

KANO ANALYSIS

The Kano model, developed by Dr. Noriaki Kano, classifies customer requirements in four ways, which illustrate how fulfilling the customer requirement affects customer satisfaction. As shown in Figures 7.7 and 7.8, delighter requirements are unexpected, so lack of a delighter requirement won't unfavorably impact customer satisfaction for a customer who doesn't expect that functionality in the first place. An unexpected upgrade to first class on a flight can be a delighter. Honda sales people could show the Honda CRV vehicle with its features, and then surprise the customers when a small picnic table was found in the back—evoking pleasant images of stopping by the side of the road for a nice picnic. Delighter requirements, if selected to help differentiate a new product, often will be considered among the "vital few" requirements.

Linear satisfier requirements improve customer satisfaction as the product does a better job of fulfilling that requirement. When gasoline prices are high, customers are happier as the miles per gallon or kilometers per liter increases. Linear satisfier requirements often are considered among the "vital few" requirements, especially if improving

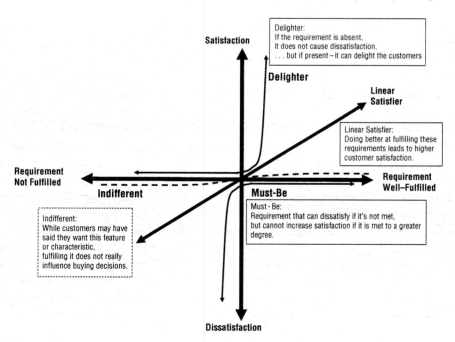

Figure 7.7 Diagram for Kano analysis; the horizontal axis represents the degree to which the requirement is fulfilled and the vertical axis describes the degree of customer satisfaction

Delighter	Linear Satisfier
The absence of a Delighter has little or no negative impact, while introducing the delighter generates excitement and satisfaction. Examples: Cupholders in cars Tamper proof tops Free drinks or entertainment on trips Portable computers	More of a Satisfier leads to more customer satisfaction, in a generally linear relationship. Examples: Gas consumption efficiency in a car Reliability Longevity
Indifferent	**Must-Be**
Customers may indicate a desire for something that turns out to have no real impact on customer satisfaction Examples: Unnecessary features (light for "radio is on") "Good for the environment"	Must-Be requirements are necessary for a product to be considered—like "table stakes" in poker. The product must meet the minimum expectation, but meeting standards above the minimum may not affect satisfaction. Examples: Overhead compartment on airplanes Biocompatibility for implanted medical devices Basic functionality requirements for cell phones

Figure 7.8 Description and examples of requirement categories for Kano Analysis

them provide a competitive advantage and involve considerable technical challenges. For example, a few years ago Motorola decided to make a surprisingly thin phone called the RAZR, based on a realization that the thickness of a cell phone could behave as a linear satisfier. However, thinning the phone involved several technical challenges, requiring special handling during product development.

Must-be requirements must be met or the customer will be dissatisfied or refuse to buy. Exceeding expectations for a must-be requirement won't impress customers. Strictly speaking, regulatory requirements (requirements that must be met in order for the product to be legally sold) are different from must-be requirements, but regulatory requirements and must-be requirements behave similarly—they must be met to some minimum level, and you get no "extra credit" for exceeding minimum requirements. Must-be requirements often end up among the "trivial many" unless those requirements involve considerable technical risk. Note that today's delighter may be tomorrow's linear satisfier and next year's must-be requirement.

VALIDATION AND PRIORITIZATION OF CUSTOMER REQUIREMENTS

The set of customer requirements filtered and structured through the KJ analysis process described in Figure 7.5 should be validated with the customers and prioritized. As shown in Figure 7.2, there are several alternative methods for obtaining numerical values for relative importance, including a customer requirements rating survey in which each selected customer is allowed to distribute 100 points among the customer requirements that emerge from KJ analysis and the results from the surveyed customers are averaged to prioritize the customer requirements (Figure 7.9).

Alternatively, conjoint analysis can provide prioritization and other valuable insights through utility indices reflecting relative priority based on customers' choices among various combinations of product features and attributes (perhaps at various prices). An efficient and comprehensive set of combinations of features can be developed using an approach such as design of experiments (DOE), as discussed in Chapter 13, and analyzed to provide the relative importance of each proposed product feature or attribute in terms of influence on prospective customers' buying decisions.[8]

TRANSLATING CUSTOMER REQUIREMENTS TO SYSTEM REQUIREMENTS: THE SYSTEM-LEVEL HOUSE OF QUALITY

Once the VOC has been gathered and organized and into a smaller set of prioritized customer requirements, the team needs a way to make these requirements measurable. Some customer requirements may provide clear mandates and measurable requirements (e.g., "the base station must weigh less than 60 kilograms," "the system downtime cannot exceed 5 minutes per year"), while other customer requirements may provide clear guidance that is not yet measurable (e.g., "the display must be readable day or night," "the police radio must be waterproof"), or rather vague requirements (e.g., "the cell phone must look cool," "the sound of the music must be crisp and clear," "the police radio must provide secure communication"). The vital few high-priority customer requirements must be converted to a set of measurable requirements to enable the team to know how it is doing and decide what it needs to do to provide a successful new product. In order to design toward meeting or exceeding customer expectations, the product development team needs to be able to measure their initial conditions, determine the impact of component and supplier decisions, and optimize the product using predictive engineering approaches (Chapters 13 and 14).

---

8. Bryan K. Orme, *Getting Started with Conjoint Analysis—Strategies for Product Design and Pricing Research,* Research Publishers, 2006.

End Customers:	Consumer-Brazil	Consumer-Japan	Consumer-USA	Consumer-Korea	Consumer-France	Average	For QFD: Max = 10
Customer Requirements (from KJ Analysis)							
Fits lightly into a pocket or purse	14	17	21	26	9	17	6
Can go 3 days without recharging	19	14	23	17	23	19	7
Can email quality photos easily & quickly	11	14	10	10	10	11	3
Support high speed internet connectivity	11	14	0	0	14	8	2
Stores reasonable number of photos	3	4	9	0	4	4	1
Few or no dropped calls	24	23	27	37	17	26	10
Provides secure m-Commerce	9	5	0	0	14	6	1
Turns on quickly	9	9	10	10	9	9	3
Total Points	100	100	100	100	100		

Figure 7.9 100-point customer requirements rating method for prioritizing customer requirements

The steps involved in gathering the VOC and then filtering, structuring, and prioritizing a set of customer requirements generally will involve and engage both the marketing and engineering organizations. The vital step of converting prioritized customer requirements to system requirements will gradually transition from marketing-led, engineering-involved efforts to engineering-led, marketing-involved efforts. In Figure 7.10, the translation of prioritized customer requirements to measurable system requirements is represented by the box labeled "House of Quality." After this step, intense and often innovative thought processes help generate alternative concepts and select a preferred concept that will facilitate the product's ability to meet the customer expectations as described in Chapter 8. The system requirements are decomposed to software and hardware requirements and distributed to the software and hardware engineering teams as described in Chapters 10 and 12.

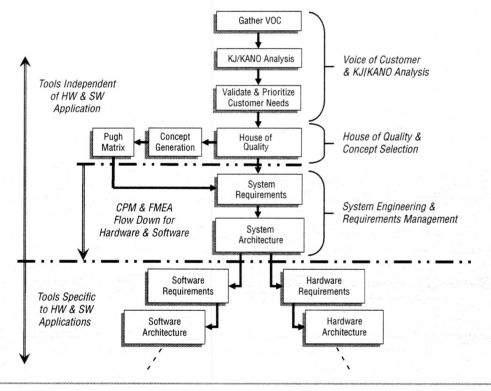

Figure 7.10 Steps for gathering the VOC, translating them to prioritized, measurable technical requirements or system requirements (represented by the House of Quality), and subsequent concept generation and selection and distribution of system requirements to the software and hardware development teams

The translation of customer requirements to measurable system requirements is most commonly done using the House of Quality—named for the resemblance of the format to a home, with several rooms and a roof. Figure 7.11 shows the use of a House of Quality to translate customer requirements from the KJ analysis (Figure 7.5), along with the prioritization (Figure 7.9), to prioritized system requirements. The House of Quality in Figure 7.11 has fewer than ten customer requirements for translation. If there are too many customer requirements, the House of Quality will become unwieldy. Figure 7.12 shows how the time required to complete a system-level House of Quality increases as the number of customer requirements increase, assuming that the number of system requirements is equal to the number of customer requirements and the team needs ten minutes to identify each system requirement, five minutes to establish specifications, two minutes to quantify the impact of each system requirement on each customer

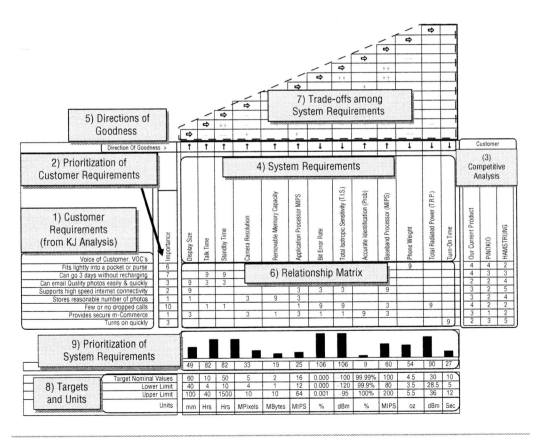

Figure 7.11 A House of Quality helps translate customer requirements into system requirements

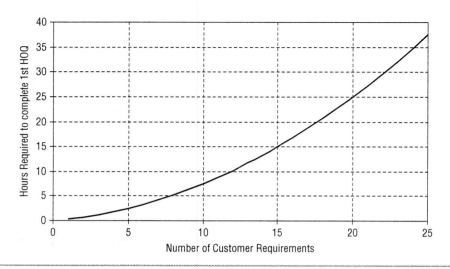

Figure 7.12 Time needed for a House of Quality depends on the number of customer requirements

requirement, and two minutes to quantify the correlation between pairs of system requirements.

QFD is especially valuable in a business with a globally dispersed team executing complex projects, as QFD also acts as a communication tool for understanding the product and its requirements—and where the requirements came from. The QFD, and the subsequent critical parameter flow-down (Chapter 10), can help new team members familiarize themselves with a project and the associated new product or service.

CONSTRUCTING A HOUSE OF QUALITY

Construction of a system-level House of Quality will involve a leader and team representing both systems engineering and marketing. The sequence of steps are shown by the numbered sections or "rooms" in Figure 7.11. The term "room" is consistent with the "house" imagery.

> **Room 1. Customer Requirements** lists the key detailed customer requirements that were derived from the VOC. These requirements have been identified, filtered, and prioritized through an approach such as KJ analysis. The customer requirements will include must-be and linear satisfier requirements that are expected to be challenging

based on the NUD criteria, along with some delighters that can help differentiate the product with the customers.

Some teams add an additional room to the left of Room 1, which indicates supporting and conflicting relationships among the customer requirements, along the lines of the arrows and crossed lines used in Figure 7.4 for KJ analysis. If used, this extra room is sometimes referred to as a "breakfast nook."

Room 2. Prioritization lists relative priorities of the customer requirements on a 1–10 scale; this prioritization may be obtained from a customer prioritization survey (Figure 7.9), or from conjoint analysis.

Room 3. Competitive analysis is optional, and generally filled out by the marketing representatives on the team. It summarizes how customers perceive your organization's performance and competitors' performance with respect to meeting the customer requirements listed in Room 1. The information might be obtained through a survey or other marketing research that collects information on customer ratings of the current generation of products—both your current products and those from key competitors. A scale of 1 to 4 is often used, and only one product (either yours or a competitor's) can receive each integer—there are no "ties." A value of four is best, one is worst. (Any perceived similarities between the names of the competitors in Figure 7.11 and actual companies are only in the reader's demented imagination.)

Room 4. System requirements can be very time-consuming. This is perhaps the most important deliverable from the House of Quality. The team discusses each customer requirement from Room 1, and goes into a brainstorming mode: "How can we measure how well we are meeting this particular customer requirement?" Each customer requirement might require one or several measurable requirements; typically between one and four measurable requirements might be needed to fulfill a customer requirement. For the system-level House of Quality, the "occupants" of Room 4 can be referred to as system requirements, technical requirements, or measurable requirements. System requirements are not solutions, and should not involve solution bias; the intent of each system requirement is to allow the team to measure the degree to which the customer requirement is fulfilled. The avoidance of solutions or solutions bias at this stage allows the team to later envision and consider a fuller range of possible solutions and concepts for the product and its subsystems and components in an unbiased way.

Brainstorming how to render a customer requirement measurable is an acquired skill. The first time a team tries to do it might be very time consuming; it might be helpful for the team to start with the simpler customer requirements, to have them become comfortable and have a few successful efforts before taking on more complex and vague customer requirements. In terms of being an acquired skill, a participant who has been involved in several of these activities may become quite adept at translating customer requirements into measurable requirements; such an experienced individual might be a welcome addition to the team developing the House of Quality.

A customer requirement is not considered to be fully translated into a set of measurable requirements until four conditions are checked and confirmed:

1. Is each of these measurable requirements necessary in order to be confident that we have fulfilled the customer requirement? Or, are some of the measurable requirements redundant?
2. If the product met this set of measurable requirements with very favorable values, would the team feel confident that the customer will be satisfied that this customer requirement has been fulfilled?
3. Are the measurable requirements solution-independent, to prevent biasing the team toward one preferred solution, concept, or architecture?
4. Is at least one—preferably all—of the measurable requirements a continuous parameter, or at least ordinal rather than binary discrete?

Examples of continuous parameters that can have any value within an acceptable range include temperatures, physical dimensions, weight, cost, delay time or latency, voltage, current, resistance, luminescence, and mean time to failure (MTTF).

Ordinal parameters can have multiple, increasing integer values; examples include the number of drops-to-failure, number of pixels in an imager or display, the number of megabytes of memory available or consumed, and the number of clicks required to get to the right screen and make a selection.

Binary discrete parameters can only have one of two values—generally pass or fail, good or bad. As discussed in Chapter 13, continuous parameters, like temperatures, lend themselves to predictive engineering, both in terms of the statistical methods available for optimization and in terms of the sample sizes required for experimentation and capability studies. Ordinal parameters with more than four levels can often be treated like continuous parameters, but binary discrete parameters have serious limitations in terms of the statistical methodologies available and the same sizes required. Oftentimes, fulfilling a customer requirement involves a combination of several measurable requirements and one or more binary discrete requirement.

For example, if the customer requirement is "the customer must be able to quickly obtain their current location," the team might come up with this set of continuous variables: accuracy in location, measured in terms of the number of yards or meters between the reported location and the true location, and acquisition time in seconds.

Using the four conditions, each of those measurable requirements are necessary—the accuracy of the reported location is needed in order to fulfill the part of the subrequirement for "obtain their current location," and the acquisition time is required to fulfill the subrequirement for "quickly."

When the team checks the second condition, whether this set of measurement requirements is sufficient to satisfy the customer, they hopefully will realize it is not. Would the customer be happy to check their portable electronic device, such as a cellular phone to find their location, and see their location reported as 22.01265 N, 159.70488 W?

In addition to accuracy in location (measured in meters) and acquisition time (measured in seconds), the device must also display the information on a map. The team might agree that the customer will be satisfied if these three measurable requirements (two continuous parameters and one binary discrete requirement) are met.

The measurable requirements seem to be solution-independent—although some on the team might have been tempted to use GPS as part of the technical requirement, they resisted the urge.

Room 5. Direction of goodness generally consumes little time (which can provide a welcome break from the intensity involved in constructing Room 4).

The team can indicate a preferred direction for the measures with arrows or symbols on each column of the House of Quality:

Up arrow ($\uparrow$) or "+": If we increase the measured value of this measurable system requirement, it will help to fulfill the customer requirement.

Down arrow ($\downarrow$) or "—": If we reduce the measured value of this measurable system requirement, it will help to fulfill the customer requirement.

Circle ($\bigcirc$) or "T": If we hit the measurement target, will that help to achieve the customer requirement?

Room 6. Relationship matrix helps in defining the strength of the fulfilling relationships between the system requirements and the customer requirements. It can be time consuming and tedious if there are a considerable number of customer and system requirements.

A strong relationship, wherein improving the system requirement (in the direction of goodness defined in Room 5) goes a long way toward fulfilling the customer requirement, is represented by the letter "H" for high or the number 9. Note that the scenarios where an adverse relationship, although improving the system requirement, works against fulfilling the customer requirement, or going in the opposite direction to that of goodness for a system requirement works favorably for fulfilling the customer requirements, are not captured in the relationship matrix in Room 6. Adverse relationships will be captured in Room 7, the roof.

A relationship that is moderately strong, that is, where improving the system requirement helps fulfill the customer requirement, is represented by the letter "M" for moderate or medium, or the number 3.

A weak relationship, that is, where improving the system requirement has some noticeable but minor favorable impact on the customer requirement, is represented by the letter "L" for low or the number 1.

If there is no relationship between the system requirement and the customer requirement, the corresponding space in the relationship matrix is left blank (or the number "0" might be entered).

An empty column may indicate that the technical requirement is not really needed because it does not correspond to any need, or it may indicate that a need was missed in the gathering of the VOC (recall that customers sometimes simply assume and therefore fail to mention "must be" needs and usually fail to articulate "delighter" needs).

Room 7. The "roof" for trade-offs among system requirements addresses the supporting and adverse relationships between the system requirements. Determine if a positive or negative relationship between system requirements exists by asking the team, "If we design to improve system requirement "m" in its direction of goodness, to what extent does that affect our ability to improve system requirement "n" in its direction of goodness, and vice versa (i.e., if we improve requirement "n," how does that affect requirement "m"?)

If improving requirement "m" is also helpful in improving requirement "n," then a symbol like "+" can be entered in the corresponding cell in the roof. If the synergistic relationship is particularly strong, then a double "++" might be entered in the cell.

If improving requirement "m" unfavorably impacts requirement "n," or vice versa, then a symbol like "-" can be entered in the corresponding cell in the roof. If this negative effect is particularly strong, then a double "- -" might be appropriate.

Generally, the Design and Optimize phases can be easier if there are lots of positives (pluses and double pluses) in the roof of the House of Quality, because the team can improve one system requirement and have a side benefit in improving other system requirements as well. Negatives (minuses and double minuses) in the House of Quality indicate trade-offs where improving one system requirement works against another system requirement. Chapter 8 introduces the TRIZ approach, which can help the team develop creative approaches to deal with trade-offs. If the team is unable to find a creative way to "have your cake and eat it, too," then the multiple response optimization approaches discussed in Chapter 14 may be useful in finding the best settings for flowed-down control factors to co-optimize measurable requirements.

Room 8. Targets and units for system requirements summarizes targets, and perhaps specification limits, for the system requirements. This room often involves considerable discussion among the marketing and systems engineering people.

The target values provide two valuable pieces of additional information that we can use. By determining from the customer how much they want of a particular measurable requirement, we identify both the target value and the units. In addition during this conversation we can often also identify the lower specification limit (LSL) or the upper specification limit (USL). Some parameters, like leakage current, bit error rate, and drops-to-failure, just have an upper specification limit, whereas other parameters, such as talk time, standby time, mean time to failure, memory capacity, and efficiency, just have a lower specification limit. Other parameters, such as timings for certain tasks, total radiated power (TRP), and contrast for a display might have both specification limits. For example, too little total radiated power from a cellular phone antenna might not allow the cellular to connect in a rural area; too high of a total radiated power value might run into regulatory or medical issues.

Room 9. Prioritization of system requirements can be performed automatically in software (Excel, Cognition Cockpit, QFD Designer), using a mathematical combination of the customer prioritization from Room 2 with the 0, 1, 3, or 9 values in the relationship matrix in Room 6.

With Excel, the "sumproduct" function can be used to calculate the importance scoring totals of each system requirement by multiplying the relationship weight (9, 3, 1) by the importance that customers assigned to the requirement in the customer requirement prioritization column. These scores are added within each column for each system requirement at the bottom of the matrix, in Room 9. Use these scores in your discussion to check your thinking and to help identify the critical requirements, as described in Chapter 9. The calculated totals should not be accepted blindly; the team should consider each system requirement and its corresponding score for prioritization, and see if it passes the "common sense test." If it doesn't, review how it was calculated—the team might have included some counter-fulfilling tendencies where a system requirement influenced a customer requirement unfavorably, or the system requirement might have a relatively low score because it only supports one customer requirement—but may be needed in order to satisfy that one very important customer requirement.

Generally speaking, a properly constructed system-level House of Quality typically takes about six to twelve hours to complete. It is recommended that a DFSS black belt or a trained facilitator lead the effort. The House of Quality should be planned in advance by identifying scope and attendees. It is highly recommended that food, snacks, and soft drinks be liberally supplied to the team during this rather intense session.

SUMMARY: VOC GATHERING—TYING IT ALL TOGETHER

Developing a successful product involves understanding customer requirements. Those requirements are often abstract and difficult to express, and it takes a disciplined, customer-focused process to capture the VOC early. Concept engineering is a proven, structured process that uses a series of steps involving identification of customers to interview, VOC gathering, KJ analysis, Kano analysis, and the system-level House of Quality.

We begin the requirements gathering process by first asking ourselves the following questions:

1. Who are the customers?
2. Which customers should we interview, in order to understand various perspectives and interests in the market or markets?
3. How can we effectively collect data on the customers needs? How can we discover "hidden needs" that the customer may not be fully aware of but that could provide our product with a competitive edge?
4. How do we condense an overwhelming number of customer voices and images into a manageable set of the customer's most important needs?
5. How can we measure how well our product might stack up against these customer requirements as we conceptualize, design, and develop the product? What are the performance targets that the design should meet to satisfy customers?
6. If we supply to an intermediate customer who sells our product or a product that uses our product to an end customer, how do we balance what the direct customer is saying about their customers' requirements with our own insights about what might be needed for the end user, and what the end user is saying?

The writers and key users of requirements should become engaged during the VOC and KJ analysis processes to fully understand the customer requirements and why they are needed.

Statements of need from VOC interviews are numerous and complex and the information from the VOC interviews appear to be unorganized thoughts, statements, and ideas. KJ analysis clarifies and condenses customer requirements from the combination of images and stated needs. Requirement analysis will identify the most significant themes from the VOC interviews and focuses on the areas where further action will have the greatest impact. Involvement through combined teams of marketing and engineering builds team knowledge of the customer that goes well beyond what the KJ diagram shows. There is a significant difference in an engineer's empathy for the customer when he or she has participated in customer visits and worked with marketing to develop customer requirements as opposed to if the engineer received a cold set of requirements tossed over the wall.

The Kano model helps the team understand the relationship between the fulfillment or nonfulfillment of a requirement and the satisfaction or dissatisfaction experienced by the customer. It is a method that allows us to sort requirements into distinct classes, where each class exhibits a different relationship with respect to customer satisfaction.

The system-level House of Quality is used by the team to translate customer needs into measurable system requirements, and assists in narrowing the focus of the project by prioritizing system requirements according to their impact on the customer. The House of Quality is a matrix that relates the customer wants to how we can determine whether the product might satisfy those wants. A goal of the House of Quality is to identify that subset of measurable requirements which the development team can focus on fulfilling in order to maximize the opportunity of meeting the customer's requirements, expectations, and perhaps a few of his or her unspoken wishes.

A structured methodology using these tools to elicit or gather requirements helps to reduce requirement churn or changes, thus reducing product development time, decreasing the risk and cost of developing requirements that don't align with customer expectations, and providing an opportunity to validate the requirements with the customers before the team decomposes the system requirements to hardware and software requirements.

Concept Generation and Selection

POSITION WITHIN DFSS FLOW

Concept generation involves a different frame of mind than most other steps involved in product development, including concept selection. Whereas most other steps involve engineering rigor and judgment, the concept generation step eschews judgment, which can impede the flow of ideas. Nonetheless, the goal of this step—to generate and subsequently select a superior concept for the product, its subsystems, or its modules and components—has repercussions throughout the rest of the new product development process, and may ultimately become the primary factor leading to the product's success or failure. Figure 8.1 highlights the position of the concept generation and selection steps, early in the new product development process.

CONCEPT GENERATION APPROACHES

Many concept generation approaches have been proposed from a variety of perspectives; rather than provide these as a smorgasbord of possible approaches, Figure 8.2 shows a detailed flowchart that describes where each of these concept generation approaches may be appropriate. Although it is possible that all of these approaches might be valuable for concept generation for a single complex product, it is more likely that a subset of these approaches will be found to be valuable for each product development effort.

Concept generation at a subsystem, module, or component level is similar but not identical to concept generation at a product or system level. Nonetheless, many complex

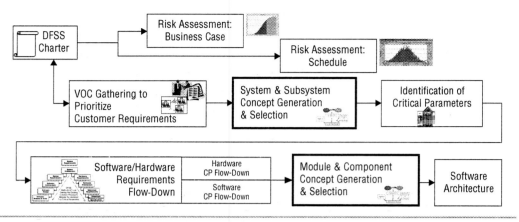

Figure 8.1 DFSS Flowchart, highlighting concept generation and selection steps

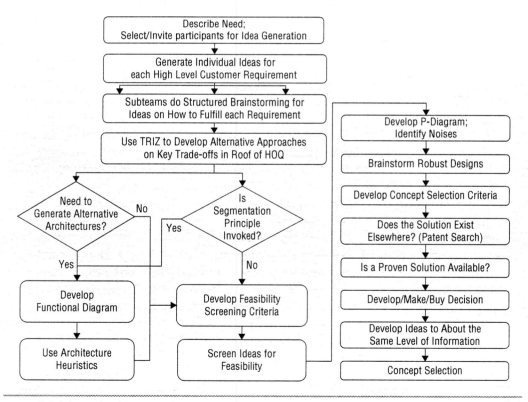

Figure 8.2 Detailed flowchart describing a proposed sequence of steps for concept generation

products involving software and hardware aggregate subsystems that in other circumstances may constitute final products. For example, a cellular phone can include a digital camera, an audio player, and removable memory—separate products or systems in other circumstances.

A primary distinction between system level and subsystem/module/component level concept generation is the source of the needs to be met by the concept. At a higher level of abstraction, particularly the product or system level, the needs are based on the VOC, as translated into measurable technical requirements via the first House of Quality. These needs primarily relate directly to the user experience—the look and feel of the product and the user interface.

For subsystems, modules, and components, many—and perhaps all—of the needs that are considered in concept generation have been flowed down and allocated from system level requirements. There are exceptions—for example, customer needs associated with the camera subsystem of a cellular phone may be derived directly from the VOC—but many needs will be less directly tied to the customer experience, and more tied to flowed down requirements of power consumption, reliability, delay times, and other aspects of the user experience, as well as technical requirements that are derived from VOB and technical challenges that the customer may not directly see or feel.

Concept generation at the system level and subsystem level may also provide a major deliverable—the architecture. The system or subsystem architecture defines how the components are arranged and interfaced to provide the functionality and support the requirements for the system or subsystem. A decision regarding the architecture influences many subordinate decisions during product development, but is (hopefully) made only once. This chapter will touch on some approaches for considering alternative hardware architectures, and Chapter 12 will discuss approaches for software architectures.

Many of the approaches discussed in this chapter can be books or courses in themselves; some consultants have developed particular expertise in some of these approaches. Although this chapter will not even attempt to explain the approaches to that depth, it is indisputable that people will not use an approach that they are totally unaware exists. Consequently, the intent of this chapter will be to introduce the approaches, show where they can be used, and provide enough insight to allow the reader to determine whether they should delve further, and perhaps read an in-depth book or contact a consultant.

Another approach is Kansei engineering; this holistic approach will be mentioned in the appendix to this chapter as an alternative to the proposed sequence of steps for concept generation and selection that constitute the remainder of the chapter.

BRAINSTORMING AND MIND-MAPPING

Brainstorming is perhaps the most common method for generating ideas; it provides a forum for a free flowing stream of ideas, sustained through contributions by participants to developing variations on a theme. For brainstorming to be most effective, it needs a shared goal or common need for the participants to focus on, a reasonably diverse set of participants with a shared goal, some ground rules to prevent judgmental actions from deflating the atmosphere of open exchange of ideas, and a leader to keep the flow going, involve the more introverted participants, and maintain the ground rules.

Brainstorming sessions can address the entire solution, or the entire user experience, or can be partitioned into subsystems (like a camera or audio subsystem for a wireless device), features (like location-based services, GPS), or sets of requirements or components. The brainstorming team can be provided with some initial information regarding the subsystem, feature, or component and the expected key functions. The team will also find a summary of the relevant customer requirements and the associated measurable technical requirements helpful.

Brainstorming is more effective if the leader clearly describes the needs, in solution-independent terminology, and is committed to sustaining the free flow of ideas. After describing the needs (functions and relevant customer-derived requirements), the leader selects a small but somewhat diverse group of two to ten participants that includes marketing and engineering. Participation of higher-level managers can stifle the free flow of ideas, transforming other participants into "yes men," and participants with a negative attitude or a predetermined "best solution" can also impede effective brainstorming. The leader should suppress attempts to screen ideas for feasibility, judge which ideas should be considered, or advocate an idea; these would impact the flow of ideas. Although brainstorming is usually very effective at generating a useful set of ideas, the forum setting may inhibit the flow of some individual ideas simply because the participants start to focus first on the ideas flowing in. The leader can ask the participants to generate their own individual lists of about five ideas beforehand, and can gather those lists to use as new starting points later in the brainstorming session when a stream of ideas starts to slow.

It is difficult (but not impossible) for the leader to serve as the recorder as well, but it is more effective to have a separate team member be the recorder. The recorder or the leader should verify with the participants that the recorded version of each idea reflects what the participants meant. Some brainstorming teams have found approaches like "mind-mapping" helpful; this approach captures the ideas in circles or boxes, and uses lines or arcs to show how some ideas flowed from other ideas, as illustrated by the mind-map in Figure 8.3 that captured a brainstorming session regarding ways to input information into a portable device. Brainstorming sessions should generate a diverse

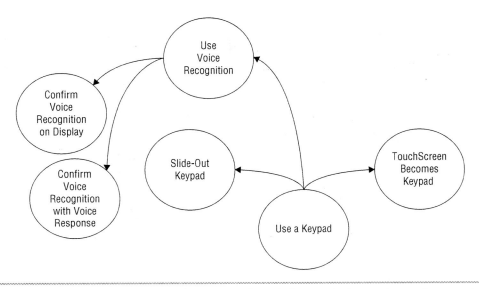

Figure 8.3 Example of mind-mapping to capture the flow of ideas in a brainstorming session regarding data entry methods for a portable device

and useful set of ideas that can be the whole set of ideas, or can be supplemented by other approaches such as TRIZ.

TRIZ

The House of Quality not only provides a summary of the customer requirements and associated measurable technical requirements but also summarizes the trade-offs among the technical requirements in the "roof" (see Chapter 7). These trade-offs can be addressed by a directed method of concept generation called "TRIZ," from the Russian phrase Teoriya Resheniya Izobreatatelskikh Zadatch,[1] also referred to as Theory of Innovative Problem Solving (TIPS).

In 1948, Dr. Genrich Altshuller wrote to Joseph Stalin to discuss problems with innovation and invention in the Soviet Union, suggesting a different approach. In response, Dr. Altshuller was arrested and spent years in the prison and Gulag system. Nonetheless, he and his friend, Rafael Shapiro, analyzed an enormous number of patents, and found

---

1. Genrich Altshuller, *The Innovation algorithm—TRIZ, systematic innovation and technical creativity,* Technical Innovation Center, 2000.

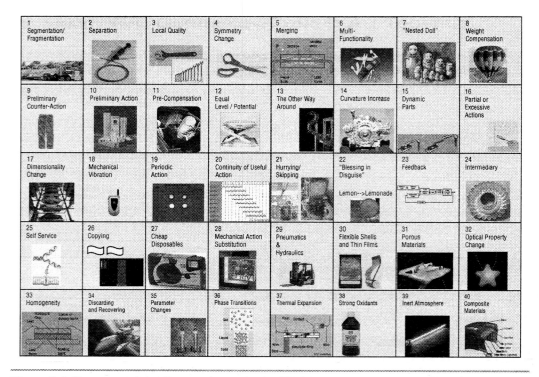

1 Segmentation/ Fragmentation	2 Separation	3 Local Quality	4 Symmetry Change	5 Merging	6 Multi-Functionality	7 "Nested Doll"	8 Weight Compensation
9 Preliminary Counter-Action	10 Preliminary Action	11 Pre-Compensation	12 Equal Level / Potential	13 The Other Way Around	14 Curvature Increase	15 Dynamic Parts	16 Partial or Excessive Actions
17 Dimensionality Change	18 Mechanical Vibration	19 Periodic Action	20 Continuity of Useful Action	21 Hurrying/ Skipping	22 "Blessing in Disguise" Lemon-->Lemonade	23 Feedback	24 Intermediary
25 Self Service	26 Copying	27 Cheap Disposables	28 Mechanical Action Substitution	29 Pneumatics & Hydraulics	30 Flexible Shells and Thin Films	31 Porous Materials	32 Optical Property Change
33 Homogeneity	34 Discarding and Recovering	35 Parameter Changes	36 Phase Transitions	37 Thermal Expansion	38 Strong Oxidants	39 Inert Atmosphere	40 Composite Materials

Figure 8.4 Illustration of the 40 principles involved in TRIZ for resolving trade-offs (© 2003 G. Tennant, Mulbury Consulting, Limited).

that the most innovative patents often were derived from identifying trade-offs among requirements. They suggested applying a subset of 40 principles (Figure 8.4) depending on the nature of the trade-off.

The TRIZ method is perhaps best illustrated with an example. During a class, some participants wanted to try to apply the TRIZ method to the trade-off posed in trying to increase the viewing area for a cell phone display (in TRIZ terms, improving the area of a stationary object: the display) without making the cell phone larger (represented in TRIZ terms as the undesired result impacting the length of a stationary object). Figure 8.5 shows the relevant subset of the 40 principles that were highlighted through using an Excel-based TRIZ Contradiction Matrix for this trade-off (http://www.sigmaexperts. com/dfss/).

The TRIZ Contradiction Matrix suggested four principles: nesting dolls, preliminary counteraction, copying, and inert environment. The participants in this TRIZ

```
Tradeoff for TRIZ:

Feature to Improve:    Area of Stationary Object

Undesired Results:    Length of Stationary Object
```
```
Relevant Principles:

        07) Nesting Dolls
        09) Preliminary Counter-Action
        26) Copying
        39) Inert Environment
```

Figure 8.5 Application of a TRIZ contradiction matrix to the trade-off for increasing the viewing area of the cell phone display without lengthening the cell phone (based on TRIZ Contradiction Matrix © 2003 G. Tennant, Mulbury Consulting, Limited).

exercise then discussed how each of these four principles could be applied to the cell phone display. Ideas included having a retracted display that could be pulled out or could spring out, based on the nesting dolls and preliminary counter action principles.

The copying principle led to the intriguing idea of using an approach in which a very small display, as on an LCD chip, could be seen through a hole and lens system, like a pinhole camera, such that the end-user could see a very large virtual image . . . but the small hold and lens, and underlying LCD chip, would take up virtually no area. The inert environment principle led to another intriguing idea—project an image from the display onto a nearby surface, like a wall or desktop. This approach was subsequently pursued on a research basis in a joint effort with Microvision[2] (Figure 8.6).

The first of the 40 principles associated with TRIZ is the principle of segmentation; this principle is closely aligned with the approaches for generating alternative architectures, which are discussed in the next section of this chapter.

ALTERNATIVE ARCHITECTURE GENERATION: HARDWARE AND SOFTWARE

In some cases, the key conceptual decision is the architecture, which involves the partitioning of the functionalities. For software, the architecture is a key conceptual

---

2. Thomas Fredrickson, *International Business Times,* July 26, 2007.

Figure 8.6 Application of TRIZ principle 39 (Inert Environment), as shown in an article regarding research into a projection display for a cell phone. From Thomas Fredrickson, *International Business Times*, July 26, 2007, with permission.

decision; for both hardware and software, the intent will be to generate several alternative architectures that align with the functional and nonfunctional requirements or constraints.

The first step of architecture generation for hardware or software would be to identify and distinguish the functional requirements from the constraints or nonfunctional requirements. From the hardware perspective, constraints are properties for the product as a whole that cannot be partitioned or allocated to a subsystem; examples include the weight of the system, the cost of the system, and the reliability of the system. Nearly everything in a complex product contributes to the weight, with the possible exception of the software; similarly for cost and reliability. From the software perspective, non-functional requirements include quality attributes such as security, usability, availability, modifiability, and testability.

The second step, after identifying and differentiating between the functional and nonfunctional requirements, is to organize the functions such that the flow of information between functions is clear. For hardware, functional modeling provides such an approach. Figure 8.7 shows a functional model for a wireless transmitter and overlays some key requirements derived from the VOC. For software, approaches such as UML (unified modeling language) provide a potentially more comprehensive

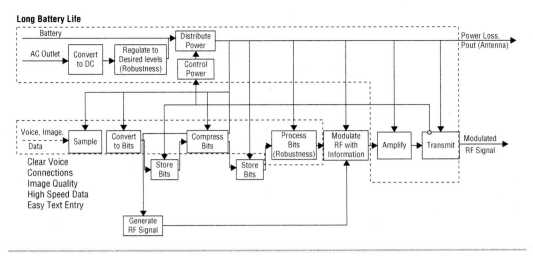

Figure 8.7 Functional model for wireless transmitter; dashed outlines highlight sections aligned with fulfilling key requirements for long battery life and for high quality voice, image, and data

approach for clarifying functions and interrelationships among functions and with the users.

The third step is to generate alternative architectures that meet the functional requirements, meet non-functional requirements or constraints, and support the VOC. The alternative architectures can be generated using heuristics—rules of thumb—to identify potential groupings of functions into modules. Some of these heuristics include:[3]

- Independence: partitioning to minimize flows between modules.
- Branching flow: partitioning before or after the information flow branches between various inputs or outputs.
- Dominant flow: partitioning to include an analog flow, a digital flow, or a material or energy flow within the module.

Some heuristics are illustrated in Figure 8.8 for command inputs for a game using either voice commands or a joystick. A much deeper dive and more examples are provided in Chapter 9. The independence heuristic is also an important aspect of the axiomatic design approach.[4]

3. Kevin Otto and Kristin Wood, *Product Design: Techniques in Reverse Engineering and New Product Development*, Prentice Hall, 2000.

4. Nam Pyo Suh, *Axiomatic Design: Advances and Applications*, Oxford University Press, 2001.

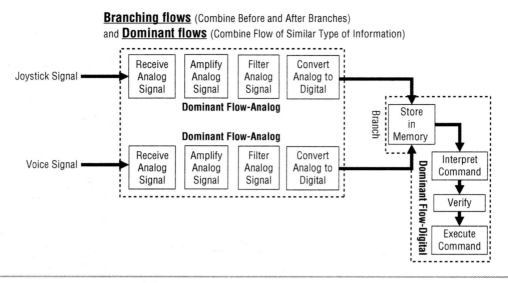

Figure 8.8 Partitioning using two heuristics for information flows for inputting commands in a game

As will be discussed in Chapter 12, Software Architecture Decisions, heuristics also can help in generating alternative software architectures. Software architectural heuristics include:

- Separation—isolating a portion of a system's functionality into a component
- Compression—removing layers or interfaces (the opposite of Separation)
- Abstraction—creating a virtual machine
- Resource sharing—sharing information among multiple independent users
- Decomposition—separating a larger system into smaller components, such as non-overlapping portions of the functionality, or specializations of the functionality

Architectural heuristics can be used to generate several alternative concepts that can stand alone, or supplement the concepts generated through brainstorming and TRIZ approaches.

GENERATION OF ROBUST DESIGN CONCEPTS

In Chapter 14, we will discuss optimization for robustness, defined as relative insensitivity to factors that you cannot control in the actual use environment. Robust design starts with a solid design concept that makes the design relatively insensitive to noises—variations in manufacturing, variations in the use environment (such as temperature and humidity), and variations in how each customer actually uses the product.

The first step in generating alternative robust design concepts is to identify the noises. These can be brainstormed or developed as part of a P-diagram, as discussed in Chapter 10. The second step is to choose which noises should be addressed through the robust design—manufacturing variation, temperature, use cases, and so on.

For each noise, the individual or team members could consider the four methods for making the design insensitive to the noise, as discussed in Chapter 14 and summarized here:

- Reducing the noise directly, such as through tightened control of manufacturing variation or immersing the product in a controlled environment.
- Determining the relationship between the noise and its impact on performance, and finding an insensitive region of that relationship, such as finding a setting for threshold voltages where the timing and output voltage levels are not very sensitive to manufacturing variation (an example is provided in Chapter 14), or, if vibration is a key noise and is a function of the thickness of a layer of material, increasing the thickness until the product is less sensitive to vibrations.
- Determining if the noise has an interaction with another factor, and if so, setting the other factor where the performance is less sensitive to the noise, such as choosing an inductor in an oscillator so frequency is less sensitive to variations in capacitor values.
- Considering ways to use correlations between factors to offset the sensitivity to a noise; this approach is used in difference amplifiers and noise cancelling methods in audio systems, and is described further in Chapter 14.

CONSIDERATION OF EXISTING SOLUTIONS

Considering existing solutions too early in the concept generation process may tend to inhibit concept generation within the design team. After the team has generated a wide range of concepts, existing solutions can be entered into the mix, to develop an even broader range of potential solutions. Some of the existing solutions will likely overlap solutions generated by the design team; other existing solutions will involve reuse of previous designs by the same design team or other design efforts within the enterprise. Other existing solutions may have been developed outside the company, and may involve patent searches and other research methods (e.g., http://www.uspto.gov to search U.S. patents, http://gb.espacenet.com to search European patents, and commercial sites such as http://patentdatabase.com and http://www.micropat.com).

A major advantage of existing solutions is that they are proven solutions, often having had the "bugs worked out," and thereby reducing risk. A major disadvantage may be the cost involved; the make or buy decision can be considered during the concept selection process.

FEASIBILITY SCREENING

Some concept generation methods, especially nonjudgmental methods such as brainstorming, are likely to produce both reasonable and not-so-reasonable ideas. The ideas should be screened to eliminate less reasonable and patently ridiculous ideas, to reduce the work required in preparing for concept selection. Table 8.1 lists a small set of potential screening criteria, along with examples for illustration—some from actual brainstorming sessions.

DEVELOPING FEASIBLE CONCEPTS TO CONSISTENT LEVELS

The concepts that survive feasibility screening should be developed to a point at which the design team can feel comfortable in the concept selection process. This development can include sketches, block diagrams, initial cost projections, and initial DFMEAs;

Table 8.1 Potential criteria for feasibility screening, along with some extreme examples

Potential Screening Criteria	Example
Cost	Using embedded diamonds to ensure a cheap (~$5) saw will cut material effectively.
Resources	Developing a personal reply and coaching network for all user concerns, requiring 10,000 people worldwide.
Schedule Impact	Relying on developing a chip set that will require 2 years to develop . . . for a product to be shipped this year.
Beyond capabilities with current or imminent technology	Developing a power supply, using room temperature fusion, that fits in your hand.
Safety issues	Providing a power supply using nuclear fission that fits in the palm of your hand.
Major mismatch in size or weight requirements	Including a cooling system (near absolute zero) for optimal operation of a supercomputer chip . . . in a 4G cell phone.

CONCEPT SELECTION

however, the concept selection process can be biased if one concept undergoes considerably more conceptual development than the others. Consequently, all of the feasible
concepts to be considered during concept selection should undergo the same preparation to a similar level of detail.

CONCEPT SELECTION

Dr. Stuart Pugh's approach for concept selection has been found to be a simple yet
powerful and effective method for triangulating in on a superior concept. A side benefit
of the Pugh concept selection process is that the design team recognizes the approach as
simple and fair, so that, once the team converges on a solution, there is general acceptance and support ("buy-in") from the team to go forward with the selected concept
with confidence that a reasonable decision was made.

> Experience gained over many projects . . . has led to the conclusion that matrices in general
> are probably the best way of structuring or representing an evaluation procedure . . . the
> type of matrix referred to here is not a mathematical matrix; it is simply a format for
> expressing ideas and the criteria for evaluation of these ideas in a visible, user-friendly
> fashion.[5]

The Pugh concept selection approach generally is applied in considering alternative
concepts for a product or subsystem or component of a product. It uses the method of
controlled convergence, which can involve several iterations of the matrix.

The matrix shown in Figure 8.9 consists of selection criteria along the left-most column, the datum and alternative concepts or solutions across the top row, and symbols
in the rows and columns between. The selection criteria include representations of the
VOC, generally in measurable form as translated into technical requirements in the first
House of Quality, plus other important criteria representing other voices, including the
VOB—criteria such as the projected cost, projected development time, and projected
reliability for that concept.

The datum is a reference concept—generally the benchmarked best-in-class current solution; it can be the concept used in the current product or in a competitor's
product. The top row can have names, descriptions, or sketches of the concepts
under consideration.

5. Stuart Pugh, *Total Design—Integrated Methods for Successful Product Engineering,* Chapter 4, Addison-Wesley
Publishing Company, 1991.

149

Iris/Retinal Looks Good	Datum (Pass Code)	Iris/ Retinal	Voice Print	Finger Print	
				Optical	Chip
Failure to Acquire		+	−	+	−
Acquisition Time		+	+	+	−
Risk of False Accept		+	−	+	+
Risk of False Reject		+	+	+	+
Robustness to Ambient Sound		+	−	+	+
Robustness to Ambient Temp, Humidity, Dust		S	+	−	−
Stability of Technology		+	−	+	+
Pluses		6	3	6	4
Minuses		0	4	1	4
Sames		1	0	0	0

A hybrid of Optical AFIS and Voice Print Looks Promising

Figure 8.9 Pugh Concept Selection Matrix for selecting a technology for security involving the verification of a person's identity. The datum is the current approach of entering a pass code.

For each cell of the matrix, the team reaches consensus on a simple rating system: for that criteria, and for that concept, is the solution better than the datum (+), or is it worse than the datum (−), or about the same (S)? After the matrix is completely filled with +, −, and S symbols, the columns representing each concept are summarized in terms of the total count of pluses, minuses, and sames. Some of the weaker concepts become apparent in terms of having few pluses and a surplus of negatives. Figure 8.9 shows a subset of a matrix for biometrics approaches for security, concepts to recognize the user as the appropriate person to engage in e-commerce. The voiceprint concept doesn't excel in comparison to the datum, a simple pass code for identity, whereas the iris/retinal scan and optical fingerprint approaches appear to have considerably more favorable aspects.

The weakest concepts are eliminated, reducing the matrix size. New concepts can be synthesized, perhaps as hybrids developed by the team looking at the weaknesses (negatives) for the better concepts and finding ways to overcome those weaknesses, perhaps by borrowing from other concepts that were strong in that aspect. The best concept can now be used as the datum or benchmark, and the matrix rerun to see if the best concept continues to excel, converging on a superior concept to pursue.

A major benefit of the Pugh concept selection approach is that it is perceived by the team to be unbiased, so that the team can achieve consensus with the feeling that the best concept won. It is simple enough that the team does not require much training to use the approach.

A common question is whether the selection criteria should be weighted, so that a favorable result for a very important criterion carries more weight than a favorable result for a less important criterion (Figure 8.10). Stuart Pugh stated that weighting " . . . implies an attempt to impart to the procedure too much precision, and thus inhibit qualitative judgements."[6] Furthermore, the determination of the relative weightings can require considerable time for debates within the team, and can inspire attempts to bias the weightings to favor a concept.

Rather than weighting, Stuart Pugh suggested "what has been found useful is the grading of criteria into three categories—important, moderately important, and desirable . . ."

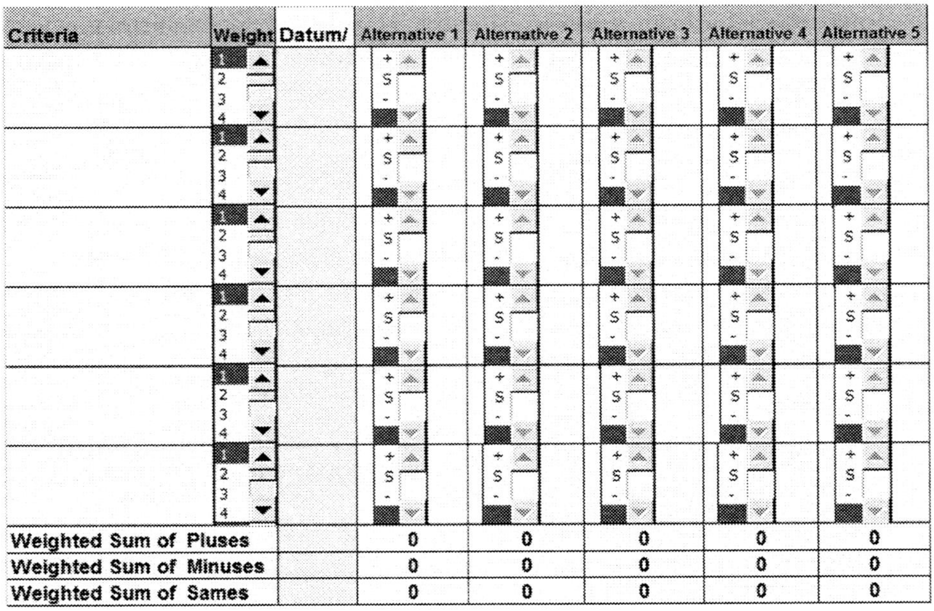

Criteria	Weight	Datum/	Alternative 1	Alternative 2	Alternative 3	Alternative 4	Alternative 5
	1 2 3 4		+ S	+ S -	+ S -	+ S -	+ S -
	1 2 3 4		+ S	+ S	+ S -	+ S -	+ S -
	1 2 3 4		+ S -	+ S -	+ S -	+ S -	+ S -
	1 2 3 4		+ S -	+ S -	+ S -	+ S -	+ S -
	1 2 3 4		+ S -	+ S -	+ S -	+ S -	+ S -
	1 2 3 4		+ S -	+ S -	+ S -	+ S -	+ S -
Weighted Sum of Pluses			0	0	0	0	0
Weighted Sum of Minuses			0	0	0	0	0
Weighted Sum of Sames			0	0	0	0	0

Figure 8.10 Weighted version of concept selection matrix, with weighting for relative importance of each criterion

SUMMARY

The selection of the concept or concepts to be used in a complex new product has tremendous influence on later stages of product development. A poorer concept may require considerable "patching" and "make-do's." The concept generation and selection processes help to ensure that a superior concept is selected by generating several concepts and selecting a superior concept—as opposed to starting with a predetermined solution, or from the best guess from one individual, or through a process of submitting to whoever shouts the loudest.

The concept generation process harnesses creativity among the individuals and the team to generate a set of concepts that will likely include or lead to a superior concept. The concept selection process engages the team in finding the superior concepts, and often developing an even better "hybrid" solution. A side benefit is the development of teamwork among the team, and buy-in for the concept they perceive to have been selected through a fair and reasonable process.

APPENDIX: KANSEI ENGINEERING

An alternative set of approaches for concept generation involves starting with a holistic view of the product, realizing that much of the purchase decision and customer satisfaction may be derived from the how the customer feels about the product. Kansei Kougaku, translated as "sense engineering," was originally developed by Professor Mitsuo Nagamachi of Hiroshima International University.

The first step is to develop a phrase that provides an emotional, sensory description of how the team hopes the customer will feel about the product. For the Mazda Miata, the phrase related to the feeling that the man and the machine were as one. For a cellular phone for teenagers, the phrase might invoke the feeling the cell phone is like a cool, helpful friend.

The team then tries to find measurable parameters that support that sensory description. Realizing that a direct flow-down (like the QFD first House of Quality) may not be applicable, the trained team develops pairs of terms or keywords that convey the range of physical parameters, and that can be correlated to the customers' feelings. Representative customers use rating scales for those ranges of physical parameters to indicate the level of correlation to the desired feeling for the product, and a multivariate method such as Factor Analysis can be used to determine the correlation between the physical parameters and associated design features to the desired feeling for the product.

There are also more advanced versions of Kansei engineering, many of which invoke expert systems approaches. More information on Kansei engineering approaches can be obtained from the Japan Society of Kansei Engineering (JSKE), http://www.jske.org/main_e.html.

Identification of Critical Parameters and FMEA

POSITION WITHIN DFSS FLOW

A vital focus of Six Sigma is to focus. For problem solving, this corresponds to focusing on a vital few within the DMAIC flow. For developing new products, the product team must be cognizant of all requirements; but to differentiate the product, manage risks, and ensure a successful new product launch, the team must focus on a vital few, measurable requirements, which are referred to as critical parameters.

This step and set of deliverables is aligned with the Requirement phase of the RADIOV DFSS process, the latter part of the Concept phase of the CDOV process, and to the transition from the Define to Measure phase of the DMADV process of DFSS. The DFSS flowchart, available for download at http://www.sigmaexperts.com/dfss/, provides a high level overview (Figure 9.1) of the sequence of steps that can be drilled down to more detailed flowcharts. Figure 9.2 is the detailed flowchart aligned with this chapter, showing the steps and methods involved in identification of critical parameters. At this stage of DFSS, there has not yet been a separation of the system's hardware and software requirements, so the steps shown in Figure 9.2 would be appropriate for a wide variety of products, including but not limited to complex electronic systems.

DEFINITION OF A CRITICAL PARAMETER

A critical parameter is a measurable requirement that has a strong impact on customer satisfaction (voiced or tacit) and also involves substantial risk to the success of the project

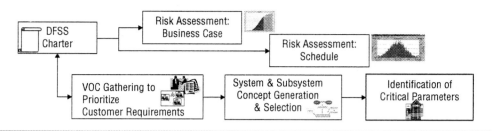

Figure 9.1 DFSS Flowchart overview highlighting step for identification of critical parameters

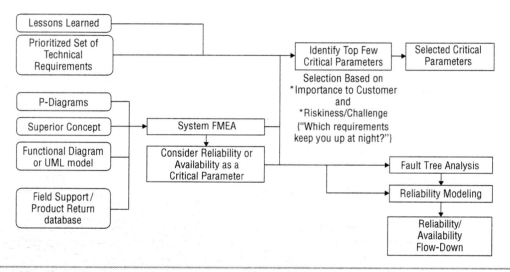

Figure 9.2 DFSS Flowchart, drilled down to detailed flowchart for identification of critical parameters

or product. The critical parameters will often consist of a prioritized subset of the technical requirements that emerge from the first House of Quality, as described in Chapter 7. Hence, a key input for this step will be the prioritized set of technical requirements, shown near the top left of Figure 9.2. Additional considerations might come from lessons learned from previous products, from competitors' products, or from products that potential customers use in conjunction with the current generation of the product.

Problems that have emerged in the past are potential risks for the next product, and the measurable parameters associated with the severity of the historical problems can benefit from the focus applied to critical parameters during development. For example, a group involved in developing cellular phones had historical problems with the ruggedness of the phones, as measured by drops-to-failure, and had once delayed a launch due to issues with total radiated power transmitted from the cell phone antenna. These were designated as critical parameters for the next-generation cellular phones under development.

CONSIDERATIONS FROM **VOB** AND **CONSTRAINTS**

The definition for critical parameters provided earlier focuses on the voice of the customer (VOC), and gently evades the topic of the voice of the business (VOB). The VOB generally involves two types of requirements: implicit VOC, and explicit success metrics for the project. Implicit VOC are requirements that were not voiced by the customer, but relate to customer satisfaction or delight—features or characteristics that will pleasantly surprise and impress the customers ("delighter" in Kano terminology), which may be the competitive edge or key requirements for this new product that the executives expect and require for the product. Implicit VOC might also include reliability or availability, which are expected by the customers but sometimes not stated explicitly during VOC gathering. Reliability and availability will be discussed in more depth later in this chapter.

Explicit success metrics for the project generally align with program risks. As described in Chapter 3 and summarized in Table 9.1, these program risks include schedule risk and financial or business case risk. Associated success metrics that are likely to be imposed upon the team are projected deviations from schedule (schedule slippage) and projected bill of material costs. Both of these are important, but are linked or impacted—at least to some extent—by decisions made by the product development team, and behave as constraints.

The distinction between functions and constraints was mentioned in Chapter 7 but bears repeating in this context. Critical parameters will generally be measurable in terms of the degree to which the product will perform a function that is desired by the customers—such as communicating information, displaying pictures, or playing sounds. The flow-down or allocation of such critical parameters will generally (although perhaps not completely) align with the hierarchical architecture of the system. For example, measurable requirements for audio quality and loudness will generally align with the architecture of the audio subsystem, consisting of amplifiers, speakers, or earphones.

By contrast, some customer requirements (and some business requirements) will translate into measurable technical requirements that do not involve functionality but, rather, properties of the system. For example, there is no subsystem that invokes the expectation of "low cost" or "lightweight"; these are properties of the total system. Moreover, the flow-down of constraints will generally involve all or nearly all

Table 9.1 Summary of new product development risks, and some associated metrics

Risks	Metrics
Business Case Risk	Forecasted Pr(Profits)
	Forecasted Costs
	Forecasted Profit or Gross Margin
	Real/Win/Worth
	Forecasted Market Share
Technical Risk	Composite Cpk
	Forecasted Composite Yield
	FMEA RPN Improvement
Schedule Risk	Pr(Schedule Slippage)
	Forecasted Completion Data
	Forecasted Schedule Slippage
	Number of Iterations
Reliability Risk	Forecasted Reliability
	Forecasted Availability
Product Delivery Risk	Forecasted Pr(On-Time Delivery)
	Forecasted Lead Time
Instability of Requirements	Number of Spec Changes during Product Development
	Percent Alignment to Expectations

hierarchical subsystems: the bill of materials (BOM) cost of a computer will essentially be the sum of the costs for each component in each subsystem and the cost to assemble them. The flow-down of weight will essentially be the sum of the weights for each subsystem, with the exception of software.

Generally speaking, measurable requirements related to functionality are preferred for critical parameters. Constraints such as cost and weight might be considered as critical parameters, or omitted from the list of critical parameters; regardless, they are frequently important (often critically important) considerations in the design process, and it is important to keep track of the value of these constraints as decisions are made during the design process.

PRIORITIZATION AND SELECTION OF CRITICAL PARAMETERS

The set of technical requirements will mostly likely exceed the capacity of the team to focus. The identification of critical parameters is both a filter and a prioritization derived from two key considerations: how important is this requirement to the customers, and how intense a focus does the team need to apply in order to have high confidence that the product will meet or exceed the customers' expectations? Note that, for an unexpected "requirement" (a "delighter" in the Kano analysis perspective), the filter and prioritization may instead be based on the importance that this requirement will have in terms of product differentiation and potential market share, and the intensity of the focus needed in order to have high confidence that the customer will be delighted by the new feature or characteristic of the product and will not be frustrated or overly mystified by the complexity (for example, of the user interface) that could undermine the delight.

The deliverables from the first House of Quality include the relative importance to the customers for each of the technical requirements, derived from the relative importance for the associated customer requirements. If there is consensus within the team and the stakeholders regarding the relative importance indices for the set of technical requirements, then the team and stakeholders would need to reach consensus on indices for added requirements based on "lessons learned," on a consistent scale.

The second key consideration is the intensity of focus required to have sufficiently high confidence that the requirement will be met, satisfying or delighting the customer and preventing problems—the level of risk or challenge involved. As shown in Figure 9.2, there are several potential sources for information to help with evaluating the risk and challenge associated with each of the measurable requirements. Particularly for requirements emerging from "lessons learned," (measurable parameters indicative of historical problems) but also for some of the technical requirements that emerged in support of meeting customer requirements, information from field support, and analysis of data in the product return database could provide the context and an estimate of the probability of occurrence associated with the risks, based on historical data and experience. The level of risk associated with each technical requirement can be assessed using a system-level FMEA, as described in the next section of this chapter.

The system-level FMEA (Figure 9.3) evaluates risks associated with many of the measurable, technical requirements; the development team could explicitly include each measurable, technical requirement in the deliberations such that the system-level FMEA would provide RPN indices that could be treated as numerical assessments of

Summary: A structured method for identifying and ranking the significance of various failure modes of the program and their effects on the product or customer

System, Subsystem or Component	Parameter at System, Sub-system, Module, Assembly or Component Level	Potential Failure Mode / Effect	Potential Effects of Failure	SEVERITY	Potential Cause	OCCURRENCE	Current Controls	DETECTION	RISK PRIORITY NUMBER (RPN)	Action	Responsibility	Due Date	Severity	Occurrence	Detection	Risk Priority
				9		7		3	189				6	5	3	90
				8		7		4	224				6	6	2	72
				9		7		3	189				5	4	2	40
				9		7		3	189				5	3	2	30
				9		7		3	189				4	2	3	24
				9		7		3	189				4	4	1	16

Failure Modes and Effects Analysis (FMEA):   Revised RPN

1.   List Functions and Interfaces or Subsystems
2.   List Potential Failure Modes.
3.   List Potential Effects.
4.   Assign Severity Rating.
5.   List potential causes.
6.   Assign Occurrence Rating. (Probability rating for Software)
7.   List current controls.
8.   Assign detection rating.
9.   Calculate Risk Priority Number.
10. Use RPN's to help decide on high priority faiulure modes.
11. Plan to reduce or eliminate the risk associated with high priority failure modes.
12. Re-compute RPN to reflect impact of action taken on failure mode.

Output:
- Ranked group of failure modes
- Impacts of failures
- Risk Priority Numbers (RPN) before and after corrective action
- Corrective actions, controls to remove or reduce the risk or impact of a failure mode

Figure 9.3 Tool summary for FMEA

risk, just as the translated relative importance from the first House of Quality could be treated as a numerical assessment of customer importance. The selection of a superior concept, described in Chapter 8, is a key starting point for the system-level FMEA effort. Other deliverables from the concept generation and selection process can include P-diagrams (which, in addition to summarizing control and noise factors and desired results, show undesirable, unintended results that relate to failure modes for FMEA) and functional models or UML (universal modeling language) models and use cases, which could also be reused as valuable inputs for the team involved in the system-level FMEA effort. For example, the functional diagram could indicate interfaces and complex information flows that could pose risks for data communication between subsystems; things often go wrong at interfaces. The error states or deviations from the ideal function (at the lower right of P-diagrams) could suggest failure modes to be included in the system-level FMEA, and the noises (at the top of the P-diagrams) could suggest potential causes for the failure modes.

An alternative to the system-level FMEA approach could be to have representatives or stakeholders (representing the teams involved in efforts to meet each technical requirement) provide a numerical assessment of the degree of risk involved, perhaps on a 1 to 10 scale, and the average or weighted average of these numerical assessments could be used as a relative index of riskiness, as shown in Figure 9.4. The assessment of riskiness would involve consideration of various types of risk: performance risk and perhaps reliability risk, but also schedule risk and cost or business case risk. The team members could assess risks in general, or each of these risks separately—the key point being that there should be consistency among the assessors as to what type of risks they are assessing and a general sense of what the range of the scale represents. The team might choose to translate all of the types into one type of risks; for example, the assessors could agree that, given sufficient time, all performance risks could be met, so each of the relative risks could be treated as schedule risk assessments—the relative impact on the schedule for meeting each requirement with high confidence.

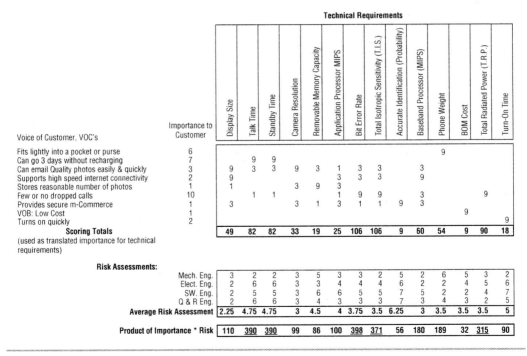

Technical Requirements

Voice of Customer, VOC's	Importance to Customer	Display Size	Talk Time	Standby Time	Camera Resolution	Removable Memory Capacity	Application Processor MIPS	Bit Error Rate	Total Isotropic Sensitivity (T.I.S.)	Accurate Identification (Probability)	Baseband Processor (MIPS)	Phone Weight	BOM Cost	Total Radiated Power (T.R.P.)	Turn-On Time
Fits lightly into a pocket or purse	6											9			
Can go 3 days without recharging	7		9	9											
Can email Quality photos easily & quickly	3	9	3	3	9	3	1	3	3		3				
Supports high speed internet connectivity	2	9						3	3	3	9				
Stores reasonable number of photos	1	1			3	9	3								
Few or no dropped calls	10		1	1				1	9	9	3			9	
Provides secure m-Commerce	1	3			3	1	3	1	1	9	3				
VOB: Low Cost	1												9		
Turns on quickly	2														9
Scoring Totals		49	82	82	33	19	25	106	106	9	60	54	9	90	18

(used as translated importance for technical requirements)

Risk Assessments:

		Display Size	Talk Time	Standby Time	Camera Resolution	Removable Memory Capacity	Application Processor MIPS	Bit Error Rate	Total Isotropic Sensitivity (T.I.S.)	Accurate Identification (Probability)	Baseband Processor (MIPS)	Phone Weight	BOM Cost	Total Radiated Power (T.R.P.)	Turn-On Time
Mech. Eng.		3	2	2	3	5	3	3	2	5	2	6	5	3	2
Elect. Eng.		2	6	6	3	4	4	4	4	6	2	2	4	5	6
SW. Eng.		2	5	5	3	6	6	5	5	7	5	2	2	4	7
Q & R Eng.		2	6	6	3	3	3	3	3	7	3	4	3	2	5
Average Risk Assessment		2.25	4.75	4.75	3	4.5	4	3.75	3.5	6.25	3	3.5	3.5	3.5	5
Product of Importance * Risk		110	390	390	99	86	100	398	371	56	180	189	32	315	90

Figure 9.4 Example of prioritization and selection of critical parameters using the product of averaged risk assessments from representatives of engineering discipline and the prioritization of technical requirements from the first House of Quality

Consistent with the definition for a critical parameter, but perhaps a bit more intu-
itive for some, is the suggestion that the team members consider the question, "What
keeps you up at night?" If there is one requirement, or a set of requirements, that
appears so difficult or risky that team members are very concerned about how or
whether that requirement can be met—that requirement or set of requirements would
be best handled using critical parameter management. For whatever reason, the "What
keeps you up at night?" criteria for selecting critical parameters, seems to resonate with
many product development engineers.

FMEA

As summarized in Figure 9.3, there are several steps involved in a system-level or subsystem-
level design FMEA; these will be described here, and will be followed by a case study.

1. **List Functions and Interfaces or Subsystems**
   For system-level or design-level FMEA, the concerns will often involve lack or loss of
   functionality. Some teams prefer to focus on the functions associated with each subsystem or component, which helps align with the team efforts, since teams tend to
   be assigned to subsystem designs. However, there may be a tendency to overlook the
   interfaces between subsystems as sources of problems; interfaces should be explicitly
   included in system-level or design-level FMEAs.
2. **List Potential Failure Modes**
   A failure mode describes how a product or service may fail in its function. As mentioned previously, these failure modes are closely related to the unintended, undesired results found at the lower left of a P-diagram, as described in Chapter 10.
3. **List Potential Effects**
   For each potential failure mode, discuss what would be the unfavorable impact if
   that mode of failure occurred.
4. **Assign Severity Rating**
   For each potential effect associated with each failure mode, the team should assign a
   value between 1 and 10 for the severity associated with that effect. A value of 10 is
   always the worst; for severity, a rating of 10 often involves injury or even death. An
   example of a severity rating scale is provided in Table 9.2. A value of 1 generally
   indicates that the effect would be hardly noticeable.
5. **List Potential Causes**
   For each failure mode, the team should brainstorm potential causes. The P-diagram
   might be helpful, as the noises might suggest potential causes in terms of manufacturing variability, environmental variations, usages and misuses, interactions with other

Table 9.2 Example of severity rating scale

Effect	Severity of Effect (Automotive Example)	Rank
Hazardous Without Warning	Very high severity ranking when a potential failure mode affects sage product operation and/or involves noncompliance with government regulation without warning.	10
Hazardous	Very high severity ranking when a potential failure mode effects sage product operation and/or involves noncompliance with government regulation with warning.	9
Very High	Primary product function is inoperable (loss of primary function).	8
High	Primary product operable, but at reduced level of performance. Customer very dissatisfied.	7
Moderate	Primary product operable, but Comfort/Convenience item(s) inoperable. Customer dissatisfied.	6
Low	Primary product operable, but Comfort/Convenience item(s) operable at a reduced level of performance. Customer somewhat dissatisfied.	5
Very Low	Fit & Finish/Squeak & Rattle item does not conform. Defect noticed by most customers (greater than 75%).	4
Minor	Fit & Finish/Squeak & Rattle item does not conform. Defect noticed by 50% of customers.	3
Very Minor	Fit & Finish/Squeak & Rattle item does not conform. Defect noticed by dscriminating customers (less than 25%).	2
None	No discernable effect.	1

subsystems or other nearby systems, and reliability degradation mechanisms. The team might find the 5-why approach (where the team repeatedly asks "Why" to try to determine the reason behind the reason) useful in getting to the potential root causes.

6. **Assign Occurrence Rating (Probability Rating for Software)**
For each potential cause, the team should assign a value between 1 and 10 for the probability of occurrence of that cause. A value of 10 implies the cause is very likely to occur—the worst situation—whereas a value of 1 indicates that the team believes that the cause is extremely unlikely. Table 9.3 shows an example of an occurrence rating scale.

Table 9.3 Example of occurrence rating scale

Probability of Failure	Possible Failure Rates (Automotive Example)	Rank
Very high: Persistent failures	≥100 per thousand functions/items	10
	50 per thousand fuctions/items	9
High: Frequent failures	20 per thousand fuctions/items	8
	10 per thousand fuctions/items	7
Moderate: Occasional failures	5 per thousand fuctions/items	6
	2 per thousand fuctions/items	5
	1 per thousand fuctions/items	4
Low: Relatively few failures	0.5 per thousand fuctions/items	3
	0.1 per thousand fuctions/items	2
Remote: Failure is unlikely	≤0.010 per thousand fuctions/items	1

7. **List Current Controls**

   For each potential cause, the team discusses controls and prevention approaches currently in place to protect the product and the customer. Prevention controls work to prevent the failure, and can include design specifications and the use of modeling. Detection controls involve detection of defects or out-of-spec conditions, and involves measurement or testing.

8. **Assign Detection Rating**

   For each potential cause, the team should assign a value between 1 and 10 for the detection or control rating of that cause. A value of 10 is again the worst case situation, and implies the cause cannot be detected in time to protect the customer, while a value of 1—the best case situation—implies that the cause can be easily detected so that the customer would be protected. Table 9.4 shows an example of a detection (control) rating scale.

9. **Calculate Risk Priority Numbers**

   The RPN is simply the product of the severity, occurrence, and detection ratings; because each can range from 1 to 10, RPN values can range from 1 to 1000.

10. **Use RPNs to Help Decide on High-Priority Failure Modes**

   Higher RPNs indicate higher risk; a potential failure that is very severe, very likely to occur, and difficult to detect (like the possibility of being involved in a car accident) is a more severe risk than a potential failure that would be very severe but is extremely unlikely (like the possibility of having a meteor crash into your house while you are

Table 9.4 Example of detection (or control) rating scale

Effect	Likelihood of Detection by Design Control (Automotive Example)	Rank
Absolute Uncertaintly	Design Control will not and/or can not detect a potential cause/mechanism and subsequent failure mode; or there is no Design Control	10
Very Remote	Very remote chance the Design Control will detect a potential cause/mechanism and subsequent failure mode.	9
Remote	Remote chance the Design Control will detect a potential cause/mechanism and subsequent failure mode.	8
Very Low	Very low chance the Design Control will detect a potential cause/mechanism and subsequent failure mode.	7
Low	Low change the Design Control will detect a potential cause/mechanism and subsequent failure mode.	6
Moderate	Moderate chance the Design Control will detect a potential cause/mechanism and subsequent failure mode.	5
Moderately High	Moderately high chance the Design Control will detect a potential cause/mechanism and subsequent failure mode.	4
High	High chance the Design Control will detect a potential cause/mechanism and subsequent failure mode.	3
Very High	Very high chance the Design Control will detect a potential cause/mechanism and subsequent failure mode.	2
Almost Certain	Design Control will almost certainly detect a potential cause/mechanism and subsequent failure mode.	1

reading this book), or a potential failure that might be very likely but not extremely severe (like the possibility of a young child falling as it learns to walk).

11. **Develop Plans to Reduce the Risk Associated with High-Priority Failure Modes**
The team brainstorms actions that could be taken to reduce the RPN for prioritized failure modes—often, these actions involve reducing the likelihood that it might occur or improving the detection.

12. **Recalculate the RPN to Reflect the Impact of Actions**

SOFTWARE FMEA PROCESS (SOFTWARE SYSTEMS, SOFTWARE SUBSYSTEMS, AND SOFTWARE COMPONENTS FMEA)

BY VIVEK VASUDEVA

Software FMEA is a proactive approach to defect prevention in the software development process. Common software FMEA acronyms are defined in Table 9.5. FMEA involves structured brainstorming to analyze potential failure modes in software, rating and ranking the risk to the software, and taking appropriate actions to mitigate the risk. This process is used to improve software quality, and reduce cost of quality (CoQ), cost of poor quality, (CoPQ), and defect density. Software FMEA roles and responsibilities are outlined in Table 9.6.

FMEA can be performed at the system level and at the network element/component level. The method of conducting an FMEA is documented in the software FMEA process phase in Table 9.7. The steps in conducting a system FMEA are outlined in the Conduct System Engineering FMEA phase in Table 9.7. The steps in conducting a network element/component-level FMEA are outlined in the Conduct Software FMEA for Network Element or Component team phase in Table 9.7.

Table 9.5 Software FMEA acronyms and definitions

Terms	Definitions
FMEA	Failure Modes and Effects Analysis. A **proactive** approach to defect prevention during software development.
VOC	Voice of Customer
FTR	Formal Technical Review
CoQ	Cost of Quality
CoPQ	Cost of Poor Quality
EDA	Escaped Defect Analysis
RCA	Root Cause Analysis
SPE	Software Process Engineer
ROI	Return on Investment

Table 9.6 Software FMEA roles and responsibilities

Roles	Responsibility
FMEA Champion	Provides resources and support. Attends some meetings. Promotes team efforts. Implements team recommendations. Generally, the senior manager who will sponsor the software FMEA activity.
FMEA Lead	Leads the software FMEA effort for the team. Maintains full team participation. The FMEA project leader.
FMEA Recorder	Maintains documentation of team's efforts. Coordinates meeting rooms and times. Distributes meeting minutes and agendas.
FMEA Facilitator	Facilitates the FMEA meetings, provides Six Sigma support such as cause and effect analysis, 5 whys. Responsible for keeping the FMEA sessions focused and productive and ensuring the process is followed.

Table 9.7 Software FMEA process phases

Phase	Description
Planning for FMEA	1. Complete FMEA team charter and obtain management approval 2. Schedule FMEA meetings 3. FMEA tools
Software FMEA Process	• Once the potential failure modes are identified, they are further analyzed, by potential causes and potential effects of the failure mode (cause and effects analysis, 5 whys, etc.). • For each failure mode, an RPN is assigned based on: 　◦ Occurrence rating, range 1–10; the higher the occurrence probability, the higher the rating 　◦ Severity rating, range 1–10; the higher the severity associated with the potential failure mode, the higher the rating 　◦ Detectability rating, range 1–10; the lower the detectability, the higher the rating • One simplification is to use a rating scale of high, medium and low for occurrence, severity, and detectability ratings: 　◦ High: 9 　◦ Medium: 6 　◦ Low: 3

continued

Table 9.7 Software FMEA process phases (continued)

Phase	Description
Software FMEA Process	• RPN = Occurrence × Severity × Detection; Maximum = 1000, Minimum = 1 • For all potential failures identified with an RPN score of 150 or greater, the FMEA team will propose recommended actions to be completed within the phase the failure was found. These actions can be FTR errors. • A resulting RPN score must be recomputed after each recommended action to show that the risk has been significantly mitigated.
Conduct System Engineering FMEA	At the system engineering level, the FMEA consists of: • Complete FMEA team charter, get management approval. Schedule meetings. • Identify and scope the customer critical and high-risk areas. • Front-end (top-down approach) analysis of system documentation. Using the system functional parameters to identify the areas of concern for system engineers and downstream development teams. FMEA will then be performed on the system requirements and subsystems identified. This phase is part of the FMEA optimization process and is not finalized.
Conduct Software FMEA for Network Element or Component team	Conduct software FMEA for network element or component team • Complete FMEA team charter, get management approval, schedule meetings. • Top-down approach, using the system engineering FMEA results. • Bottom-up approach, using history of previous releases to identify areas of concern in the current software architecture. • Perform FMEA analysis • Box Requirements phase • Box high-level design and low-level Design phase • Box low-level Design phase • Box Coding phase (If required) • Collect FMEA metrics and ROI

The high-level process steps for performing software FMEA are:

1. Planning for system software FMEA
2. Train and familiarize the team with traditional FMEA process
   a. Cause and effect analysis
   b. Identifying potential failure modes
   c. Assigning original RPN ratings pre-risk mitigation
   d. Assigning resulting RPN ratings post-risk mitigation

3. Conduct software system, software subsystem (network element level), or software component (sub-subsystem level) FMEA, as required
4. Collect appropriate metrics to analyze return on investment (ROI) on the software FMEA effort

A complete software FMEA process document is included in the appendix to this chapter. A software FMEA tracker example is shown in Figure 9.5, and a software FMEA presentation template example is shown in Figure 9.6.

Software FMEA results in significant cost savings, by detecting defects early that would have otherwise been detected in the test phases or by the customer. A software defect cost model from Motorola showed that the later that a defect is detected, the more the cost; a defect detected by the customer can cost up to $70,000. However, the argument may continue as to how one can measure the benefit of software FMEA effort. It can be considered a "chicken and the egg" type problem because issues identified early are not looked on as severely as defects; defects by definition are issues identified after a test phase, so the true measure of an FMEA activity would require a comparative analysis on the software system or subsystem, comparing typical defect density, testing costs, productivity, in a software FMEA-centric release versus a non-software FMEA release.

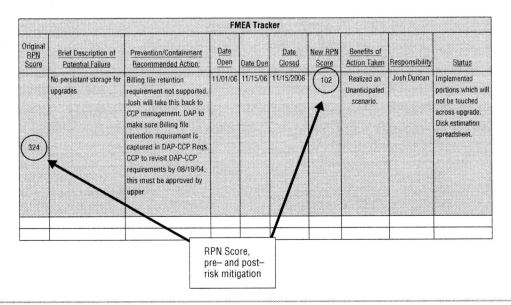

Figure 9.5 Software FMEA tracker example

Software FMEA Project Dashboard
<team> <target release> <Prevention or Containment>

Milestones	Plan	Fcst/Act	FLT	
Prevention/Containment FMEA Scope	01/01/06	01/01/06	0	FMEA Session Planned/Completed to date
Complete FMEA Team Charter and obtain management approval. Prevention or Containment FMEA	01/01/06	01/01/06	0	Planned = ? / Completed = ? Staff Hours Planned/Actual to date Planned = ? / Actual = ?
Identify areas of concern in existing product/component/application via trend analysis	01/01/06	01/01/06	0	Number of Resources to date Planned = ? / Actual = ? Total number of Actions taken ?
Identify components/tasks associated with critical parameters for new features (if applicable)	01/01/06	01/01/06	0	FMEA Process Issues/Risks Issue-1 Issue-2
FMEA Weekly Meetings	01/01/06	01/01/06	0	Risk-1 Risk-2
Analysis of existing defects and new requirements to develop an initial list of potential failures	01/01/06	01/01/06	0	Return on Investment from FMEA Metric-1 Metric-2
Weekly review of failure modes until all risks have been mitigated. This includes prioritizing all resulting action items and monitoring them to completion.	01/01/06	01/01/06	0	
FMEA Closure Activities	01/01/06	01/01/06	0	
Deployment of resulting Process Improvements logged into the DDTS PI database	01/01/06	01/01/06	0	
Perform FMEA Post Mortem	01/01/06	01/01/06	0	

Figure 9.6 Software FMEA presentation template example

Case studies at Motorola have shown that there is an extremely high ROI for each software FMEA activity; the return ranges from 10 to 25 times. One way to look at the software FMEA ROI is in terms of a cost avoidance factor—the amount of cost avoided by identifying issues early in the life cycle. This is accomplished relatively easily by multiplying the number of issues found in a phase by the software cost value from the software cost table.

The main purpose of a software FMEA is to catch software defects in the associated development phases: catching requirements defects in Requirements phase, design defects in the Design phase, and so on. This ensures reliable software, with significant cost and schedule time savings to the organization. Earlier detection of defects is a paradigm change, but may not be obvious to software managers or leaders; the software FMEA subject matter expert may need to convince senior leaders and management to commit to this effort.

The quantitative benefits of software FMEA are

- Software that is more robust and reliable
- Software testing cost is significantly reduced (measured as cost of poor quality)
- Productivity of the organization increases, in terms of developing reliable and high-quality software in a shorter duration
- Improvement in schedule time

We have all heard that real estate is all about location, location, and location—similarly, developing highly reliable software is all about FMEA, FMEA, and FMEA!

SOFTWARE FMEA IMPLEMENTATION CASE STUDY

The objective of this software FMEA implementation case study, provided courtesy of Vivek Vasudeva, a software quality manager in Motorola's Networks Group, was to lower two types of software defects based on the root cause analysis of customer data over several software releases. The organization had several software root cause categories, but the top two cause categories, persistent over several releases, were lack of robustness (issues such as retry mechanisms, error checking, validations) and unanticipated scenarios (issues such as timing, unexpected messages, missing scenarios), as shown in Figure 9.7. The cost was estimated at almost $2 million over the prior three software releases. Attempts at process

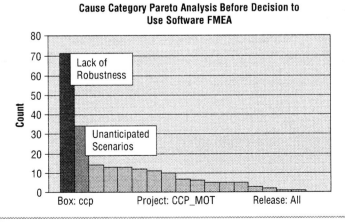

Figure 9.7 Pareto analysis of root cause categories before use of software FMEA

improvements and additional reviews did not help. A decision was then made to try using software FMEA.

An FMEA team was formed consisting of those responsible for the middleware functional area that was most prone to lack of robustness and unanticipated scenarios:

- FMEA consultant
- Technical software lead
- Two software developers
- Component integration test lead

For the pilot study of using FMEA for a software release, there were significant changes in the middleware layer; hence, software FMEA was performed for the complete software feature, and middleware layer issues were identified through root cause analysis.

With tight schedules, weekly software FMEA meetings were limited to one hour. The FMEA effort was initiated in the software development Architecture phase (after the first formal review of the architecture document), and were continued until the team believed all of the major risks had been mitigated. Activities during the weekly, one-hour software FMEA meetings consisted of:

- Analysis of all the root cause analysis data, identified the traditional "soft spots."
- Analysis of new requirements and consideration of "overlap areas"—where new requirements overlapped with the "soft spots."
- The team then asked a series of questions in order to anticipate the potential software failure modes in the Design/Architecture phase (as the team had already completed the first formal review of the architecture document before initiating the FMEA effort): What if . . .? What can go wrong?
- The team then used the P-diagram approach by analyzing the state machine for the software and developed a transition table with the starting state, event state, and ending state, as well as error scenarios. This helped make the software extremely robust; some missing states in the software were identified, along with many "unexpected" state changes.
- The inclusion of the test lead was an excellent decision; the test lead made his living by testing and breaking the code—experience which helped the team analyze potential failures in the architecture and design phases.
- It took the team about three to four weeks to reach the "performing" state from the "storming state" of team dynamics (the four stages of team dynamics are forming, storming, norming, and performing), after which the team increased the frequency of software FEMAs to twice a week.
- In addition to anticipating many failure modes, the team added 38 percent more test cases to the software regression test suite.

- The team anticipated 25 high-severity failure scenarios in approximately eight KLOC software size, along with four critical failures.
- The software FMEA effort lasted about four months—with a total of 18 FMEA sessions.

The results of the pilot study are shown in Table 9.8, and the Pareto analysis after using software FMEA is shown in Figure 9.8.

Table 9.8 Results from pilot study on the use of software FMEA

FMEA Pilot Goals	FMEA Pilot Results
Reduction of Lack of Robustness Defect Categories	18 Lack of Robustness category prevention Action Items.
Reduction of Unanticipated Scenarios Defect Categories	7 Unanticipated Scenario category prevention Action Items. Added 38% more test cases to contain all categories of defects.
Improve Documentation/Add Documentation	Low level Detailed Design Document done for CCP-MW team as part of the pilot. Internal ICD containing all internal CCP messages to be documented.
Improve Tracebility	Adding End to End Tracebility for CCP in the E-SPMP. Tracing System Requirements to Box Requirements to Architecture to Design to Code and Test Cases.

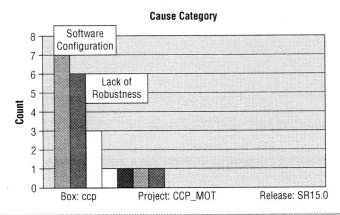

Figure 9.8 Pareto analysis of root cause categories after six months of using software FMEA

The results from the decision to use software FMEA were very encouraging:

- Cost savings of over US $250K
- Eliminated unanticipated scenario type defects
- Significantly reduced lack of robustness type issues
- Cost of the effort was estimated at US $15K

As a result of this pilot study, software FMEAs were institutionalized for the Architecture phase for all software projects in the organization. Documentation for this software FMEA process is provided in the Appendix to this chapter.

CONSIDERATIONS OF RELIABILITY AND AVAILABILITY

As shown in Figure 9.2, prioritization can directly provide the selected set of critical parameters that are the deliverable for this part of the Design for Six Sigma process. However, an additional requirement should be considered: reliability or availability. Whether or not explicitly stated by customers, there is an implicit expectation that any product involving electronics continue functioning for a sufficient period of time after acquisition.

Reliability can be defined as the probability that the system or product will operate, without failure, for a given time in a given environment. Availability can be defined as the percent of time that the system or product is functioning and providing the service or functions expected by the customer. Studies have indicated that, when customers cite reliability as an expectation, their expectations often align better with the measurable definition for availability described in this paragraph rather than the measurable definition for reliability. Although availability is more commonly associated with infra-structure systems such as servers for the Internet or base stations for cellular networks than for consumer products, I'll use a consumer product example—a cell phone—to illustrate this point.

One day, my cell phone stopped working. The proximal cause was the fact that I had dropped it one too many times—but the important thing to me was not where the fault lay, but the inconvenience I now faced with no means to contact people while on the go. I immediately drove to a storefront for my service provider, but they were un-able to repair my cell phone and—because it was a business phone—they were not willing to replace it. Instead, I needed to call a phone number, during normal business hours—from a working phone. This delayed my ordering of a new phone for the rest of the weekend, and then an additional week as the new phone was shipped to me by the slowest transportation available. As a customer, I was greatly annoyed by the inconvenience and the impact on my need to communicate as part of my job. In other words, I was annoyed by the amount of time that I was without the communication service.

Imagine, by contrast, that the consumer system (the cell phone) had the high availability expected from a network system. As soon as my cell phone stopped functioning, the rest of the network would have noticed the gap and transmitted information regarding my loss of cell phone functionality to the local service center. Within minutes, a car could have driven up to my house and a gentleman with a smiling face would have handed me a new, functioning cell phone. I would have been delighted!

Between these two scenarios, it hopefully is clear that customer satisfaction is more clearly linked to availability (the time that I was without the service or functionality) than to reliability (the probability that the device would continue to function without interruption). Availability and reliability are related measures, but although reliability only relates to the probability of failure, or the mean time to failure (MTTF) or mean time between failures (MTBF), availability is related both to the mean time to service interruption (MTTF or MTBF) and the mean time to service restoration (mean time to repair or replace).

There are two ways to use critical parameter management with reliability or availability. It can be treated as a critical parameter in itself, and later allocated or flowed down to reliability measures for each subsystem, module, and component or each function and subfunction. Alternatively, measurable aspects for each of the key functions expected by the customer can be linked to critical parameters, and reliability for that function can be treated as a noise factor in flowing down the functionality.

Allocating or flowing down reliability of the system or electronic product can be handled by developing a maximally acceptable failure rate—the reciprocal of the mean time to failure. Assuming an exponential model for system reliability, the failure rate of a system, treated as a series reliability model (in which the system is considered to have failed if any major subsystem or subfunction fails), is simply the sum of the failure rates of the major subsystems or subfunctions. Alternatively, the flow-down and allocation of reliability can be achieved in terms of allocating the overall probability of failure to the subfunctions using an approach referred to as fault tree analysis (FTA), as summarized in Figure 9.9.

Allocating or following down availability is deceptively simple in concept. The team sets an expectation level for availability that can be translated into time per year without service. For example, a "5 nines" level of availability, 99.999 percent availability, corresponds to no more than five minutes of downtime per year. Once the availability is expressed in terms of time, the maximum tolerable downtime for the system can be allocated to maximum tolerances for downtimes for the subsystems or subfunctions.

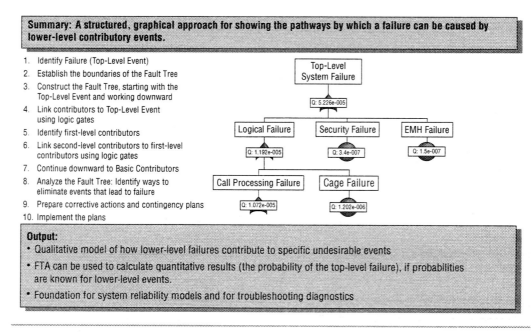

Summary: A structured, graphical approach for showing the pathways by which a failure can be caused by lower-level contributory events.

1. Identify Failure (Top-Level Event)
2. Establish the boundaries of the Fault Tree
3. Construct the Fault Tree, starting with the Top-Level Event and working downward
4. Link contributors to Top-Level Event using logic gates
5. Identify first-level contributors
6. Link second-level contributors to first-level contributors using logic gates
7. Continue downward to Basic Contributors
8. Analyze the Fault Tree: Identify ways to eliminate events that lead to failure
9. Prepare corrective actions and contingency plans
10. Implement the plans

Top-Level System Failure
Q: 5.226e-005

Logical Failure
Q: 1.192e-005

Security Failure
Q: 3.4e-007

EMH Failure
Q: 1.5e-007

Call Processing Failure
Q: 1.072e-005

Cage Failure
Q: 1.202e-006

Output:
- Qualitative model of how lower-level failures contribute to specific undesirable events
- FTA can be used to calculate quantitative results (the probability of the top-level failure), if probabilities are known for lower-level events.
- Foundation for system reliability models and for troubleshooting diagnostics

Figure 9.9 Tool summary for fault tree analysis (FTA)

EXAMPLES OF CRITICAL PARAMETERS

As illustrated in Figure 9.2, the process described in this chapter results in the establishment of a set of selected critical parameters, after appropriate consideration of customer expectations, risks, and tacit expectations such as reliability or availability. Table 9.9 shows some examples of critical parameters for a personal computer, a base station, a communication device for public safety officers, and an audio player.

An example of a communication device for secret agents (spies) can involve the prioritization of the critical parameters using the product of the importance to the ultimate customers, secret agents, times the average of the technical challenge scores from the lead mechanical, electrical, and software engineers. Recall that the importance to the ultimate customers was based on the ratings from a representative sample of the customers, and translated to relative importance for the associated technical requirements through the first House of Quality (Chapter 7). The technical requirements with the highest product of customer importance and technical challenge become the critical parameters that will be flowed down to the subsystems and components for the

Table 9.9 Examples of critical parameters for various electronic systems

System	Customer Requirements	Translated Technical Requirements	Software/ Hardware	Importance	Technical Challenge	C.P. Priority
Personal Computer	Rapid Reboot	Start-up time	Software	7	4	28
Base Station	High System Availiabilty	Forecasted downtime per year	Both	9	9	81
	Small size	Footprint	Hardware	6	4	24
	Support multiple users	System access time at high loading	Both	7	7	49
Public Safety	Secure Communication	Cipher Effectiveness: Time required to defeat cipher	Software	7	7	49
Communicator	Water Proof	Submersion depth where water intrusion begins	Hardware	9	6	54
Audio Player	High quality of sound	Mean Opinion Score (MOS) for Perceived Audio Quality	Both	9	7	63
		Frequency Response	Hardware	7	6	42
	Easy to select song	Number of clicks to select song from list	Both	8	5	40

communication device at the end of the next chapter. The selected critical parameters for the communication device for secret agents are

- Provides secure communication   Customer Importance 7 × Average Risk 8 = 56
- Small   Customer Importance 8 × Average Risk 4 = 32
- Unobtrusive   Customer Importance 8 × Average Risk 5 = 40
- Durable   Customer Importance 8 × Average Risk 7 = 56
- Long battery life   Customer Importance 5 × Average Risk 6 = 30
- Water resistant   Customer Importance 4 × Average Risk 7 = 28

SUMMARY

It may not be an overstatement to claim that the identification of the critical parameters is the most critical step of DFSS and, indeed, for the success of the new product. This step serves to focus the team for the rest of the new product development effort—to design and optimize the subsystems, modules, and components to have confidence that the product will provide features and functionality that will ensure a successful product launch and ongoing customer satisfaction.

With the selection of a superior concept and with the identification of the set of critical parameters essential to success for the product, the Requirement phase of RADIOV, (corresponding to the Concept phase of CDOV or the Define and Measure phases of DMADV) has been completed. Subsequent phases will build on these key deliverables.

APPENDIX: SOFTWARE FMEA PROCESS DOCUMENTATION

BY VIVEK VASUDEVA

Table 9.10 describes all the phases and their details, as well as a set of "RASCI roles" for the software FMEA process. RASCI roles are defined as:

R—Responsible, typically the team lead, who runs the software FMEA effort.
A—Approve, typically senior management, who approve the software FMEA effort.
S—Support, typically middle management, who support and provide the resources.
C—Consult, typically the software FMEA expert, such as a Six Sigma black belt.
I—Inform, typically software developers, or software testers who are conveyed the findings of the software FMEA effort.

Table 9.10 Software FMEA process documentation

Roles				Sequence of Activities
FMEA Champ	**FMEA Lead**	**Recorder**	**FMEA Facilitator**	**Planning for FMEA**
I	R	S	C	1. Complete FMEA team charter and obtain management approval.

       • Form FMEA team charter, which includes an estimate for the ROI (Return on Investment) for FMEA effort. The ROI estimate for an FMEA effort should be in terms of CoPQ Dollars (US$).

       • The system FMEA team should consist of system engineers who have expertise in the new features or subsystems.

       • The software FMEA team should consist of an architect, software development leads, one integration test or box test expert, a software process engineer (SPE), and software developers. The size of the team should generally be around three to seven people.

       • Software requirements FMEA team: system engineer expert (recommended, not required), box architect, SPE, development technical leads, and a test expert.

       • Box software architecture and high-level design team: architect, SPE, development technical leads, software developers (if applicable), and test expert.

       • Box software component architecture and high-level design: architect, SPE, development technical leads, test expert, and software developers (if required).

       • Software component low-level design: architect, SPE, development technical leads, test expert, and software developers.

       • Software component code: architect, SPE, development technical leads, test expert, and software developers.

continued

Table 9.10 Software FMEA process documentation (continued)

Roles				Sequence of Activities
FMEA Champ	FMEA Lead	Recorder	FMEA Facilitator	Planning for FMEA
I	R	S	C	2. Scheduling FMEA meetings.

- Rule of thumb estimate for software FMEA is three percent of total staff months.
- System FMEA meetings have to be planned based on system FMEA effort estimate; hourly meetings, once-a-week per subsystem.
- The Software FMEA tends to left shift the development paradigm, thus the effort spent up front will in return save on the development effort. The net impact of doing an FMEA is usually 1.5 percent of total staff months.
- FMEA meetings have to be planned based on the FMEA effort estimate.
  - Generally weekly meetings scheduled for the team starting with the Requirements phase and through Coding phase if the standard development (waterfall) model is followed.
  - In case of Agile development the FMEA meetings have to scheduled during the Elaboration phase and during construction (high-level and low-level Design phases). Careful planning of FMEA is required when following the Agile model for development.
- FMEA meetings are usually scheduled from Requirements phase through coding, once a week and for an hour for box/component teams.
- Schedule should be flexible, where meetings are added or removed based on the project priorities.

I	S	S	R	3. Select FMEA Tool.

- Choose one of the following:
  - The legacy FMEA tool to capture FMEA brainstorming sessions is an Excel-based spread sheet, see example below.
  - There are professional FMEA tools such as Relex. Please see http://www.relex.com.

Table 9.10 Software FMEA process documentation (continued)

Roles				Sequence of Activities
FMEA Champ	**FMEA Lead**	**Recorder**	**FMEA Facilitator**	**Planning for FMEA**
				• FMEA project status reporting template is used for reporting FMEA project status and results monthly to senior management. See example.
FMEA Champ	**FMEA Lead**	**Recorder**	**FMEA Facilitator**	**FMEA Process Phase**
I	R	S	C	1. Cause and effects analysis of potential failure modes. • Once the potential failure modes are identified, they are further analyzed by potential causes and potential effects of the failure mode. (Please reference DSS Tools, Cause and Effects Analysis, 5 Whys, and so on, for details on cause and effects analysis/)
I	R	S	C	2. Calculate the potential failure modes risk priority number (RPN). • Occurance rating, range 1–10, the higher the occurance probability, the higher the rating—O. • Severity rating, range 1–10, the higher the severity associated with the potential failure mode, the higher the rating—S. • Detectability rating, range 1–10, the lower the detectability the higher the rating—D. • RPN = O × S × D Maximum value = 1000, Minimum value = 1.

I	R	S	C	3. Setting FMEA ratings.
				
				• The current recommendation for ALL is to use a rating scale of High, Medium, and Low for all of the above ratings:
				  ※ High: 9
				  ※ Medium: 6
				  ※ Low: 3

continued

Table 9.10 Software FMEA process documentation (continued)

Roles				Sequence of Activities
FMEA Champ	**FMEA Lead**	**Recorder**	**FMEA Facilitator**	**FMEA Process Phase**
				• We can always modify the ratings guidelines based on typical software failures identified. This is usually an ongoing process, as we learn more about potential failures in our software the granularity of the ratings is a logical progression.
				• For all potential failures identified with an RPN score of 150 or greater the FMEA team will propose recommended action to be completed within the phase the failure was found. These actions can be FTR errors or wisdom records. All CRs/DRs resulting from software FMEA will have FMEA in the wisdom record title.
I	R	S	C	4. Computing the resulting RPN score after risk mitigation action. • A resulting RPN score **has to be computed** for each potential failure mode after each recommended action to show that the risk has been significantly mitigated.
FMEA Champ	**FMEA Lead**	**Recorder**	**FMEA Facilitator**	**Conduct System Engineering FMEA**
I	R	S	C	1. Perform system FMEA **planning phase** activities noted above.
A	R	S	C	2. Rate and rank the features based on the VOC. • Rank all the features based on input from customer to identify the high priority requirements and subsystems. The ranking of features is discretionary to the team performing FMEA. • The ranking is based on customer priority, complexity of implementation for down stream teams, and deficiencies in current system design.
I	R	S	C	3. Front-end analysis consists of identifying the system requirements and subsystems to perform FMEA.

Table 9.10 Software FMEA process documentation (continued)

Roles				Sequence of Activities
FMEA Champ	**FMEA Lead**	**Recorder**	**FMEA Facilitator**	**Conduct System Engineering FMEA**
				• System engineering team will identify all the requirements for FMEA analysis.
				• System engineering team will identify all the subsystems for FMEA analysis.
				• System engineering team will perform **FMEA process**, as shown earlier, on the requirements and subsystems identified.
				• Network element (box) teams will use the system FMEA analysis results to focus on the components further as required.
FMEA Champ	**FMEA Lead**	**Recorder**	**FMEA Facilitator**	**Conduct Software FMEA for Network Element or Component Team**
I	R	S	C	1. Perform network element/component FMEA **planning phase** activities noted above.
				2. Perform front-end analysis for the box software based on results from system FMEA.
I	R	S	C	• Front-end analysis consists of utilizing results from system engineering FMEA analysis, implying:
				◦ Focus is limited to areas identified by the system engineering teams.
I	R	S	C	3. Back-end analysis involves:
				• Root cause analysis trends: The root cause analysis trends will help determine the historically weak areas of the network element.
				• Analysis of defects that escape to the field.
				• EDA analysis. EDA helps determine typical defects escaping the given testing phases.
				These targeted FMEA efforts will help reduce the internal failure time and improve the cost of poor quality. (CoPQ)
	R	S	C	4. Perform FMEA **process phase**.

continued

Table 9.10 Software FMEA process documentation (continued)

Roles				Sequence of Activities
FMEA Champ	**FMEA Lead**	**Recorder**	**FMEA Facilitator**	**Conduct Software FMEA for Network Element or Component Team**
	R	S	C	5. Conduct software FMEA at box requirements level.
				• Software FMEA is initiated during the box software requirements phase.
				• Upon completion of the initial version of requirements, post the first formal technical review.
				• Software requirements can be scoped down based on FMEA analysis by the system engineering team.
				• Generally software requirements are scoped by complexity of implementation and risks to the software development.
				• The potential failure modes are to be identified by the Software FMEA team by trying to "anticipate" and asking the question "what can go wrong." Initial potential failure scenarios can be developed by analysis of the scoped requirements and history of failures from previous releases, phase sourced to requirements. Please follow the FMEA process overview once potential failure modes are identified.
				• If the box team utilizes Use Cases, they should develop "abuse cases" for software FMEA.
	R	S	C	6. Conduct software FMEA at high-level Design phase
				Most of the effort for software FMEA is at this phase, the Design phase, high-level and low-level design.
				• The software FMEA should be done after the initial version of the high-level design exists, that is, after the first high-level design FTR.
				• Upon completion of the initial FTR the high-level design can be scoped to the areas where the software failure analysis will be performed based on input from system engineering FMEA and back-end analysis.

Table 9.10 Software FMEA process documentation (continued)

Roles				Sequence of Activities
FMEA Champ	**FMEA Lead**	**Recorder**	**FMEA Facilitator**	**Conduct Software FMEA for Network Element or Component Team**
				• The effort to scope usually takes a few meetings. This is based on complexity, risk to the down stream phases, and historical failures sourced to high-level design.
				• The potential failure modes are to be identified by the software FMEA team by trying to "anticipate" and asking the question "what can go wrong." Initial potential failure scenarios can be developed by analysis of the scoped high-level design and history of failures from previous releases, phase sourced to high-level design. Please follow the FMEA process overview once potential failure modes are identified.
	R	S	C	7. Conduct software FMEA at low-level Design phase.
				Most of the effort for software FMEA is at this phase, the Design phase, high-level and low-level design.
				• The software FMEA should be done after the initial version of the low-level design exists, that is, after the first low-level design FTR.
				• Upon completion of the initial FTR the low-level design can be scoped to the areas where the software failure analysis will be performed based on input from system engineering FMEA and back-end analysis.
				• The potential failure modes are to be identified by the software FMEA team by trying to "anticipate" and asking the question "what can go wrong."
				• The software FMEA team should try and identify all possible unintended results from normal scenarios based on current inputs to the software design.

continued

Table 9.10 Software FMEA process documentation (continued)

Roles				Sequence of Activities
FMEA Champ	**FMEA Lead**	**Recorder**	**FMEA Facilitator**	**Conduct Software FMEA for Network Element or Component Team**
				• Initial potential failure scenarios can be developed by analysis of the scoped low-level design and history of failures from previous releases, phase sourced to low-level design.
	R	S	C	8. Conduct software FMEA during the Coding phase.
				Software FMEA can be performed at the Coding phase.
				• During the Coding phase, the failure modes identified in the Design phases are usually implemented.
				• The coding of high-risk areas should have been identified by the requirements and high-level and low-level design FMEAs.
				• Some times a further analysis might be required at the Coding phase because of complexity or unanticipated issues. The potential failure modes are to be identified by the software FMEA team by trying to "anticipate" and asking the question "what can go wrong" based on abnormal inputs to the code. Goddard notes can be helpful during the Coding phase.
				• Any potential failures identified at the Coding phase should be addressed immediately and implemented.
	R	S	C	9. Collecting FMEA metrics for return on investment (ROI).
				• Number of defect reports filed
				• Number of change requests filed
				• Number of test cases added during the FMEA effort for the system software release
				• Number of defects found in Testing phase by the FMEA test cases
				• Improvement in defect density

Table 9.10 Software FMEA process documentation (continued)

Roles				Sequence of Activities
FMEA Champ	**FMEA Lead**	**Recorder**	**FMEA Facilitator**	**Conduct Software FMEA for Network Element or Component Team**
				• Development cycle time reduction
				• Reduction of CoQ
				• Reduction of CoPQ
				FMEA prevention effort ROI is usually measured in terms of CoPQ dollars. This represents the dollars saved in Testing phases as issues are identified in earlier phases utilizing FMEA.

Requirements Flow-Down

POSITION WITHIN DFSS FLOW

Requirements flow-down is aligned with the early part of the Design phase of the RADIOV process, which corresponds to the transition from the Measure to Analyze phases of the DMADV process of DFSS or from the Concept to Design phases of the CDOV process. The DFSS flowchart, which can be downloaded from http://www.sigmaexperts.com/dfss provides a high-level overview (Figure 10.1) of the sequence of steps that can be drilled down to detailed flowcharts, and further drilled down to summaries for key tools and deliverables within each detailed flowchart. Figure 10.2 is the detailed flowchart aligned with this chapter, showing the steps and methods involved in flowing down system requirements.

For a system involving both hardware and software, the flow-down for system requirements will result in software and hardware requirements, evolving to subsystem requirements and to subassembly requirements and requirements for components (software components and hardware components). The sequence of steps in the flow-down process is iterative, in the sense that the anticipation of potential problems, measurement system analysis, and initial design capability analysis will be first performed at the system level, then at the subsystem/subassembly level, and then at the component level, as illustrated with the iterative nature of the flowcharts in Figures 10.2 and 10.3.

Figure 10.3 starts with a set of high-level system requirements. The process of "systems design" consists of turning these requirements into a specification of the system. First, a concept or architecture must be specified at the system level that identifies the subsystems, the interfaces between them, and any interfaces to the outside

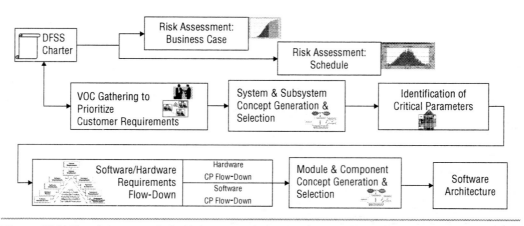

Figure 10.1 Flowchart overview highlighting step for flow-down of critical parameters

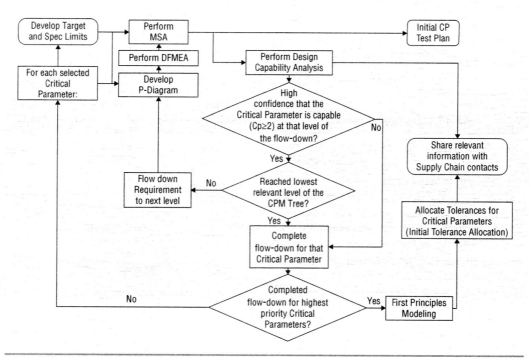

Figure 10.2 Flowchart, drilled down to detailed flowchart for flow-down of critical parameters

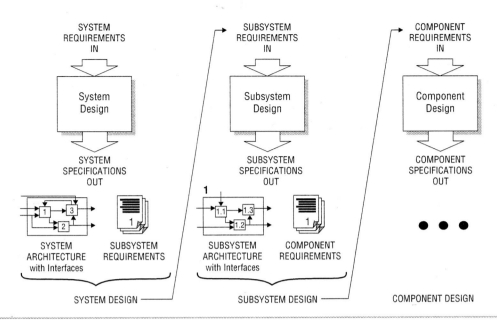

Figure 10.3 Evolution of system requirements

of the system—including the user interface and interfaces to or interactions with other systems. For electronics systems, interactions with other systems can be intentional, as in data communication linkages, or unintentional, such as with EMI (electromagnetic interference—unwanted disturbances caused by electromagnetic radiation emitted to or from another electronic system).

- For each subsystem, define the behavior and performance with subsystem requirements. Identify other subsystems and systems with which this system might interface or interact, and define the requirements for the interfaces.
- The team responsible for a subsystem is expected to not only deliver that subsystem so that it meets all of the requirements when isolated, but to meet all of the requirements when it is integrated with the other subsystems.
- The design for a subsystem requires considerations and decisions for how to meet all of the requirements jointly. This design should be documented in a subsystem specification that also contains the architecture for the subsystem, consisting of the components within the subsystem, the interfaces between these components, and the interfaces to other subsystems or other systems with which it can interact (see Chapter 12 for software architecture examples). Furthermore, each component has

requirements that define the behavior and performance required for the component to work with the other components and meet requirements jointly. This process continues down to low-level design.

FLOW-DOWN FOR HARDWARE AND SOFTWARE SYSTEMS

When developing a system comprised of both hardware and software, the flow-down of requirements to measurable technical requirements (subordinate y's, x's, and noises) might lead to three situations:

- Critical parameters that only involve hardware aspects
- Critical parameters that only involve software aspects
- Critical parameters that involve both hardware and software aspects.

For a hardware intensive system, or a system that involves little or no software, the critical parameters flow down to subsystems and components that are electrical or mechanical. The technical challenges for some products or some subsystems might fall entirely within one engineering discipline. For example, a team of mechanical engineers might flow down or decompose the critical parameters for a lawnmower to subsystem and component requirements. Other hardware intensive products might require a team composed of electrical and mechanical engineers, or the development organization might be structured so that these teams are separate, but a cross-functional team would handle the electrical-mechanical interactions and interfaces.

The flow-down for some requirements of a cell phone is particularly relevant for electrical engineers who are knowledgeable about radio frequency (RF) performance. Figure 10.4 shows part of the system-level House of Quality example discussed in Chapter 7. Two critical parameters, total radiated power (TRP) and turn-on time, are highlighted. In Figure 10.5, a second House of Quality focused on the RF sections of the cell phone indicates that TRP flows down to some measurable requirements for the antenna and for the transmitter. Figure 10.6 shows the flow-down, juxtaposed with some images of the physical layout within the cell phone.

Figure 10.7 shows the flow-down or decomposition of the critical parameter, cellular phone turn-on time, to hardware and software requirements. A team of system engineers, software engineers, and hardware engineers discussed this flow-down, and developed a simple mathematical model for the turn-on time, which showed that the delays in phone turn-on caused by the hardware requirements such as phase locked loop (PLL) lock time were negligible. The critical parameter for turn-on time then became a software development team focus.

Direction Of Goodness >	Importance	Display Size ↑	Talk Time ↑	Standby Time ↑	Camera Resolution ↑	Removable Memory Capacity ↑	Application Processor MIPS ↑	Bit Error Rate ↓	Total Isotropic Sensitivity (T.I.S.) ↓	Accurate Identification (Prob) ↑	Baseband Processor (MIPS) ↑	Phone Weight ↓	Total Radiated Power (T.R.P.) ↓	Turn-On Time ↓
Voice of Customer, VOC's														
Fits lightly into a pocket or purse	6											9		
Can go 3 days without recharging	7		9	9										
Can email Quality photos easily & quickly	3	9	3	3	9	3	1	3	3		3			
Supports high speed internet connectivity	2	9						3	3	3	9			
Stores reasonable number of photos	1	1			3	9	3							
Few or no dropped calls	10		1	1				1	9	9	3		9	
Provides secure m-Commerce	1	3			3	1	3	1	1	9	3			
Turns on quickly	3													9
Scoring Totals		49	82	82	33	19	25	106	106	9	60	54	90	27
Target Nominal Values		60	10	50	5	2	16	0.000	−100	99.99%	100	4.5	30	10
Lower Limit		40	4	10	4	1	12	0.000	−120	99.9%	80	3.5	28.5	5
Upper Limit		100	40	1500	10	10	64	0.001	−95	100%	200	5.5	36	12
Units		mm	Hrs	Hrs	MPixels	MBytes	MIPS	%	dBm	%	MIPS	oz	dBm	Sec

Figure 10. 4 System-level House of Quality, focusing on two critical parameters, total radiated power from the transmitter and antenna, and turn-on time for the cell phone

Critical parameters that can involve both software and hardware aspects require that initial combined team approach. If software or hardware is totally dominant, then the effort can be handed off to the appropriate team as was the case for the turn-on time for the cellular phone. If neither software nor hardware dominates to such an extent, the effort on the critical parameter can either continue to be addressed by a team consisting of system engineers and software and hardware engineers, or the software aspects can be handed to the software team and the hardware aspects can be handed to the hardware team. In the latter instance, the interfaces and interactions between hardware and software risk "falling in the crack," so an additional effort is required to consider these interactions and to integrate the hardware and software aspects. In many cases, emulation can be used to evaluate the software aspects without the final version of the hardware but, rather, an existing hardware platform modified to behave like and substitute for the hardware.

The second House of Quality, as shown in Figure 10.5, is one of several methods to flow down requirements. Other methods can use a P-diagram, as discussed in the next section, or brainstorming session with a set of engineers, including system engineers, to

System Requirements	Importance Score	Antenna Gain	Antenna NF	Antenna Interface Attenuation	Antenna Cost	Antenna Weight	Receiver Gain	Receiver Cascaded Noise	Receiver Cascaded	Receiver Current Drain	PLL Lock Time	Receiver Cost	Receiver Weight	Transmitter Gain	Transmitter Output Power	Transmitter Power Added	Transmitter Current Drain	Transmitter Cost	Transmitter Weight
Display Size	4																		
Talk Time	8	M		M			M			H	L			H	H	H	H		
Standby Time	8	L	L							H	L			H	H	H	M		
Camera Resolution	2																		
Removable Memory Capacity	1																		
Application Processor MIPS	2																		
Bit Error Rate	10	M	M	M			H	H	M					M					
(T.I.S.)	10	H	H	H			H	H	L										
Accurate Identification (Prob)	0																		
Baseband Processor (MIPS)	5																		
Phone Weight	5					M							M						M
BOM Cost	0				M							M						M	
Total Radiated Power (T.R.P.)	8	H		M										H	H	M			
Turn-On Time	2								M		H								
Scoring Totals		224	128	168	0	15	204	186	40	144	34	0	15	246	216	168	96	0	15
Normalized Scores		9	5	7	0	1	8	8	2	6	1	0	1	10	9	7	4	0	1
Target Values																			
Units		dB	dB	dB	$	g	dB	dB	dBm	mA	sec	$	g	dB	dBm	%	mA	$	g

Figure 10.5 Second or subsystem-level House of Quality, focusing on the radio frequency (RF) subsystems including the antenna, receiver, and transmitter for a cell phone

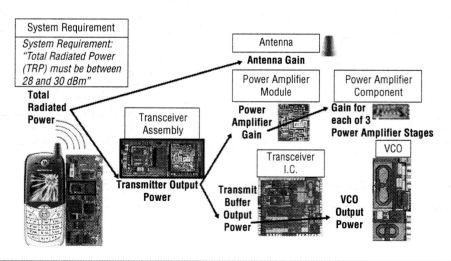

Figure 10.6 Flow-down of the total radiated power requirement for a cell phone to measurable requirements on the antenna and for the transmitter within the transceiver assembly

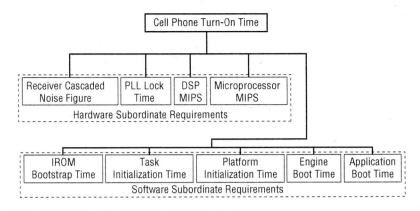

Figure 10.7 Flow-down of cell phone turn-on time to hardware and software requirements

identify indirect or intermediate requirements (subordinate y's), control factors (x's), and noises (n's) that affect the critical parameter, as discussed in the flow-down section later in this chapter.

ANTICIPATION OF POTENTIAL PROBLEMS: P-DIAGRAMS AND DFMEA

System requirement flow-down also involves anticipation of potential problems. At the system level, a P-diagram and FMEA can be part of the concept generation and selection process for the system, as described in Chapter 8, and the system-level FMEA is included in the flowchart for identification of critical parameters, as described in Chapter 9.

As the flow-down proceeds iteratively, similar anticipation of potential problems should be subsequently applied for each critical parameter, or at the subsystem/ subassembly level and the component level—and possibly at the manufacturing process level. Some perspectives should underline the importance of this anticipation as the flow-down proceeds: the system level flow-down will naturally involve a bird's-eye view of failure modes, and will involve a broader cross section of expertise for this purpose—but anticipation of failure modes and mechanisms at subsystem and module/ component levels will involve a more focused set of experts to dissect the potential problems involved at that deeper, more detailed level. Essentially, at these subsequent iterations, the subsystem and component under consideration becomes "the system" for

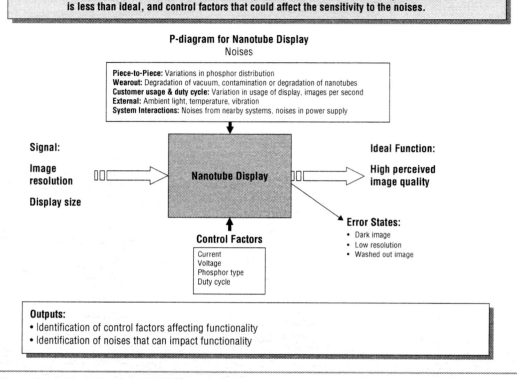

Figure 10.8 Tool summary for P-diagram

the team. It is worth noting that many of the subsystems for complex electronic products could literally *be* the "system" or product in other situations. For example, many cellular phones include digital cameras—but digital cameras are separate products or systems for camera manufacturers. Many cell phones ("music phones") incorporate music players, which also exist as separate products. In turn, many music players as products include flash drives, and some companies sell flash drives as products.

Either as an integrated subsystem, or as a separate product, anticipation of potential problems is a vital step toward prevention of problems. The P-diagram (Figure 10.8) and DFMEA (Figure 10.9) can be useful for the module and component-level concept generation and selection process described in Chapter 8, and also valuable in anticipating and preventing problems with aspects that were not selected as critical parameters (or flowed-down from critical parameters) but that could impact the success of the product if not adequately addressed.

> **Summary: A structured method for identifying and ranking the significance of various failure modes of the program and their effects on the product or customer**

Failure Modes and Effects Analysis (FMEA):													Revised RPN			
System, Subsystem or Component	Parameter at System, Subsystem, Module, Assembly or Component Level	Potential Failure Mode / Effect	Potential Effects of Failure	SEVERITY	Potential Cause	OCCURRENCE	Current Controls	DETECTION	RISK PRIORITY NUMBER (RPN)	Action	Responsibility	Due Date	Severity	Occurrence	Detection	Risk Priority
				9		7		3	189				6	5	3	90
				8		7		4	224				6	6	2	72
				9		7		3	189				5	4	2	40
				9		7		3	189				5	3	2	30
				9		7		3	189				4	2	3	24
				9		7		3	189				4	4	1	16

1. List Functions and Interfaces or Subsystems.
2. List Potential Failure Modes.
3. List Potential Effects.
4. Assign Severity Rating.
5. List Potential causes.
6. Assign Occurrence Rating. (Probability rating for Software)
7. List current controls.
8. Assign detection rating.
9. Calculate Risk Priority Number.
10. Use RPNs to help decide on high priority failure modes.
11. Plan to reduce or eliminate the risk associated with high priority failure modes.
12. Re-compute RPN to reflect impact of action taken on failure mode.

> **Output:**
> - Ranked group of failure modes
> - Impacts of failures
> - Risk Priority Numbers (RPN) before and after corrective action
> - Corrective actions, controls to remove or reduce the risk or impact of a failure mode

Figure 10.9 Tool summary for design failure modes and effects analysis (DFMEA)

A P-diagram (Figure 10.8) offers several benefits. It can help with the development of the DFMEA, in which the error states or deviations from the ideal function (at the lower right of P-diagrams) could suggest failure modes to be included in the DFMEA, and the noises (at the top of the P-diagrams) could suggest potential causes for the failure modes.

As will be seen later in this chapter, the team approach for identifying control and noise factors used in developing the P-diagram can be leveraged in flowing down requirements to the next level. The control factors portion of the P-diagram generally are the *x*'s in the flow-down, and the noises in the P-diagram obviously are the noises in the flow-down. The missing pieces are the subordinate y's—subrequirements that can be flowed down to other subordinate y's, x's, and n's. If the critical parameter does not involve subordinate y's, then the P-diagram can be used for the flow-down. However, many critical parameters cannot be directly flowed down to the final control factors with one P-diagram, and the P-diagram just provides a good start.

The P-diagram can also prove useful in generation and subsequent evaluation of alternative concepts for the subsystem, module, or component, particularly in terms of considering the noises that can affect performance when brainstorming potentially robust design approaches—the relative insensitivity of the alternative concepts to those noises can and should be considered in selecting a superior concept for the subsystem, module, or component.

The P-diagram can also prove valuable during transfer function determination (Chapter 13), in terms of initializing the identification of control and noise factors to use in an experimental design approach. The P-diagram will also prove valuable during optimization (Chapter 14), for evaluating and optimizing robustness against the noises. Some of the noises from the P-diagram can also be used as stress factors or for verification of reliability (Chapter 17).

FMEA (including system FMEA and design FMEA or DFMEA) has been discussed in Chapter 9, but it will be briefly reviewed here. The objective of DFMEA (summarized in Figure 10.9) is to consider the ways a product, subsystem, function, or interaction can fail, then analyze the risks, and take action where warranted. Typical applications include preventing defects, improving processes, identifying potential safety issues, and increasing customer satisfaction. It can be applied throughout the development life cycle. To be more effective, the DFMEA should relate to the nature of the development process itself. In either case, it considers overall architecture and functionality problems while at the same time addressing process problems. Therefore, DFMEA is an effective engineering tool for evaluating systems at a number of stages in the design process.

DFMEA evaluates risks posed by potential failure modes by considering the severity of the impact if the failure mode occurred, the probability that the failure mode could occur (based upon the probabilities for occurrences of potential causes of the failure mode), and the possibility that the problem would be detected in time. These three aspects of the risk are rated on a scale of 1 to 10, and then multiplied to provide RPN indices (on a scale of 1 to 1000) that can be treated as numerical assessments of risk. DFMEA and associated assessments are performed in a team setting, the atmosphere for which can become rather intense. It has been suggested that the DFMEA process be broken into two to three shorter sessions, during which the team is locked in a meeting room, and necessities (drink, raw meat . . .) are tossed over the wall.

There are systems that are heavily software oriented and that could benefit from a software DFMEA effort. The objective of a software DFMEA is to identify all failure modes in a software artifact. Its purpose is to identify all catastrophic and critical failure probabilities so they can be minimized as early as possible. For example, a common problem in software involves memory leaks. A memory leak is an unintentional memory consumption by a computer program where the program fails to release memory when no longer needed. Memory is allocated to a program, and that program subsequently loses the ability to access it due to program logic flaws. A memory leak can diminish the performance of the computer by reducing the amount of available memory. Eventually, too much of the available memory may become allocated and all or part of the system or device stops working correctly, the application fails, or the system slows down unacceptably. For example, code that has a "*malloc*" (a subroutine for dynamic memory allocation) or a "*new function constructor*," which is evaluated each time it is encountered, can increase the risk of creating a memory leak. Memory leaks can corrupt and misalign pointers (which reference values stored elsewhere in memory), and may cause part or all of the system to go down; the system may have difficulty recovering, and in severe cases, key data may be lost.

DFMEA and P-diagrams can be used and reused through many of the subsequent steps of DFSS. This continuing value is realized because DFMEA and P-diagrams, in concert, help the team conceptualize and share an understanding of the risks by assessing the risks in terms of the severity or impact, the probability of occurrence, and the opportunities for errors. The team can also gain insight into noises as potential sources of variation, stresses, and failures.

TARGET AND SPEC LIMITS

Target values or specification limits for the critical parameters might have been developed as part of the QFD/first House of Quality effort, as discussed in Chapter 7. The specification limits are involved in the calculation of the P/T ratio in the measurement system analysis, the design capability analysis, and the tolerance allocation topics discussed in later sections of this chapter.

If the critical parameter is a lower-is-better type parameter, then it will generally just have one specification limit, the maximum. Examples of such one-sided parameters include leakage currents, defects or defect densities, costs, weight, delay times, and power consumption. The target in this situation could be half of the specification limit or maximum, or perhaps an achievable low value that would represent a value considered desirable for the customers.

Similarly, if the critical parameter is a higher-is-better type parameter, then it will have one specification limit corresponding to the minimum. Examples include battery life, drops-to-failure, mean-time-to-failure (MTTF), efficiency, and resolution. Some of these examples are bounded on both sides by the nature of the metric or by physics; for example, percent efficiency is bounded by 0 and 100 percent, even though it is considered a higher-is-better type parameter. The target in this situation could be twice the lower specification limit, or an achievable high value that would be considered desirable for the customers.

If the critical parameter is a target-is-best type parameter, then it will have both an upper and a lower specification limit. Examples could include total radiated power (TRP) for a transmitted signal and some timing requirements in a clocked system constrained by issues such as race conditions. Generally, for two-sided limits, the target will be midway between the upper and lower specification limits; however, there will be exceptions to this, such as situations where the critical parameter is believed to follow a lognormal distribution, in which case the target might be the geometric average of the upper and lower specification limits (that is, the square root of the product of the upper and lower specification limits). Alternatively, the target is an achievable value that would be considered desirable by most of the customers; ideally, if the manufacturer could produce all parts with exactly that value, the customers should be satisfied (if not downright ecstatic).

Companies are rife with examples of problems with measurement systems analysis (MSA), capability indices, SPC, and customer issues that trace back to specification limits set arbitrarily, such as to some target ± 10 percent. The specification limits should be based on what is needed to meet the customers' expectations—and, subject to that consideration, the spec limits should be as wide apart or as generous as reasonable for the design team. This enables the design team to have the best chance of success in meeting the specifications with high confidence, and creates a high likelihood that the customers will be satisfied by the result of the design teams' innovation, optimization, and robust design of the product.

In some instances, appropriate specification limits may be hard to pin down. One possible cause for this fuzziness might be that different customers or sets of customers may have different expectations. There are at least three alternative approaches that can be used to deal with this issue: the best and widest-spaced compromise can be selected to satisfy the largest groups of key customers, the product can become multiple products each tuned to the expectations of different customers, or the characteristic can be designed to be tunable, programmable, or selectable by the customers.

MEASUREMENT SYSTEM ANALYSIS

Once the critical parameters have been selected, and specification limits have been set, it seems reasonable that the next steps might be to set things up so that progress towards achieving expectations can be monitored. As discussed in Chapters 7 and 9, the critical parameters have been defined in measurable terms. The next logical step is to set up the measurement systems and determine whether each is capable of measuring appropriate critical parameters.

Figure 10.10 summarizes the purpose, results, and outputs from measurement system analysis, focusing on MSA for critical parameters that are continuous rather than discrete. There are several indices used to determine if the measurement system is adequate for the purposes of optimization and validation of the critical parameters, including assessments of stability, linearity, accuracy, and measurement error. MSA is discussed further in Chapter 16.

The assessment and estimate of measurement error is a key, recurring topic in DFSS, and this is an appropriate point to begin that discussion. The measurement error is one of several "noises" that can be flowed down, as discussed later in this chapter, and that will be encountered along the way as the design team uses approaches such as design of experiments (DOE) and response surface methodology (RSM).

This aspect of the flow-down process is illustrated in Figure 10.11, which starts with the concept of squared deviation from the target. If the target is the desired value, as discussed in the previous section, then one can define a statistical index, the second moment about the target, which can represent the degree of customer satisfaction.

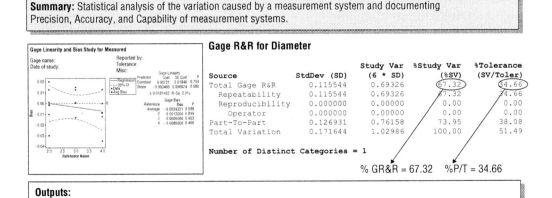

Summary: Statistical analysis of the variation caused by a measurement system and documenting Precision, Accuracy, and Capability of measurement systems.

Gage R&R for Diameter

Source	StdDev (SD)	Study Var (6 * SD)	%Study Var (%SV)	%Tolerance (SV/Toler)
Total Gage R&R	0.115544	0.69326	67.32	34.66
Repeatability	0.115544	0.69326	67.32	34.66
Reproducibility	0.000000	0.00000	0.00	0.00
Operator	0.000000	0.00000	0.00	0.00
Part-To-Part	0.126931	0.76158	73.95	38.08
Total Variation	0.171644	1.02986	100.00	51.49

Number of Distinct Categories = 1

% GR&R = 67.32   %P/T = 34.66

Outputs:
- Assessment of Stability
- Estimate of Accuracy (Bias)
- Estimate of Linearity
- Estimate of Measurement error, Std dev of Repeatability (within same conditions) and Reproducibility (operator-to-operator)
- Assessments of Precision: Precision-to-Tolerance (P/T) Ratio and Gage R&R

Figure 10.10 Summary for measurement system analysis

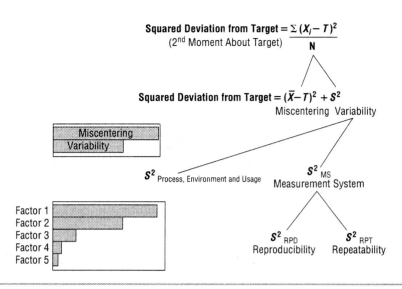

Figure 10.11 Partitioning of squared deviation from the target, including variance associated with the measurement system

The ideal case, in which every product is exactly on target with no variation, would have a value of zero for this statistical index. As further illustrated in Figure 10.11, this squared deviation from (or second moment about) the target corresponds to the Taguchi Loss Function for a target-is-best situation. A useful aspect of this equivalence is that the deviation from the ideal situation can be partitioned into two parts: the degree to which the deviation is a result of the average being off-target, and the variance about the mean. This variance can be further partitioned into variance as a result of the measurement system (discussed here) and variance as a result of manufacturing variation and variations in usage and environment (including system interactions).

MSA for continuous parameters provides an estimate for the variance caused by the measurement system, and compares it to the tolerance in terms of the precision to tolerance ratio (P/T ratio), and to the total observed variance in terms of the GR&R ratio (gauge repeatability and reproducibility). The P/T ratio is defined as six times the standard deviation of the measurement system divided by the difference between the upper and lower specification limits. The GR&R ratio is defined as the standard deviation of the measurement system divided by the total observed standard deviation, combining sources of variation including measurement error, variation from manufacturing, variation from how the customers use it, variations from the environments where the product will be used, and variations in how the interactions among the subsystems and the product with other systems affect the parameter.

If the measurement variance consumes too much of the tolerance window, or obscures the ability to assess the other sources of variation, then the measurement system is not acceptable. For many situations, the rule of thumb is that both the P/T ratio and the GR&R ratio should be less than 30 percent; for other situations, a rule that both should be less than 10 percent is imposed. Acceptable values for the P/T ratio derive from statistical analyses that indicate that a P/T ratio more than 30 percent corresponds to a very high risk of incorrectly passing bad parts or incorrectly rejecting good parts.

The measurement system for critical parameters at the system or product level will generally link to the test and evaluation plan for the product, as illustrated by the arrow to the deliverable initial critical parameter (CP) test plan in Figure 10.2. This deliverable is a starting point for the verification of capability discussed in Chapter 16, and summarized in Figure 10.12. Clearly, the preparation of the measurement systems to be used

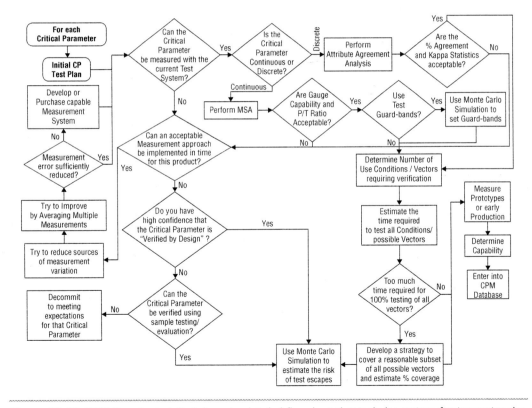

Figure 10.12 DFSS flowchart, drilled down to detailed flowchart that includes actions for improving the measurement system if MSA results are unacceptable

for verification do not need to wait, and should not wait, but should be initiated with the initial measurement systems analysis effort.

If the GR&R or the P/T ratio, or both, fail to meet acceptable guidelines, then there are a series of actions for improving the measurement system that are summarized in Figure 10.12 and discussed in Chapter 16.

CAPABILITY ANALYSIS

The next step shown in Figure 10.2 involves preliminary assessment of design capability. The design capability is a predicted capability, as opposed to the measured capability assessed on existing products or processes in the DMAIC process improvement flow. The preliminary assessment is performed in the Design phase of RADIOV (corresponding to the Measure phase of DMADV); if it is inadequate, then later steps (largely in the Optimize phase of RADIOV or Analyze through Design phases of DMADV) will improve the capability and a new assessment of the design capability will presumably be reflected in an improved value for the design capability indices. Later, in the Verify phase, the actual capability will be assessed on prototypes or early production samples, as discussed in Chapter 16.

As a predicted capability, the design capability might be assessed using predictive methods such as Monte Carlo simulation or a method referred to as the propagation of errors or system moments method in some situations and the root sum of squares method in other situations. These predictive engineering methods are discussed in Chapter 14.

There are two key indices used to assess design capability: the Cp (also known as Pp) and the Cpk (also known as Ppk). Equations for these two indices are given here:

$$Cp = \frac{USL - LSL}{6s} \tag{10.1}$$

$$Cpk = \min\left[\frac{USL - xbar}{3s}, \frac{xbar - LSL}{3s}\right] \tag{10.2}$$

Six Sigma performance is defined as having a Cp greater than or equal to 2 and a Cpk greater than or equal to 1.5. It is possible that the initial design capability assessment will forecast Cp and Cpk values that meet Six Sigma performance expectations at the get-go. If the team has confidence in this initial estimate, the design team can breathe a sigh of relief, celebrate, party, and paint the town red as appropriate to their personalities and local laws and customs. In addition to this emotional reaction, the design team need not expend any further effort on this critical parameter unless something changes that would jeopardize this pleasant state of affairs. Consequently, the detailed flowchart in Figure 10.2 shows that the flow-down for that critical parameter can be considered complete, and the design can move on to the efforts for the next critical parameter.

In those cases in which the initial assessment of the design capability do not provide sufficient confidence that the initial design is capable, the next step would entail flow-down or decomposition, as discussed in the next section.

FLOW-DOWN OR DECOMPOSITION

If the initial design capability analysis does not provide high confidence that the critical parameter will reside comfortably and consistently within the specification window, robust against noises ranging from manufacturing variation through variations in use conditions, environments, system interactions and measurement error, then the team will need to engage in robust design and optimization efforts that will generally be performed at the subsystem, module, subassembly and/or component levels. Consequently, the next steps involve identifying the parameters at these levels that are affecting the performance of the critical parameter at the system level. This is referred to as the critical parameter flow-down process.

A valuable tool for critical parameter management in general, and for this critical parameter flow-down and the later process for critical parameter flow-up (Chapter 14) is called Cognition Cockpit (http://www.cognition.us). This software tools provides an easy-to-use, Web-based interface that handles virtually all aspects of critical parameter management and provides interfaces to other software commonly used in DFSS and in product development.

The critical parameter flow-down is a team activity involving the appropriate expertise to identify x's, n's, and subordinate y's that affect the performance of the system level critical parameter. The second House of Quality can help with this flow-down. The approach used previously in the system-level or first House of Quality, described in Chapter 7, would be used again at the subsystem level or the next level down in the product hierarchy, but with the measurable system-level parameters along the left side and subordinate measurable technical parameters for the subsystem described across the top, as in Figure 10.5. Although this approach tends to provide a useful set of subordinate y's for each subsystem, the control factors (x's) and noises (n's) are a bit more difficult to obtain from this method.

The term "x's" refers to factors that are under the design engineers' control: design choices, component choices, or settings of continuous variables, like the choice of a resistor or capacitor value or for a voltage-controlled oscillator or the setting on a voltage supply.

The term "n's" refers to noises: factors that will not be under the design engineers' control when the product is operating out in the field among customers. Noises like environmental temperature might be controllable in the lab environment, which will be useful for evaluation purposes, but cannot be controlled once it leaves the controlled environment—a customer may use the product during a summer in Phoenix, Arizona, or Riyadh, Saudi Arabia, and the same or a different customer may use the product during winter in Alaska or Sweden.

The term "subordinate y's" refers to measurable parameters at a lower level in the flow-down that affect the system-level performance for the critical parameter and are in turn affected by other factors and parameters at an even lower level in the flow-down. A mechanical example of this might be the water resistance of the system being flowed down to subordinate y's representing the water resistance of various inserts, holes, and user interfaces that cannot be affected directly but can be affected indirectly through the choices of O-rings and dimensions that can ensure acceptable water resistances for those subordinate y's. An electronic example might be the total isotropic sensitivity, corresponding to the weakest signal strength that the system can dependably handle, which can be flowed down to subordinate y's representing the antenna gain and the LNA (low noise amplifier) gain at the component level, which cannot be directly affected but can in turn be affected by decisions about the design of the antenna and selection of the LNA part or components in the LNA.

The flow-down effort can be facilitated by the use of the P-diagram (Figure 10.3) discussed earlier in this chapter, which could already have identified the noises and may simply require differentiation between the subordinate y's and the x's. Alternatively, the team can use a second House of Quality approach or participate in a meeting to brainstorm the factors that affect the critical parameter and subsequently differentiate the factors as noises, x's, or subordinate y's, with an additional step to further explore the subordinate y's to complete the flow-down to x's and n's.

The process described here has proven very efficient at quickly generating a more thorough first-pass flow-down, which can subsequently be refined and expanded or "fleshed out." It also can be used to quickly generate P-diagrams and subsystem or component Houses of Quality or entered into the Cognition Cockpit database.

PROCEDURE FOR CRITICAL PARAMETER FLOW-DOWN OR DECOMPOSITION

1. Ask the critical parameter owner to describe the critical parameter, how it's measured, current estimates about its most likely value and possible distribution, and any progress that's already been made towards developing confidence that the critical parameter will be capable.
2. Discuss with the team:
   Is the critical parameter clearly measurable as-is? How would it be measured?
   If the measurement approach is clearly defined, would meeting that measurable

requirement fulfill customer and business expectations? If not, develop an operational definition, a measurable definition for the critical parameter.

3. Brainstorm subrequirements with the team, first pass. The template shown in Figure 10.13, available for download at http://www.sigmaexperts.com/dfss/, can help with this process.

4. Classify the subrequirements into subordinate y's, x's, and noises.

Subordinate y's are measurable, but not directly controllable (i.e., there is not a "knob" to change the value of the subordinate y to a selected value).

Control factors or x's are directly controllable and affect the value of the critical parameter or a subordinate y to the critical parameter.

Noises or n's are factors that affect the critical parameter or a subordinate y, but that the team does not control in normal usage (although the team might be able to control a noise like temperature in a lab).

Subordinate y's (Subsystem, Module, or Component)— Measurable Requirements:	Priority	Units	y, x or N	Continuous or Ordinal Subrequirements:	Binary or Obligatory Subrequirements:	Is this Necessary to Fulfill the Requirement?	Is this Set Sufficient?
						Yes	
						Yes	**Yes**
						Yes	
						Yes	
						No	**No**
						Yes	
						Yes	
						No	**Yes**
						No	

Figure 10.13 Template for critical parameter flow-down process and for associated P-Diagrams; this template can be downloaded at http://www.sigmaexperts.com/dfss/chapter10flow-down

5. Classify the subordinate y's into continuous requirements, ordinal requirements, and binary discrete or obligatory requirements (pass/fail or meets/doesn't meet requirements).

   Continuous requirements have a full range of possible values—like a voltage measurement; ordinal requirements have integer values, where higher or lower is better—like a score of 1 to 7 on a Likert survey form, or the number of drops to failure, or the number of clicks to get to a certain screen.

   Binary requirements are either acceptable or unacceptable—like whether an Excel-compatible table of data is output from the software or not.

6. For each subordinate y, ask the team—is each subrequirement necessary?

   If this set of subrequirements was satisfactorily met, would that provide sufficient confidence that the product will meet expectations for the critical parameter? If not, brainstorm what additional subordinate y's are needed to have sufficient confidence that the customers will be satisfied that this critical parameter has been fulfilled.

7. For each necessary subordinate y that is continuous or ordinal, the team should discuss their confidence. If the team is highly confident that a subordinate y will be satisfactorily achieved, then it need not be flowed down further, but otherwise the team would brainstorm other lower level subordinate y's, control factor (x's) and noises (n's) that affect that subordinate y. Continue until the lowest level of the flow-down consists of only x's, n's, and subordinate y's that the team is highly confident will be satisfactorily achieved.

8. For each necessary subordinate that is binary or obligatory, ask the team: What are the goals? Should we consider a Pr (success) metric (as discussed in Chapter 14)? Would fault tree analysis be helpful for this obligatory requirement? (Fault tree analysis is discussed in Chapter 9.)

9. Put the results of the flow-down of the critical parameter into a diagram, with the critical parameter or Y placed at the top or left side of the diagram, and the subordinate y's, x's, and n's linked by lines or arrows. If appropriate, capture this flow-down in a database such as the critical parameter management database associated with Cognition Cockpit. If appropriate, capture part or all of the flow-down as a P-diagram.

10. Ask the team or assign team members to obtain goals/preliminary spec limit(s) for the continuous and ordinal requirements.

FLOW-DOWN EXAMPLES

The flow-down described in this chapter can be applied to a variety of parameters, including mechanical, electrical, and software parameters. In Figures 10.14, 10.15, and 10.16, qualitative flow-down will be applied to three critical parameters for a set of

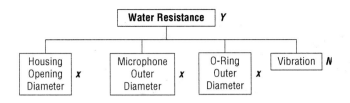

Figure 10.14 Critical parameter flow-down of water resistance for "Simon" communication device

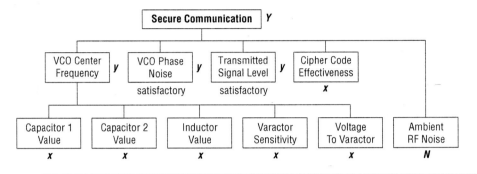

Figure 10.15 Critical parameter flow-down of secure communication for "Simon" communication device

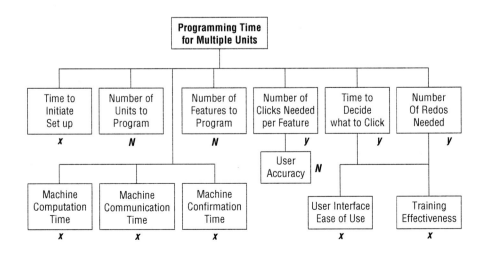

Figure 10.16 Critical parameter flow-down of programming time for multiple units

communication devices for secret agents; the device is code-named "Simon." These critical parameters are water resistance, secure communication, and programming time for multiple units. By sheer coincidence, the first critical parameter is primarily a mechanical engineering example, the second is a combined software and electrical engineering example, and the third is primarily a software example.

Water resistance is largely dependent on the materials used in the housing (outer shell) of the communication device and the effectiveness of the seals involved in the opening in the housing to accommodate a microphone. The mechanical engineering design team is confident that the housing itself is impervious to water intrusion, and the primary risk is seal for the microphone. The team has identified the x's as the housing opening diameter, the microphone outer diameter, and the outer diameter of an O-ring that must not exhibit excessive compression. The design team is also concerned with vibration as a noise. The ultimate customers (spies, secret agents, and informers) are prone to considerable vibration in the usage environment, as the team ascertained through exhaustive research (watching James Bond movies; popcorn optional). This example is similar to an actual DFSS project, which is shared as an example for optimization and flow-up in Chapter 14.

The design team flowed-down secure communication (measured by a secure communication effectiveness metric) to a software parameter (cipher code effectiveness) and several subordinate y's, including the center frequency of the voltage-controlled oscillator (VCO). This subordinate y was flowed down to parameters associated with the two capacitors, an inductor, a varactor, and the voltage applied to the varactor, as shown in Figure 10.15.

The team also flowed down the requirement for the time required to program a set of communication devices, as shown in Figure 10.16.

INITIAL TOLERANCE ALLOCATION

After the flow-down has been completed in a qualitative respect and specification limits and a target value for a system-level critical parameter have been set, the next step is to allocate tolerances from the critical parameter to the subordinate y's and x's involved in the flow-down.

Initial tolerance allocation is the quantitative part of the critical parameter flow-down. Tolerances may be available from suppliers for some of the subordinate y's and x's. For others, the initial tolerance may be the start of communication with the suppliers and assembly and manufacturing areas involved in the supply chain. If the supplier already has a proposed tolerance, it could be helpful for the engineering team to compare the tolerances proposed by the suppliers to a baseline to ascertain whether the suppliers' tolerances align with reasonable or expected tolerances.

	Units	LSL	Target	USL	kmin	kmax	
Frequency	MHz	2.4	2.5	2.7	0.07	0.13	

Subordinate y's	Units	Slope	Target		Allocated LSL	Target	Allocated USL
Capacitor 1	pF	−0.24793	2		1.9	2.0	2.3
Capacitor 2	pF	−0.24793	1		0.9	1.0	1.1
Varactor Sensitivity	pF/V	−0.48868	1		0.9	1.0	1.1
Inductor	nH	−0.15582	8		7.5	8.0	9.0
Voltage	V	−0.24793	2		1.9	2.0	2.3

Figure 10.17 Excel worksheet for quantitative flow-down of allocated tolerances

The approach described in this section can be used for a variety of situations:

- **Schedule allocation:** Starting with best case/most likely/worst case durations for developing features, and allocating these to subtasks required to develop the feature.
- **Timing/delay allocation:** Starting with a range or mean and standard deviation for overall timing for a feature or function, and allocating the total timing or delay to the individual tasks.
- **Mechanical tolerance allocation:** Allocating tolerances that are additive, like tolerances for components in a gap analysis.
- **Electrical tolerance allocation:** Allocating tolerances for a function to its subfunctions when the transfer function is not necessarily additive and some of the subordinate y's or x's might be in different units than the Y (for example, the Y may be frequency in MHz and the x's might be capacitance in pF and inductance in nH).

A step-by-step approach for allocating tolerances to the subordinate y's and x's is provided here. The Excel template shown in Figure 10.17 can be downloaded to assist with these calculations. The subordinate y's and x's are both referred to as subordinate y's and treated the same in this approach. If the transfer function is a simple additive or sum of terms function, then the slopes will be set to unity. If the transfer function is a simple multiplicative or product of terms function, then a logarithmic transform could allow the same approach to be used, with the slopes similarly set to unity.

1. Determine Tolerance for Y: Target, USL and LSL.
2. Determine Target or Most Likely Values for subordinate y's: $Y_{T,i}$ and slopes:
   $b_i = dY/dy_i$.

3. Estimate the constant percent tolerance for each subordinate y_i:

$$k_{max} = \sqrt{\frac{(USL - T)^2}{\sum (b_i y_{T,i})^2}}$$

4. For each subordinate y, set $USL(y_i) = y_{T,i} (1 + k_{max})$.
5. If the Tolerance for Y is symmetrical, set $LSL(y_i) = y_{T,i} (1 - k_{max})$.
6. If the Tolerance for Y is not symmetrical:
   a. Determine the constant percent tolerance for each subordinate y to its lower limit:

$$k_{min} = \sqrt{\frac{(T - LSL)^2}{\sum (b_i y_{T,i})^2}}$$

   b. Set $LSL(y_i) = y_{T,i} (1 - k_{min})$.

As illustrated in Figure 10.2, these allocated tolerances should be shared with suppliers and manufacturing and assembly engineers, or with supply chain experts who can work with suppliers and manufacturers.

The quantitative aspect of the flow-down described in the previous section will be applied through an Excel worksheet set up as a template, as shown in Figure 10.17. Figure 10.15 includes a subordinate y called "center frequency." The quantitative flow-down for the subordinate y of center frequency for the VCO is illustrated in Figure 10.17, using the template that can be downloaded from http://www.sigmaexperts.com/ dfss/chapter10allocation. The transfer function for the center frequency is not additive; it is a constant divided by the square root of the product of the inductor value and the sum of the capacitances for the two capacitors and the varactor. This transfer function was evaluated to obtain slopes for frequency versus each factor for use with the spreadsheet template.

SUMMARY

The system requirements flow-down is the beginning of the Design phase of the RADIOV DFSS process, initiating activities for predictive engineering: to design and optimize the flowed-down parameters at the subsystem, subassembly, module, and component level in order to provide high confidence that expectations for the system-level critical parameters will be successfully met. The systems (software and hardware) requirements flow-down also initiates focused activities in the systems/field test and supply chain to help develop confidence in meeting the flowed-down requirements.

Software DFSS and Agile

Chapter 10 discussed the flow-down of requirements, including the flow-down from system requirements to hardware and software requirements. However, requirements often change during the course of development—software requirements seem to be notoriously unstable in this respect. The intent of this chapter is not to teach Agile or even provide a step-by-step process for Agile methods—these are clearly beyond the scope of this book, and an argument can be made that Agile methods are only really learned by experience. Instead, the intent of this chapter is to discuss the combination of software DFSS with Agile methods. For example, Agile provides approaches to handle changing or evolving requirements; in the ideal situation, a knowledgeable customer works with the product development team, helping to clarify requirements and answer questions, and also providing feedback. Furthermore, the iterative nature of Agile software development provides an environment for responding in a later iteration if a requirement changes—in that respect, Agile almost embraces changing requirements.

Although the term "Agile" usually is applied only to software development, the methods have a historical connection with Toyota approaches ("lean"); Mary and Tom Poppendieck were involved in the Agile Alliance, to be discussed shortly, and had considerable experience with lean methods and had written two books about applying lean methods to software. Agile methods for software development bear a resemblance to set-based concurrent engineering, which is part of the Toyota new product development process that is applied primarily on the hardware aspects.

In early 2001, motivated by the observation that software teams in many corporations were stuck in a quagmire of ever-increasing process complexity, a group of industry experts met to outline the values and principles that would allow software teams to work quickly and respond to change. They called themselves the Agile Alliance. Over the next several months, they developed a statement of values.[1] The result was *The Manifesto of the Agile Alliance.*[2]

Manifesto for Agile Software Development
We are uncovering better ways of developing software by doing it and helping others do it. Through this work we have come to value

- Individuals and interactions over processes and tools
- Working software over comprehensive documentation
- Customer collaboration over contract negotiation
- Responding to change over following a plan

The Agile Alliance states that, while there is value with items on the right, they value the items on the left more. In DFSS, we value both, especially the need for collaboration, critical change management, and iterative development.

A software development lifecycle (SDLC) addresses lower-level software-specific aspects, like software third-party suppliers, release management, and configuration management, which are not part of DFSS. Agile can be considered a software development process built on the foundation of iterative development. Iterative development is a cyclical process of repeating a set of development activities to progressively elaborate and refine a complete solution (see Figure 11.1). The "unit" of iterative development is an "iteration," which represents one complete cycle through the set of activities. An iteration addresses the requirements first, then the design, and then the code is written as part of the implementation. Each iteration (after the last creation phase) produces intermediate deliverables. Often, each iteration includes testing of the software as an intermediate deliverable, where the testing confirms that the software meets the requirements associated with that iteration; this is often referred to as test-driven development (TDD). In effect, each iteration could involve a subset of the RADIOV process, with the TDD aspect linking the Requirements phase of the iteration with the Verification phase of the iteration.

DFSS can be considered a sequence of events within the iterative development life cycle, from requirements through system integration. It can be a framework in which to drive measurement, forecasting or predictive engineering, quality, and downstream control of business results (some of which might be considered outside the scope of SDLCs). Iterative development should especially be used for large, complex systems, instead of the one-pass waterfall approach that commonly leads to unreliable, rigid, and fragile software.

---

1. Agilealliance.org.
2. Robert C. Martin, *Agile Software Development Principles, Patterns, and Practices*, Prentice-Hall, 2002.

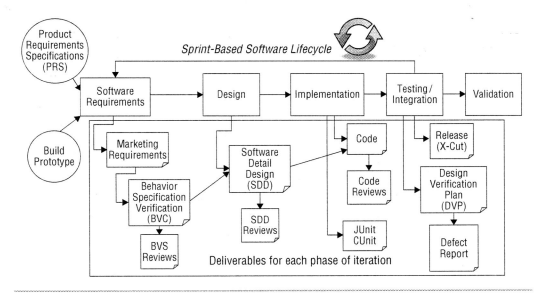

Figure 11.1 Agile software development lifecycle (SDLC)

Many Agile practitioners advocate working on the most challenging, highest risk requirements, such as complex features, in the earliest iterations; the VOC gathering, prioritization, and translation into system requirements (Chapter 7) and flow down (Chapter 10) provides that prioritization.

The Agile process introduces an iterative approach to feature development. A feature is decomposed into a set of iterations, in which every iteration has a set of requirements that must be implemented and test cases that must be validated for that iteration to be complete. All iterations, while short on functionality, are developed as a working version of the final deliverable. Therefore, all iterations must be organized, developed, and tested as if that software was to be delivered and used as a functioning product. This approach increases collaboration and accountability and reduces risk.

The appropriate duration for a planned iteration will vary from feature team to feature team, but most Agile methodologies prescribe between two and six weeks per iteration. An iteration is called a Sprint. For the Motorola Agile process, there are no restrictions on an iteration's length. Instead, the process recommends a duration of three to four weeks, but also provides flexibility to adapt the length of feature iterations up to six weeks based on the needs and circumstances of a feature team.

During Agile activities, especially feature development, the operational aspects of critical parameters should be used to define some of the feature sets. Conversely, having

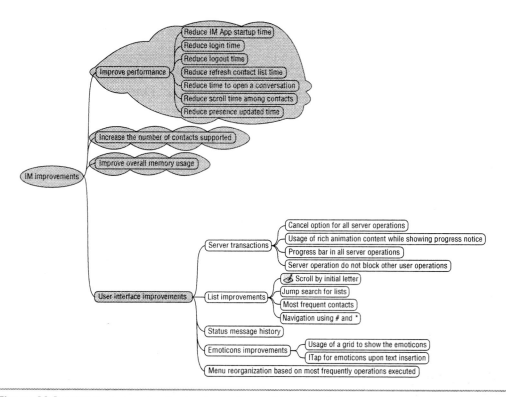

Figure 11.2 VOC Brainstorming Session for an Instant Messaging feature

the customer involved in Agile development can provide another source of VOC or VOB requirements that flow into critical parameter management. Capturing requirements is difficult and VOC in DFSS can help (see Chapter 7). See Figure 11.2 for an example of a brainstorming session that was held to get VOC on an instant messaging feature.[3]

During the requirement gathering phase the team use the DFSS VOC to understand the requirements and identify any derived requirements. It is important in Agile development to have an understanding of the requirements as soon as possible. Thus, rather than competing with an SDLC such as Agile, DFSS actually supports any life cycle by bringing useful tools and data to it, facilitating effective decisions and adjustments to manage risk and cycle time.

The RADIOV model can provide a baseline to guide the deployment of the DFSS methods and is used as the baseline requirements for DFSS projects. Throughout the

---

3. Motorola DFSS Instant Messaging Feature Performance Project.

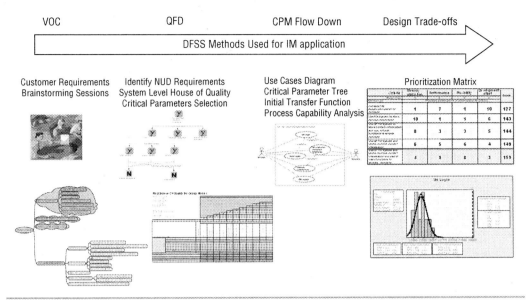

Figure 11.3 Example of applications of DFSS tools during early stages of an Agile project to develop an instant messaging feature

iteration cycles of requirements, architecture, design, implementation, and testing the development team can and should use DFSS tools such as FMEA/FTA, Rayleigh model, design patterns, performance modeling, Monte Carlo simulation, multiple Y optimization, and so on.

It is well documented that software development projects often don't deliver on time, or within budget or with the function that was agreed when they started. These DFSS tools can assist in reducing cycle time and preventing defects from escaping to the customer. *The fundamental concept of DFSS is to predict the final quality level earlier (even while still in the Design phase), evaluate design trade-offs, and make decisions to address potential low-quality issues.* Agile must go through the Design phase within weeks (maybe days) and by using DFSS the project can make design trade-offs quickly and effectively. Figures 11.3 and 11.4 illustrate an example of how the developers of the instant messaging team used DFSS tools throughout the Agile project.[4]

DFSS can provide an understanding of the importance of requirements modeling and systems analysis, introducing various modeling and analysis methods during the Requirements phase. The initial transfer functions in DFSS can relate critical parameter

---

4. Motorola DFSS Instant Messaging Feature Performance Project.

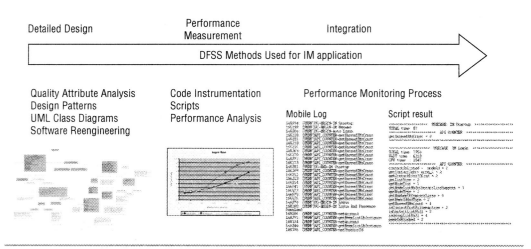

Figure 11.4 Example of applications of DFSS tools during later stages of an Agile project to develop an instant messaging feature

requirements to design parameters. This allows the Agile design team to know if they are capable of meeting the customer specification as early as possible. The most tragic situation that can occur on your Agile development project is to have the application features completed by the third iteration only to find out that you are not able to meet your lower-level requirements specification on the fifth iteration. DFSS transfer function concepts can help determine the capability of meeting lower-level requirements.

Suppose a team is developing a new feature for an ATM (automatic teller machine): authentication using fingerprints. After obtaining the requirements, the team obtains a sensor and some off-the-shelf software in the first iteration. The team verifies that, under ideal conditions, the software and sensor will, in fact, correctly identify a perfectly clean fingerprint.

In the second iteration, the team develops a P-diagram (or parameter diagram, see Figure 11.5). The customer representative on the team clarifies the customer requirements and expectations: the risk of false authentication involves legal issues like identity theft, and must be extremely low—consistency with error rates of less than 3.4 errors per million could be considered acceptable. The risk of incorrect rejection and inconclusive results requiring retests involves customer irritation; customers would accept no more than a five percent risk of incorrect rejection, and less than one percent would be preferable. Inconclusive results could be handled with automatic retesting, but the customer would be willing to wait no more than five seconds for authentication.

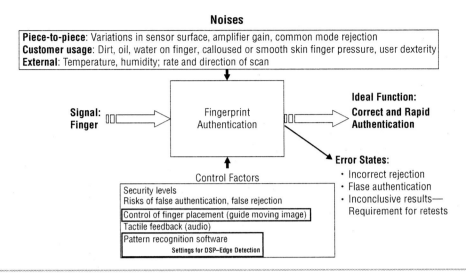

Figure 11.5 P-Diagram for authentication using fingerprints

The Agile feature team considers the impact of the noises in Figure 11.5. Noise factors are parameters that:

- *cannot* be controlled during use, or
- are *difficult* to control, or
- are much too *expensive* to control.

Reviewing the P-diagram, the software team invites in some hardware engineers, and they jointly design an experiment in which the control factors are settings for edge detection for the digital signal processor (DSP) that handles pattern recognition in an off-the-shelf fingerprint identification software package (purchased in the first iteration), and the mechanical design of a guide to help ensure consistent finger placement relative to the sensor. The experimental design takes about two hours, and the next day a subteam is ready to begin running the experiment using a mock-up of an ATM. The experiment includes the noises, and the fingers run across the sensor with various guides and various settings for edge detection including clean fingers, dirty fingers, and fingers that look like someone just changed their own oil for their car.

After two days running the experiment, the data is analyzed using a logistic regression approach similar to the approach used in the signature recognition example described by Figure 13.8 in Chapter 13. The team finds optimal settings for edge detection for the software controlling the DSP and an optimal design for the mechanical guide. By the end of the third iteration, the team is highly confident that the ATM fingerprint authentication is robust against the noises from the P-diagram and will meet customer expectations in terms of both low probability of false authentication (identity theft risk) and low probability of incorrect rejection or retests (customer irritation). The team also adds a screen—in cases where the customer's fingerprint is not recognized due to noises, the new screen gently and very politely advises the customer to try to clean or wash their hands and then try again.

Verification is involved as part of the deliverable for each iteration when Agile methods are used. As part of the test-driven development aspect of Agile, a planned iteration generally includes a set of test cases that will thoroughly cover all aspects of that iteration. These test cases encompass all layers and features that will be created or modified to address the functionality dictated by that iteration. It may take several iterations and many test cases to completely cover a single set of requirements. Iteration is said to be completed when all planned test cases for that iteration have passed.

Using DFSS testing techniques, as described in Chapter 18, fewer test cases are generally required for verification. A test strategy called pair-wise testing can be used to reduce the number of exploratory tests the team creates. Pair-wise testing takes advantage of the "sparsity of effects" principle and assumes that most failures will occur because of interaction of no more than two factors. Orthogonal arrays are used to identify the minimum number of cases to assure all two-way interactions are tested (see Chapter 18).

Measuring the Agile Design

The team should use similar measurements for the Agile development as in other SDLC, but the main question is how can the team know if the software design using Agile is good? There must be a measurement system for the design to verify and validate that the design met your customer requirements.

For Agile Development, the following software measures are important:

a. On Time: Planned vs actual iterations
b. On Scope: Planned vs actual test execution
c. Defect Discovery Rate: Number of defects discovered in iterations
d. Iteration Containment Effectiveness: The number of defects passed to the next iteration
e. Percent of Feature Complete per Iteration: Index for the number of features not completed in the prior iteration

This knowledge makes the design process more efficient and helps to prevent wasted efforts that are unaffordable in an Agile environment. With DFSS, a design verification plan is created to help clarify what should be measured and how to measure it. The verification plan forces the design team to consider how to test while in the Design phase of that iteration; during design, the team may plan for monitoring methods, checkpoints, or test points, before completing the Design phase. The Agile feature design team is responsible for specifying how the design performance will be verified. Therefore, the team should create a data collection plan as part of their verification plan. Here is an example of a data collection plan for a prototype from the Motorola design team in Brazil.

DATA COLLECTION PLAN FOR VIEWHOME PROTOTYPE[5]

1. **Establish objectives**

   I need to collect data on *Refresh Rate on ViewHome* in order to *prove the product is able to handle the required refresh rate.*

2. **Determine specific measures for defined test sample**

   What Y or X data is needed to meet the objective outlined in step 1?

A verification plan also provides a ***clear direction*** and ***method*** for achieving the objectives for a prototype or a deliverable from an iteration. An organized approach

	Measure	Operation Definition
Y	Frame rate	Number of frames per minute
X_1	Available memory	Available memory during frame transfers
X_2	Network throughput	Network capability during frame transfers
X_3	Concurrent applications	Applications running at the same time
X_4	Image quality	Quality of the image being transferred

---

5. Motorola's Brazil DFSS green belt ViewHome project.

3. Determine how to measure

	Measure	Data Type	Tool
Y	Frame rate	Continuous	Log with frame timestamp
X_1	Available memory	Discrete	Log (memory logging)
X_2	Network throughput	Continuous	Internet speedometer
X_3	Concurrent applications	Discrete	Manual testing
X_4	Image quality	Discrete	Manual testing

3.1 Sampling strategy

	Measure	When?	Where?	How Many?	Measurement Method
Y	Frame rate	Upon image reception	Equipment under test	5 measures of 5 minutes	Count number of frames per minute
X_1	Available memory	Before image reception	Equipment under test	1 measure before each test	Log free memory during frame transfer
X_2	Network throughput	During the test execution	Similar equipment	5 measures of 5 minutes	Measure throughput in the similar equipment
X_3	Concurrent applications	Before image reception	Equipment under test	1 measure before each test	Test performance with 0–5 apps running at the same time
X_4	Image quality	After image reception	Equipment under test	1 measure after each test	Vary the quality and verify the changes on frame rate

220

ensures that the prototype runs smoothly and that the *data collected is valid*. The process of creating a plan for executing the prototype helps ensure that the team will:

- Thoroughly test the critical parameters and transfer functions under the proper operating conditions and with an appropriate number of samples
- Consider the risks associated with the design
- Look for and address problems and opportunities that surface
- Use resources efficiently
- Verify the validity and completeness of data collection

SUMMARY

DFSS assists Agile development projects to achieve the following objectives:

- The project team learns how to measure the design and identify critical parameters early in the Agile iterations.
- Improvement and guarantee of the quality of the design.
  - The completeness of the results to be delivered.
  - Defined interim results to make early assessment of the architecture and design possible.
- Checking the costs for the entire lifecycle via DFSS scorecards:
  - The generation of relevant project-specific development measurements and its assessment will be simplified.
  - Makes the cost calculation more transparent. Any risks in connection with the costs can be recognized better.
  - Allows for the reduction in the use of resources.
- Undesirable developments are recognized at an earlier stage using modeling and simulation.
- Improvement in the communication between the different parties and reduction in misunderstandings between all parties involved.

Software Architecture Decisions

Architecture has emerged as a crucial part of the design process and encompasses the structures of large software systems. The architectural view of a system is abstract, refining away details of implementation, algorithm, and data representation and concentrating on the behavior and interaction of "black box" elements.[1]

The focus of this chapter is software architecture trade-off analysis using DFSS tools and methodologies. Architecture imposes a structure for a system or subsystems and it affects the structure of the development of the software. An archetypal software architecture can be divided into layers. The layers are hierarchical, with each layer providing services to the layers immediately above and below it in the hierarchy, with which it interfaces. In some systems, every layer has access to parts of all other layers.

For example, the architecture for an application processor can be divided into these layers:

Layer 1: System services: This consists of the operating system (OS), the kernel (which manages the system resources, especially the communication between hardware and software components), and the drivers file system support (support for computer software that allows higher-level programs to interact with a hardware device like a printer through a computer bus or communication system). Layer 1 contains the core system services required to support the application layer. In a Linux architecture, Layer 1 would be composed of open source Linux components and include system

---

1. Len Bass, Paul Clements, and Rick Kazman, *Software Architecture in Practice, Second Edition*, Addison-Wesley, 2003.

libraries and daemons (software that runs in the background for handling systems logs, responding to network requests, or running schedule tasks) such as the clib (C libraries for communications between a personal computer and a real time processor) and Internet networking daemon, system utilities, such as /init, device drivers, and the kernel.

Layer 2: Application support services: This layer consists of middleware components (software that connects components or applications, consisting of a set of services that allow multiple processes running on one or more machines to interact for interoperability) like databases, messaging, and browsers. These components publish well-defined APIs and other functionalities used by applications, and interact with underlying hardware and OS.

Layer 3: Application framework and user interface: framework (support programs, code libraries, and scripting language to glue together different components of software), application management.

Layer 4: Applications: This consists of all QT-based applications (QT is a cross-platform framework for GUIs [graphical user interfaces]). All user interactions are handled by Layer 4.

The description of the architecture of the system would show the external and internal interfaces and the architectural units that make up the system. It is not always easy to structure a system in layers. The software development team must ensure that changes to the external and internal interfaces are done with considerable forethought and not done unintentionally. Unintended or poorly defined changes can lead to serious defects. There are several architectural styles an architect can choose from (e.g., computational schemes, data elements, control loop, process control, client-server, etc.) DFSS tools and methods can help make smart decisions about which architectural style is best.

SOFTWARE ARCHITECTURE DECISION-MAKING PROCESS

Through architecture trade-off analysis, software developers can predict the impact of software architecture decisions. The ability to make decisions is a vital skill, both in personal and professional lives.

- **What am I going to have for lunch?**
- **What kind of computer should I buy next?**
- **What display and keypad should Motorola use to meet a customer's need for a robust design?**

Making a decision involves knowing what the possible choices are and what criteria distinguish the choices. In Figure 12.1, the architecture team surely did not consider the impact of the location of the ATM on its customers trying to access the ATM.

Decisions about architecture will impact future features and application of the product line and affect the ability to fulfill customers' requirements for future systems. Figure 12.2 summarizes the steps, activities, and deliverables involved in software architecture decisions, and provides an overview of this process that aligns with both the critical parameters that are derived from the voice of the customer (VOC) and with key quality attributes.

The input into the architecture decision-making process is the selection of critical parameters (see Chapters 7 and 9). These critical parameters are a set of customer requirements and features that have been prioritized, analyzed, and decomposed (Chapter 10). The architecture must satisfy functional requirements while meeting non-functional requirements such as quality attributes; for example, availability, usability, security, performance, and maintainability. Functionality of the system includes basic requirements regarding the system's behavior, capabilities, operations, and services. Quality attributes determine the quality of the system and how the design will affect the overall architecture, implementation, quality, and business goals.

Figure 12.1 Example of the impact on customers of an inferior design decision (ATM location)

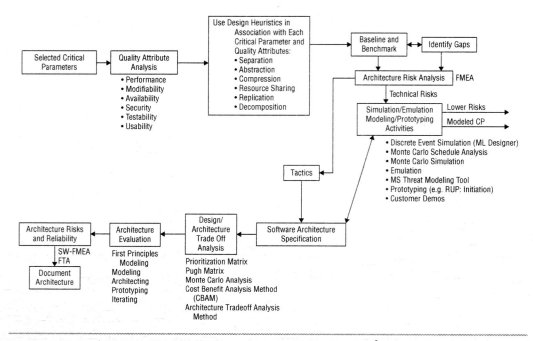

Figure 12.2 Detailed flowchart for software architecture decision making[2]

Bass, Clements, and Kazman[3] describe a list of system quality attribute scenarios that are quality-attribute-specific requirements. These scenarios consist of six parts.

- **Source of stimulus.** This is some entity (a computer system, or any other actuator) that generated the stimulus.
- **Stimulus.** The stimulus is a condition, input, or change that needs to be considered.
- **Environment.** The stimulus occurs within certain conditions. The system may be in an overload condition or may be running when the stimulus occurs.
- **Artifact.** Some artifacts are stimulated; this may involve the whole system or some pieces of it.
- **Response.** The response is the activity undertaken after the arrival of the stimulus.
- **Response measure.** When the response occurs, it should be measurable in some fashion so that the requirement can be tested.

---

2. Excerpt from DFSS flowchart developed by Dr. Eric Maass.

3. Len Bass, Paul Clements, and Rick Kazman, *Software Architecture in Practice, Second Edition*, Addison-Wesley, 2003.

Table 12.1 Performance quality attribute scenario

Portion of Scenario	Possible Values
Source	One of a number of independent sources, possibly from within system
Stimulus	Periodic events arrive; sporadic events arrive; stochastic events arrive
Artifact	System
Environment	Normal mode; overload mode
Response	Processes stimuli; changes level of service
Response Measure	Latency, deadline, throughput, jitter, miss rate, data loss

Table 12.1 is an example of a performance quality attribute scenario: Performance is concerned with how long it takes the software system to respond when an event occurs.

Not every system-specific scenario has all of the six parts. The parts that are necessary are the result of the application of the scenario and the types of testing that will be performed to determine whether the scenario has been achieved.

Making a system-specific scenario helps the team to decide which features will interact with other features and which scenario will have an overall impact on the customer's basic needs and delighters. It is important for the organization to perform a gap analysis on the quality attributes and the system-specific scenario in order to understand the greatest benefit to cost and schedule.

USING DESIGN HEURISTICS TO MAKE DECISIONS

Design heuristics state that there is a best practice (technique, method, process, or activity) that is more effective at delivering a great design than any other technique, method, process, and so on. The idea is that with using a few key concepts at code, verification, and validation testing we will have with fewer defects and unforeseen complications.

Architectural design heuristics can also be defined as the most efficient and effective way to architect a product or platform based on repeatable techniques that have proven themselves over time to work. The architects who apply design heuristics and principles (such as those listed here) make better decisions about their architecture than those who do not, therefore, they achieve the task of satisfying the functional critical parameters while meeting the nonfunctional critical parameters.

COMMON DESIGN HEURISTICS AND PRINCIPLES

Separation
- Isolates a portion of a system's functionality into a component. For example, isolate the user interface of the entertainment system from the other applications in the system. So if something goes wrong with the entertainment system it will not spill over to the other applications that could be more critical.

Abstraction
- The operation of creating a virtual machine and hiding its underlying implementation.

Compression
- Removing layers or interfaces (i.e., the opposite of separation). Sometimes, this is done to remove unnecessary layers in your architecture. For example, if designed correctly your system might not need to have an application layer and a user interface layer.

Resource sharing
- Encapsulation of either data or services.
- Sharing among multiple independent end-users, consumers, or customers.

Replication
- Operation of replicating a component. Offers a way to achieve the high performance or reliability quality attribute by deploying replicas on different processors.

Decomposition
- Separating a large system into smaller components.
- Part-whole: each subcomponent represents nonoverlapping portions of the functionality.
- Is-a: Each subcomponent represents a specialization of its parent's functionality.

USING ARCHITECTURE TACTICS TO MAKE DECISIONS

Tactics are design decisions that you make about how you will achieve the quality attributes. For example, the customer's need for high performance may depend on high availability, which depends on having some form of redundancy in either data or code. Both the performance requirements and the redundancy that is needed by availability generate additional considerations for the architect (such as considering resource demand). Each of these are referred to as a tactic and become a design option for the architect to consider.

For each of the six quality attributes (availability, modifiability, performance, security, testability, and usability) listed here, there are tactical approaches for achieving them. The organization of tactics is intended to provide a path for the architect to search for appropriate tactics and make a good decision.[4]

Examples of tactics include:

- Performance
  - LIFO
  - Leaky bucket
- Security
  - Trusted computing base
  - Authenticate users
  - Authorize users
- Usability
  - Parameter hiding
  - Undo
  - Clearly marked exits
- Availability
  - Trusted computing base
  - LIFO
  - Leaky bucket
  - Garbage collection
- Modifiability
  - Abstract common services
  - Anticipate expected changes
  - Runtime registration
- Testability
  - Record/playback

Many of the tactics we discussed are available within standard execution environments such as operating systems, application servers, and database management systems. It is still important to understand the tactics used so that the effects of using a particular one can be considered during design and evaluation.

Let's say that we're developing architecture for an online Internet store and the gap analysis of the quality attributes revealed that the *security* response of the Internet store is a critical parameter.

---

4. Pugh matrix courtesy of Vivek Vasudeva.

- Desired response: Keep mean-time-to-detect within five minutes (i.e., to detect an attack). What design decisions can we make to achieve this?
  - Leverage a tactic called "trusted computing base."
- Design decision:
  - Choose an architectural boundary, within which the data is trusted.
  - Note that this decision could result in a degradation of performance response. Tactics will often support one attribute, but at the expense of another.

USING DFSS DESIGN TRADE-OFF ANALYSIS TO MAKE DECISIONS

The objectives for design trade-off analysis are to be able to assess the risks of changing legacy architecture to accommodate the new features and to be able to characterize economic factors that influence architecture decisions (e.g., value weighting of scenarios). We also need to be able to quantify the impact of an architecture strategy in terms of the response on the quality attributes. *Changing a legacy architecture is one of the riskiest development activities that a team can undertake.*

The key to making the right design trade-off is the following:

- Keep the effort under control by making changes in small steps.
- Understand the risks that derive from how well you know your architecture, requirements, and current implementation:
  - Risk of insufficient documentation of legacy products
  - Risk of insufficient domain experience/understanding
  - Risk of insufficient architecture skills on team
- Make the investment that is needed (i.e. long-term Architecture phase).

Design trade-off tools provide a formal method of evaluating possible decision choices against key selection criteria. There are many such tools, all of which are aimed at revealing the best concept. Common steps in design trade-off tools are:

1. List the choices, options, or alternatives.
2. List the selection criteria (e.g., critical parameters).
3. Score each choice against each criterion.
4. Score the choices against each other.
5. Document which choice scores the highest.

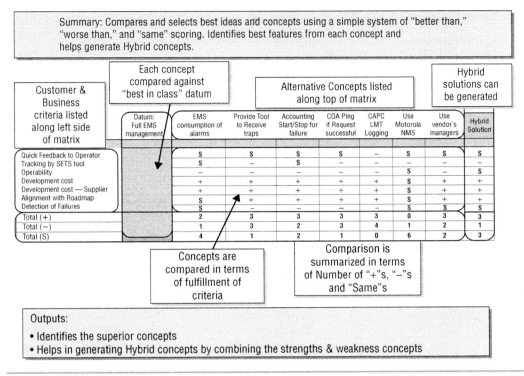

Figure 12.3 Example of Pugh matrix for concept selection[5]

The RADIOV process offers unique tools such as the Pugh and prioritization matrices, to help an architecture team to resolve architecture issues and weigh their options for the best design or architecture concept.

The Pugh matrix helps the architecture team to choose a best concept based on a standard, widely known as the datum. The datum is the concept that is measured against every other concept idea. Use a Pugh matrix when you need to:

- Do a quick qualitative assessment.
- Determine if there is an obvious winning concept.
- Get a quick feel for concept comparison.
- Down-select from many concepts to the most promising few.

Figure 12.3 is an example and summary of a Pugh concept selection matrix.

---

5. Pugh matrix courtesy of Vivek Vasudeva.

Criteria	Memory Utilization	Performance	Flexibility	Development Effort	Score
Weightings	10	9	4	5	
Alternatives	*Positive correlation of alternative to criteria*				
Increase the maximum number of contacts	1	7	1	10	127
Use of file system to store contact information	10	1	1	6	143
Use of file system and cache to store contact information and use of hash functions to access contacts	8	3	3	5	144
Use of file system and cache to store contact information	6	5	6	4	149
Use of file system and cache to store contact information and use of hash functions to access contacts	4	8	8	3	159

Figure 12.4 Mobile phone contact prioritization matrix example[6]

The prioritization matrix (Figure 12.4) can be used to help narrow down options through a systematic approach of comparing choices by selecting, weighting, and applying criteria. It will quickly focus the team on the best things to do given the critical parameters and quality attributes that the customer wants. It limits hidden agendas by surfacing the criteria as a necessary part of the process. It also increases the probability of completion because consensus is sought at each step in the process from criteria to conclusion.

Follow the following steps to complete a prioritization matrix:

1. List the alternatives.
2. Establish the criteria.
3. Weight the criteria (on a 1–10 scale).
4. Rate each alternative against each criteria (1–10 scale).
5. Rank the alternatives by score.
6. If no clear winner emerges, consider adding criteria.

Another example of a design trade-off matrix is a simple method called cost-benefit analysis method (CBAM).[7] It studies how much a solution will cost to create,

---

6. Excerpt from Cristina Enomoto and Wendel Assis, Motorola SDFSS green belt project.

7. Len Bass, Paul Clements, and Rick Kazman, *Software Architecture in Practice, Second Edition*, Addison-Wesley, 2003.

beforehand. CBAM does not make decisions for the stakeholders. It assists in the elicitation and documentation of the costs, benefits, and uncertainty of a "portfolio" of architectural investments and gives the stakeholders a framework within which they can apply a rational decision-making process that suits their needs and their risk aversion.

The idea behind CBAM is that architectural tactics affect the quality attributes of the system and these in turn provide system stakeholders with some benefit. SEI refers to this benefit as utility. Each architectural strategy provides a specific level of utility to the stakeholders. Every stimulus-response value pair in a scenario provides some utility to the stakeholders, and the utility of different possible values for the response can be compared. For example, a very high security in response to performance might be valued by the stakeholders only slightly more than moderate security. But high availability might be valued substantially more than moderate availability. Each also has cost and takes time to implement.

CBAM can aid the stakeholders in choosing architectural strategies based on their ROI. It is the responsibility of the architecture team to determine the architectural strategies for moving from the current quality attribute response level to the desired or even best-case level. The ROI value for each architectural strategy is the rationale of the total benefit, Bi, to the Cost, Ci, of implementing it. The cost is calculated using a model appropriate for the system and the environment being developed.

$$Ri = \frac{Bi}{Ci}$$

Using this ROI score (Figure 12.5), architectural strategies can be rank-ordered. This rank ordering can then be used to determine the optimal order for implementation of the various strategies. The ROI score shows that the biggest ROI will be in the choice to

Architecture Strategies	Cost ($K)	Benefit (Utility)	ROI	Rank
Move to faster, fault tolerant platform	500	2700	5.4	3
Move to Google Android platform	300	50	0.2	4
Introduce 3rd party software for messaging	600	−80	−0.1	5
Implement iterative development lifecycle for architecture design standards	125	850	6.8	2
Partner with advanced technology company to build hardware abstract layer	60	850	14.2	1

Figure 12.5 Architecture strategies and ROI

partner with a leading edge advance technology company to build the hardware abstraction layer.

Using Design Patterns, Simulation, Modeling, and Prototyping for Decisions

How does one know whether the application of the tactics will really pull the critical parameters within the customer specification limits? Early modeling, simulation, emulation, or even prototyping can help with making good decisions while saving time and money. An architecture team can create executable models of its architecture specification to simulate or develop a prototype that implements parts of the architecture.

Prototypes are created as models presented in a format that is immediately recognizable to the users and therefore is beneficial for making judicious decisions about the design. Normally, a mock-up is a user interface model that may or may not be skeletal in terms of functional capability but could have the quality attributes (i.e., security and performance) that are needed to help with the decision regarding which architecture to use. Prototypes are created through the use of specification languages that are directly machine-interpretable (e.g., flash) and often are developed by using a high-level language that is application oriented. An iterative development lifecycle such as Agile will enable a team to develop prototypes on critical parameters faster.

After a prototype is produced and helps with discovering attributes associated with a concept and with making an architecture decision, the results from the prototypte should be discarded after the architecture selection is completed, lest someone (like a well-intentioned manager) suggest using the prototype as an evolutionary design.

We can model the behavior and relationship of a system by using various software architecture design patterns to come up with a solution.

The design patterns (see Chapter 18 for more information and examples of design patterns used in software architecture design) used on your project can be integrated with your overall DFSS solutions. The architect uses the prioritization or Pugh matrix to select the correct approach, followed by an analysis of the quality attributes, and then creates the design patterns using diagrams that can be easily implemented by the design teams. This technique provides a specific practice to address the need for selecting a sound technical solution to an architectural design.

Summary

Software architecture decisions cast a long shadow through the remainder of the development effort. This chapter discussed the considerations involved in these

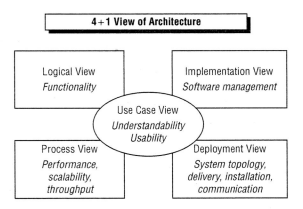

Figure 12.6 The 4+1 view model of architecture

decisions, and the process for making these decisions. The process by which the architects prescribe the architecture is vital to the future product's modification and maintainability. The famous 4+1 view model (Figure 12.6) of architecture, by Phillipe Kutchen, can be used to validate the architecture structures after the project has decided which architecture offers the best design for meeting the customers' critical parameters.

However, it is crucial to understand that decisions made in architecture can have a resounding effect on the implementation and testing of the software units. Your architecture team is best served by having tools such as the Pugh and prioritization matrices to help select the best architecture styles. After making the decisions, the team can use design patterns to document the design for evaluation. Also, the DFSS FMEA tool (see Chapter 10) will allow the team to evaluate the architecture that they decided on and find failure modes in the architecture design. This process of making decisive and clear decisions improves the quality attributes, especially maintainability and testability.

Predictive Engineering: Continuous and Discrete Transfer Functions

The critical parameter flow-down provides insight into *what* affects each critical parameter. Predictive engineering is related to an essential aspect of engineering—developing an equation or a predictive model or performing experiments to understand and represent *how* the critical parameter is affected.

Figures 13.1 and 13.2 show what might be considered a rather daunting flowchart. They provide guidelines for navigating through the maze of potential methods to develop transfer functions, which mathematically represent the relationships between the critical parameters and the subordinate y's, x's, and n's that affect them:

$$Y_i = f(x_1, x_2, \ldots, x_i, y_1, y_2, \ldots, y_i, n_1, n_2, \ldots, n_i) \tag{13.1}$$

where:

x_i represents factors that are under the design engineers' control, such as design choices, component choices, or settings of continuous variables.

y_i represents subordinate y's, measurable parameters at a lower level in the flow-down that affect the critical parameter and are in turn affected by other factors and parameters at a lower level in the flow-down.

n_i represents the noises, which are factors that will not be under the design engineers' control after the product has been delivered to the customers.

These transfer functions are equations that can range from simple, additive functions, to polynomials, to complex functions (meaning complicated . . . but some RF performance situations may literally involve complex functions).

The first decision point in Figures 13.1 and 13.2 is whether a transfer function is needed for the parameter. If the team has high confidence that the requirement will be

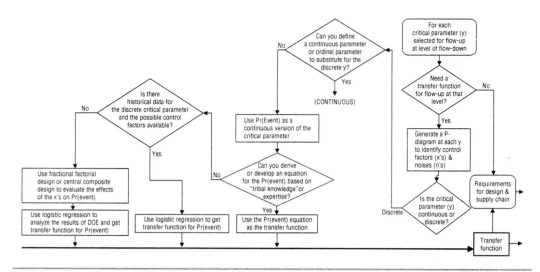

Figure 13.1 DFSS flowchart for selecting an approach for developing a transfer function for a discrete critical parameter, in preparation for predictive engineering, critical parameter flow-up, and optimization

met without the need for transfer function or optimization efforts, then the team can skip this step. This is not trivial—at some point in a flow-down for a critical parameter, the team may have confidence that the subordinate y requirement at that level will be met by the supplier for the component, not requiring further effort in determining a transfer function.

DISCRETE VERSUS CONTINUOUS CRITICAL PARAMETERS

The next decision point is whether the parameter is continuous or discrete. Throughout Six Sigma methods, there is a strong preference for continuous parameters—and this preference also holds for DFSS: continuous parameters are preferred, but if not continuous—then multivalued ordinal parameters are usable. Discrete parameters suffer from at least two major disadvantages, as illustrated in Figures 13.3 and 13.4. Figure 13.3 shows a decision tree for statistical methods. In the left side of the figure, the statistical methods appropriate for discrete parameters are highlighted, whereas statistical methods appropriate for continuous parameters are highlighted on the right side. Comparing these two sides, it should be very clear that there is a much richer palate of statistical methods available for working with continuous parameters. (The statistics decision tree can be found at http://www.sigmaexperts.com/stattree.)

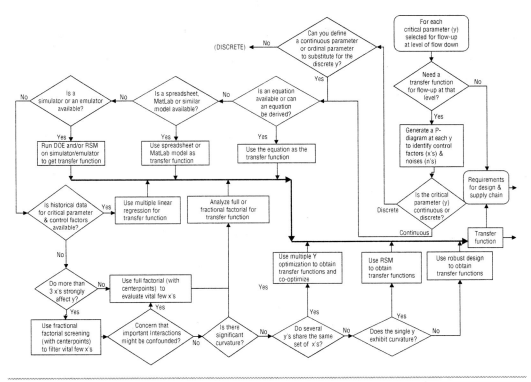

Figure 13.2 DFSS flowchart for selecting an approach for developing a transfer function for a continuous critical parameter, in preparation for predictive engineering, critical parameter flow-up, and optimization

The second disadvantage relates to sample sizes. Imagine that you have a parameter, such as power consumption, that can either be treated as a continuous parameter or a discrete (pass/fail) criteria. If the design team thinks that it has a design approach that will approach or achieve the Six Sigma goal, what sample size is needed to provide 95 percent confidence that the improvement to a Five Sigma level or a Six Sigma level have been achieved? A Six Sigma level corresponds to a Cpk of 1.5 as a continuous parameter, or 3.4 ppm failures as a discrete parameter; a Five Sigma level corresponds to a Cpk of 1.17 or 233 ppm failures, and a Four Sigma level corresponds to a Cpk of 0.83 or 66,807 ppm failures. As shown in Figure 13.4, the sample size required for the discrete version of the parameter could be almost 12,000 for showing improvement from a Five Sigma to Six Sigma level, and about 650 for showing improvement from a Four Sigma to a Five Sigma level. By contrast, the sample size required for the continuous version of the parameter would generally be less than 100 and possibly as low as 30, depending on the type of

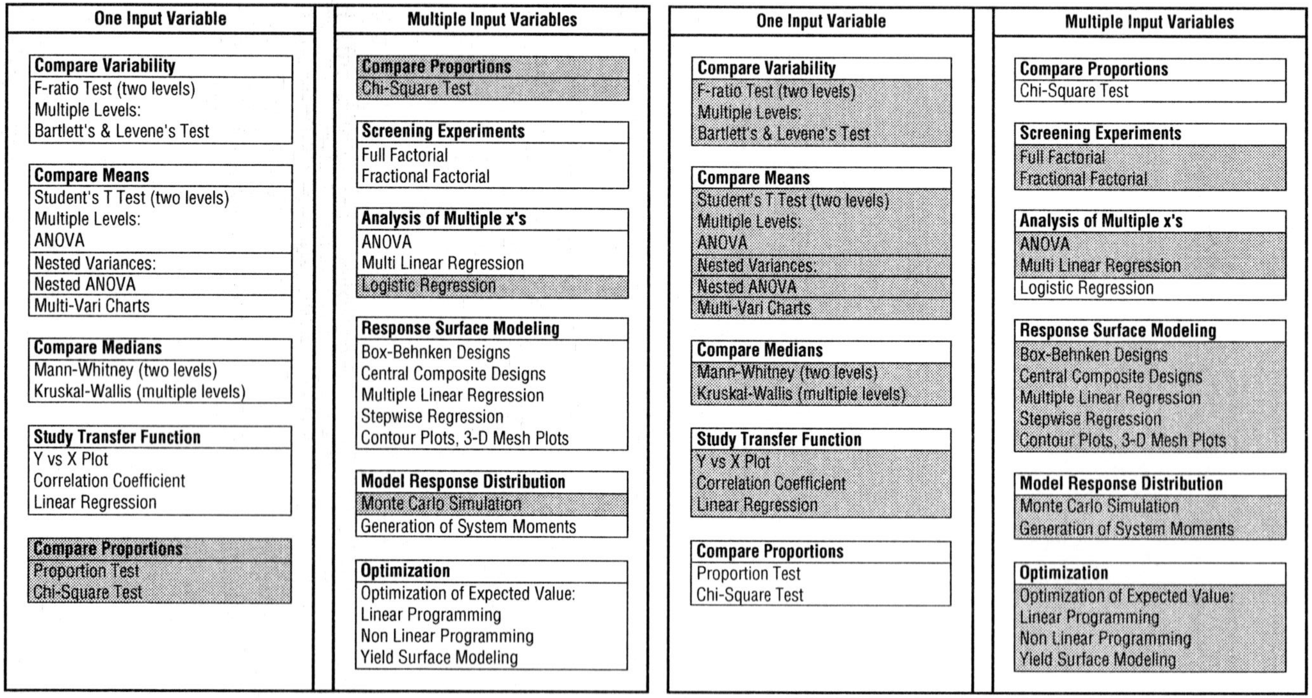

One Input Variable

Compare Variability
F-ratio Test (two levels)
Multiple Levels:
Bartlett's & Levene's Test

Compare Means
Student's T Test (two levels)
Multiple Levels:
ANOVA
Nested Variances:
Nested ANOVA
Multi-Vari Charts

Compare Medians
Mann-Whitney (two levels)
Kruskal-Wallis (multiple levels)

Study Transfer Function
Y vs X Plot
Correlation Coefficient
Linear Regression

Compare Proportions
Proportion Test
Chi-Square Test

Multiple Input Variables

Compare Proportions
Chi-Square Test

Screening Experiments
Full Factorial
Fractional Factorial

Analysis of Multiple x's
ANOVA
Multi Linear Regression
Logistic Regression

Response Surface Modeling
Box-Behnken Designs
Central Composite Designs
Multiple Linear Regression
Stepwise Regression
Contour Plots, 3-D Mesh Plots

Model Response Distribution
Monte Carlo Simulation
Generation of System Moments

Optimization
Optimization of Expected Value:
Linear Programming
Non Linear Programming
Yield Surface Modeling

One Input Variable

Compare Variability
F-ratio Test (two levels)
Multiple Levels:
Bartlett's & Levene's Test

Compare Means
Student's T Test (two levels)
Multiple Levels:
ANOVA
Nested Variances:
Nested ANOVA
Multi-Vari Charts

Compare Medians
Mann-Whitney (two levels)
Kruskal-Wallis (multiple levels)

Study Transfer Function
Y vs X Plot
Correlation Coefficient
Linear Regression

Compare Proportions
Proportion Test
Chi-Square Test

Multiple Input Variables

Compare Proportions
Chi-Square Test

Screening Experiments
Full Factorial
Fractional Factorial

Analysis of Multiple x's
ANOVA
Multi Linear Regression
Logistic Regression

Response Surface Modeling
Box-Behnken Designs
Central Composite Designs
Multiple Linear Regression
Stepwise Regression
Contour Plots, 3-D Mesh Plots

Model Response Distribution
Monte Carlo Simulation
Generation of System Moments

Optimization
Optimization of Expected Value:
Linear Programming
Non Linear Programming
Yield Surface Modeling

Figure 13.3 Comparison of the palate of statistical methods available for analyzing discrete parameters (left) versus continuous parameters (right)

**Power and Sample Size: Detect Difference
between 5 Sigma and 4 Sigma Levels**
Test for One Proportion

Testing Propotion = 0.000233 (versus not = 0.000233)   **[Sigma Level = 5]**
Alpha = 0.05

Alternative Proportion	Sample Size	Target Power	Actual Power	
0.00621	646	0.95	0.950106	**[Sigma Level = 4]**

**Power and Sample Size: Detect Difference
between 6 Sigma and 5 Sigma Levels**
Test for One Proportion

Testing Propotion = 0.0000034 (versus not = 0.0000034)   **[Sigma Level = 6]**
Alpha = 0.05

Alternative Proportion	Sample Size	Target Power	Actual Power	
0.000233	11886	0.95	0.950002	**[Sigma Level = 5]**

Figure 13.4 Sample sizes required for providing confidence that the sigma level has improved from Four to Five Sigma (top) or from Five to Six Sigma (bottom), if the parameter is discrete

continuous distribution and whether the improvement involves a change in the mean or a reduction in the variance or both.

In many cases, a continuous variable can replace or substitute for the proposed discrete parameter. For example, a pass/fail reliability test can be replaced by the continuous parameter, time to failure (perhaps under accelerated testing). If a continuous parameter can not be found, then perhaps a multivalued ordinal parameter can substitute for the discrete parameter. For example, a pass/fail test for whether a device functions after 15 drops can be replaced by a count of the number of drops to failure.

METHODS FOR DERIVING A TRANSFER FUNCTION FOR A DISCRETE CRITICAL PARAMETER

If all efforts to find a continuous parameter or ordinal value that can represent or substitute for the discrete parameter fail, then there are a few statistical methods that can be used to deal with the discrete variable, as shown in Figure 13.3. If there is only one factor or "x" that affects the discrete parameter, and that factor only has two levels, then a proportion test may be useful; however, as noted earlier in this chapter, large sample sizes might be required. If there are more than two levels and/or more than one factor or "x," then the chi square test may be used to compare the observed discrete results to those expected from a model. A useful and effective method for predictive engineering purposes is logistic regression, as described in the next section of this chapter.

LOGISTIC REGRESSION FOR DISCRETE PARAMETERS

Logistic regression provides an approach that effectively converts a discrete parameter from "pass/fail" to the probability of "pass," through regression using the logistic function:

$$P = \frac{\exp(\beta_0 + \beta_1 X)}{1 + \exp(\beta_0 + \beta_1 X)} \tag{13.2}$$

where:
  P is the probability of a "pass," passing the discrete pass/fail criteria
  X is a factor affecting the probability
  β_0 and β_1 are coefficients

The coefficients in equation 13.2 can be obtained by transforming equation 13.2 into linear form (equation 13.3), by setting y equal to the natural logarithm of the odds, defined as the ratio of the probability of a pass to the probability of a fail:

$$(\beta_0 + \beta_1 X) = \mathrm{Ln}\left(\frac{P}{1-P}\right) \tag{13.3}$$

Logistic regression can handle discrete and continuous factors or x's, as described in a few examples that follow. In Figure 13.5, binary logistic regression was applied to a mechanical issue to find an equation for how the degree of a visually inspected issue, a factor measured on a Likert scale, affected the probability that the casing would crack.

In Figures 13.6 and 13.7, binary logistic regression was applied to discrete and continuous factors that might affect the probability of software code passing an inspection:

- Experience (Continuous, in months)
- SW type of work (Nominal discrete: enhancement, new release, urgent bug fix)
- Day of week (Nominal discrete: Monday through Friday)
- Development platform (Nominal discrete: platform 1, 2, or 3)
- Home/office (Binary discrete: work from home or at office)
- Inspector experience (Continuous, in months)

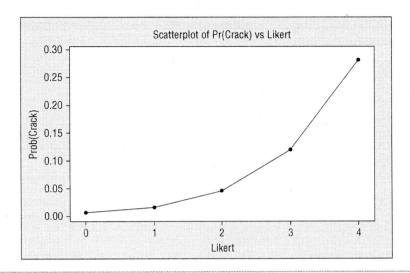

Figure 13.5 Scatterplot of the probability of a crack versus a Likert scale visual inspection of a battery, as evaluated through binary logistic regression

Logistic Regression Table

Predictor	Coef	SE Coef	Z	P	Odds Ratio
Constant	-12.1718	0.795929	-15.29	0.000	
Experience	0.541602	0.0333121	16.26	0.000	1.72
SWTypeWork					
New_Release	-1.27166	0.271096	-4.69	0.000	0.28
Urgent_Bug_Fix	1.85183	0.301608	6.14	0.000	6.37
DayofWeek					
Monday	0.0212215	0.210551	0.10	0.920	1.02
Thursday	0.142624	0.231336	0.62	0.538	1.15
Tuesday	0.267547	0.227429	1.18	0.239	1.31
Wednesday	0.204093	0.231994	0.88	0.379	1.23
DevtPlatform					
2	0.338570	0.181646	1.86	0.062	1.40
3	0.0355954	0.174062	0.20	0.838	1.04
Home/Office					
Office	0.168296	0.158969	1.06	0.290	1.18
InspectorExperience	0.0015825	0.0033399	0.47	0.636	1.00

Log-Likelihood = -599.027
Test that all slopes are zero: G = 1577.814, DF = 11, P-Value = 0.000

Figure 13.6 Results from binary logistic regression for software inspection test results

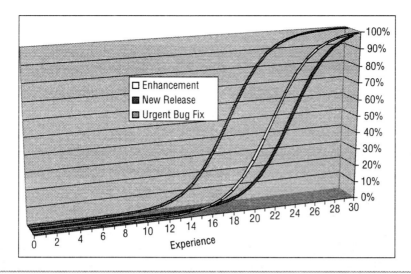

Figure 13.7 Illustration of the effect of experience and software type of work on the probability of passing the inspection, based on the reduced logistic regression equation

As illustrated in Figure 13.6, some of these factors were significant. Figure 13.7 shows the probability of passing the inspection based on the equation derived from binary logistic regression with a reduced model obtained by removing the terms that were not significant. The steps involved in this example are shown and explained in more detail at http://www.sigmaexperts.com/dfss/chapter13transfer.

Figure 13.8 shows a case of applying ordinal logistic regression combined with design of experiments (DOE) to help with decisions for software involved in recognizing signatures. The DOE systematically varied aspects of the handwritten signatures, providing two responses: the percent of the signature area that was dark, and a Likert scale ordinal discrete response in terms of the quality of the image that resulted from subsequent image processing. This combination of DOE and ordinal logistic regression provided an optimal setting for the threshold that led to substantial improvement in accurate signature recognition.

Methods for Deriving a Transfer Function for a Continuous or Ordinal Critical Parameter

Figure 13.2 can help with navigating through the relatively rich set of approaches and statistical methods available for deriving a transfer function for a continuous parameter. Most of these methods may also have some applicability for ordinal

Reduce "Blank Signatures"
DOE & Logistic Analysis to optimize Threshold

Legend:

Current Status only 1%

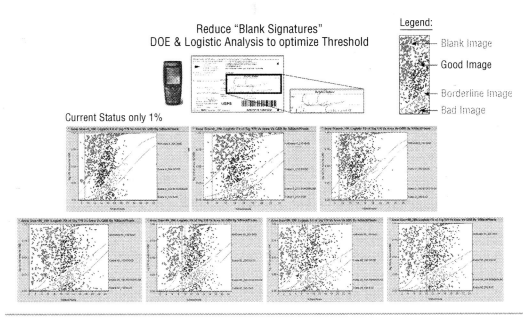
— Blank Image
— Good Image
— Borderline Image
— Bad Image

Figure 13.8 Application of ordinal logistic regression to recognition of signatures. A designed experiment directly varied parameters relating to a signature, and provided two key outputs: the ability to recognize the signature, and the percent dark area. The combination of logistic regression and DOE allowed the team to set an optimal threshold in the software for recognizing signatures (courtesy of Felix Barmoav).

parameters—parameters that range over a set of integer values, like the number of drops to failure for the ability of a portable product to withstand dropping onto a hard surface and the number of clicks required to get to a certain screen. The appropriate method for obtaining a transfer function for a continuous or ordinal parameter largely depends on the level of knowledge about the relationship between the parameter (Y) and the control factors (x's), noise factors (n's), and subordinate parameters (y's) that affect it; this can range from having an established and proven equation available, through having a model, simulator, emulator, or spreadsheet that represents the relationship, to starting from low initial knowledge and needing to develop an empirical model through experimentation. The next chapter will show how this transfer function can be used for optimization and critical parameter flow-up.

EXISTING OR DERIVED EQUATION (FIRST PRINCIPLES MODELING)

There are many situations in which there already is a mathematical model for the relationship between a parameter and the factors that affect it. In some cases, the engineer can find the equation by means of a search through a book or reference work or the published literature. Examples in electrical engineering include Ohm's Law ($V = IR$), the resonant frequency for an LC oscillator circuit using capacitors and inductors (Frequency = $1/(2\pi((L \times C))$), and the diode equation ($I = I_s \times e^{qV/kT}$). Examples in mechanical engineering include Newton's Law ($F = ma$), Hooke's Law ($F = -kx$), and additive equations that might be used in tolerance analysis. Software examples might include equations for software complexity models, software availability models, or equations for the time it takes to complete a set of tasks:

$$\text{Duration for Set of Tasks} = \sum_{i}^{N} (\text{duration of serial task}_i) + \max(\text{durations of tasks in parallel}) \qquad (13.4)$$

In other cases, the engineer can derive an equation using algebra or calculus, such as determining that the value of a parameter depends on the rate of change of another parameter. A variation of this approach could include developing an electrical or mechanical analogy and using algebra or calculus based on established models for that analogous situation. For example, thermal resistance and heat capacity can be represented with an electrical analogy of resistors and capacitors. A mechanical system with springs and dampers can be represented with an electrical network analogy with springs represented by inductors and dampers represented by capacitors. Reliabilities of systems dependent on the reliability of subsystems and components can be represented with an electrical network analogy of series and parallel resistors.

MODELING WITHIN A SPREADSHEET, MATHEMATICAL MODELING SOFTWARE, OR SIMULATION SOFTWARE

In many cases, a parameter can be modeled within a spreadsheet such as Excel, using mathematical modeling software (such as MatLab, Mathematica, and MathCad) or simulation programs. Examples of the simulation programs for electronics include SPICE for electronic circuits and ADS for RF (radio frequency) circuits. Examples for mechanical simulators include Albaqus, ProE, or other simulators using finite element analysis. Approaches for software include emulation or programs such as MLDesigner for discrete event simulation of latency and timings associated with software. A case study applying predictive engineering for availability of a system composed of both hardware and software is provided in Chapter 17.

To some extent, the capabilities and speed of the software involved in the spreadsheet, mathematical modeling software, or simulator determines how this approach can be used for optimization and the flow-up described in the next chapter. Stochastic optimization and flow-up require the ability to predict the variability of the output or response parameter; in some cases, the software has approaches such as Monte Carlo simulation or discrete event simulation capabilities built in. In other cases, the software is so quick that the user can develop a macro to vary the inputs and noises and quickly predict the variability of the response parameter.

If the software is much slower, then optimization and critical parameter flow up may require the development of an empirical model abstracted from the spreadsheet or simulator through the use of approaches such as design of experiments (described later in this chapter); the use of these approaches with a simulator is often referred to as DACE (design and analysis of computer experiments), as described in the next chapter.

EMPIRICAL MODELING USING HISTORICAL DATA: REGRESSION ANALYSIS AND GENERAL LINEAR MODEL

The top part of Figure 13.2 reflects situations where a model can be developed based on theory and represented either as an equation (first principles modeling) or as a model within a spreadsheet or simulator. The bottom part of Figure 13.2 represents situations in which no theoretical model exists, so a transfer function would need to be developed through empirical methods. Equations or transfer functions developed this way are often represented as a linear or quadratic equation.

If the factor or factors (x's) are continuous, then an equation for a continuous parameter (y) can be developed using multiple linear regression analysis (if one or more of the factors are discrete, then the general linear model [GLM] can be used instead). Using available historical data, regression analysis provides coefficients of a linear equation that provide the best fit equation for the historical data. The best fit is usually determined based on the least squares criteria: the best fit coefficients for the equation minimize the sum of the squared differences between the predicted values from the equation (using the coefficients) and the observed values.

Figure 13.9 shows the fitted line plot that results from a simple linear regression model, with one factor (x) and one response (y). R-squared describes the percentage of the variance of the response values that can be explained by the equation combined with the variance of the factor values.

Somewhat more complicated functions can be fit using linear regression through transformations of the factor, x, or the response, y, with examples in Table 13.1. Applying a transform for the capacitance example improves the R-squared value, as shown in Figure 13.10.

If the historical data includes information for several factors (x's), then multiple linear regression can provide both a linear equation and tests for statistical significance for each term, in terms of p-values (probability that the coefficient could appear to be non-zero

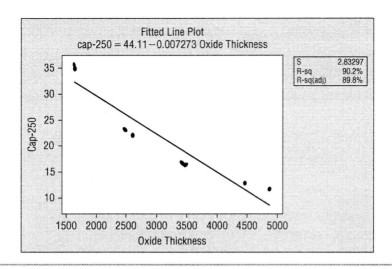

Figure 13.9 Fitted line plot for capacitor values as a function of the oxide thickness. The capacitor consists of two conductors separated by a silicon dioxide layer that behaves as the dielectric. The R-squared indicates that about 90 percent of the variance of the capacitance values can be explained through the linear equation combined with the variance of the thickness values.

Table 13.1 Examples of transformations for linear regression

Untransformed Equation	Transformed for Linear Regression
$Y = A \times e^{Bx}$	$Ln(Y) = ln(A) + Bx$
$Y = A \times x^{B}$	$Ln(Y) = ln(A) + B \ln(x)$
$Y = A + B/x$	$Y = A + B \times (1/x)$
$Y = 1/(A + Bx)$	$1/Y = A + Bx$
$Y = x/(A + Bx)$	$1/Y = B + A\,(1/x)$

due to chance alone) that can be compared to a criteria such as 5 percent alpha risk (risk of incorrectly assuming that the coefficient is nonzero). Figure 13.11 shows the regression analysis for the capacitor values as a function of both oxide thickness (thickness of the dielectric in the capacitor) and the area of the capacitor. Figure 13.12 shows an improved R-squared value when each term is replaced with its logarithmic transform; the equation obtained from the logarithmic transform can be restated (using the inverse transform) to

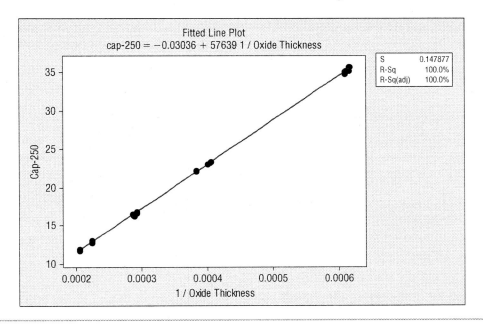

Figure 13.10 Fitted line plot for capacitor values as a function of the reciprocal of the oxide thickness. The R-squared indicates that nearly all of the variance of the capacitance values can be explained through the transformed linear equation.

an equation of the form Capacitance = Constant × Area/dielectric thickness—an equation that corresponds to the theoretical or first principles equation, Capacitance = $\varepsilon_0 \varepsilon_{\text{dielectric}}$ Area/dielectric thickness.

Historical data can be messy, and there can be correlations among the factors (x's). For example, the data may consist of a few batches of components, with similarities within each batch manufactured together. The similarities within each batch, and differences between batches, may be reflected in correlations among some properties that are treated as factors. Figure 13.13 illustrates a variation on the capacitor example in which data from capacitors were selected to develop a correlation between the factors of oxide (dielectric) thickness and area. The correlation of the factors impeded the ability of the statistical software to differentiate between the effects of the two confounded factors, such that the regression results showed no significant effect of oxide dielectric thickness on the capacitance.

A key set of approaches to provide clear differentiation of the effects for each factor is orthogonal design experimentation, in which the factors are orthogonal (i.e., mutually independent or uncorrelated). Orthogonal design is also known as design of experiments (DOE), described in the next section.

The regression equation is
Capacitance = 146 – 0.0469 OxideThickness + 0.0840 Area

Predictor	Coef	SE Coef	T	P
Constant	145.61	21.67	6.72	0.000
OxideThickness	–0.046855	0.006353	–7.37	0.000
Area	0.084007	0.003623	23.19	0.000

S = 70.4316   R-Sq = 86.4%   R-Sq(adj) = 86.1%

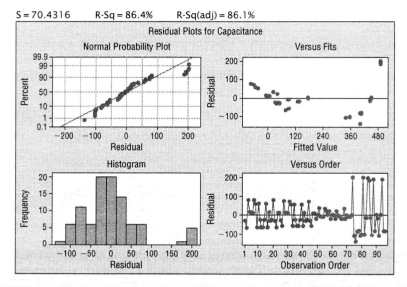

Figure 13.11 Results from multiple linear regression for two factors (top), and residual plots (bottom). The p-values from regression analysis indicate that both factors are statistically significant using a 5 percent alpha risk criteria. In the residual plots, curvature in the residuals versus fitted value plot indicates the need for a transform.

The regression equation is
logCapacitance = 5.50 – 1.00 logOxThickness + 0.991 logArea

Predictor	Coef	SE Coef	T	P
Constant	5.49792	0.05087	108.08	0.000
logOxThickness	–1.00108	0.00625	–160.25	0.000
logArea	0.991207	0.001624	610.42	0.000

S = 0.0238682   R-Sq = 100.0%   R-Sq(adj) = 100.0%

Figure 13.12 Results from multiple linear regression for two factors, after logarithmic transform of both factors and the response

Correlations: OxideThk1, Area1, Capac1

```
              OxideThk1       Area1
Area1             0.977
                  0.000

Capac1            0.974       0.995
                  0.000       0.000

Cell Contents:  Pearson correlation
                P-Value
```

The regression equation is
Capac1 = 10.3 + 0.00059 OxideThk1 + 0.0390 Area1

Predictor	Coef	SE Coef	T	P
Constant	10.265	3.618	2.84	0.016
OxideThk1	0.000593	0.002145	0.28	0.787
Area1	0.038977	0.005415	7.20	0.000

S = 2.13613   R-Sq = 99.1%   R-Sq(adj) = 98.9%

Figure 13.13 Variation on the capacitor example in which the areas and the oxide thicknesses were selected to be correlated (top), impeding the ability of the regression analysis to distinguish the effects of the two factors. The regression analysis provides a p-value that (inappropriately) indicates the oxide thickness does not have a statistically significant effect on the capacitance.

EMPIRICAL MODELING USING DESIGN OF EXPERIMENTS

Design of experiments[1] provides at least two primary benefits—the ability to develop a transfer function for the response (y) as a function of the factors (x's), and the ability to find a subset of the factors that have a statistically significant effect on the response. Screening experiments generally use fractional factorial designs to efficiently identify significant factors with a relatively small number of experimental runs. This section will provide an introduction and an overview of design of experiments; a "deeper dive," with examples, is provided at http://www.sigmaexperts.com/dfss/ and the book *Design and Analysis of Experiments* is highly recommended for its readability and value in understanding the topic and concepts for DOE.

Figure 13.14 shows these applications of DOE in the relevant context; variations of this figure will be used in the next chapter to illustrate optimization and flow-up. Starting from the right side, the Requirements phase of RADIOV, the Concept phase of CDOV, or the Define and Measure phases of DMADV/DMADOV assist in focusing the development team on some critical parameters, Y's that *are* important to the customers,

---

1. D.C. Montgomery, *Design and Analysis of Experiments,* John Wiley and Sons, 2004.

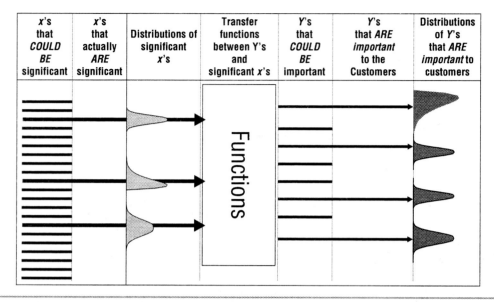

x's that COULD BE significant	x's that actually ARE significant	Distributions of significant x's	Transfer functions between Y's and significant x's	Y's that COULD BE important	Y's that ARE important to the Customers	Distributions of Y's that ARE important to customers

Figure 13.14 Illustration of the flow of predictive engineering in predicting the distributions of important Y's based on the distributions of significant x's

from among a larger set of technical requirements, or *Y*'s that *could* be important. The goal of predictive engineering is to predict distributions of the important *Y*'s to provide confidence that the *Y*'s will meet or exceed expectations. The critical parameter flow-down identifies the factors or *x*'s that could be significant, at the left of the figure. Screening experimentation can be used to identify which of these *x*'s actually have a significant effect on the *Y*, as illustrated by the filtering process emerging from the left side of the figure.

DOE provides a systematic approach for exploring the effects on the response or *Y* from combinations of the factors or *x*'s at various levels. For example, Figure 13.15 shows a full factorial design in which there are three factors, each at two levels, represented by −1 and +1. The table to the left of Figure 13.15 lists the eight runs (which, for the digitally inclined, relates to counting from 0 to 7 in binary, with −1's substituted for 0's and the sequence of columns reversed). The figure to the right of Figure 13.15 shows a graphic representation of those eight runs, with each of the three axes representing a factor and each of the corners of the cube representing one of the eight runs.

Run	X1	X2	X3
1	−1	−1	−1
2	+1	−1	−1
3	−1	+1	−1
4	+1	+1	−1
5	−1	−1	+1
6	+1	−1	+1
7	−1	+1	+1
8	+1	+1	+1

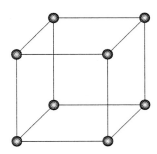

Figure 13.15 Full factorial design with three factors at two levels (−1 and +1). The table (left) lists the set of levels for each of the eight runs, which can also be viewed as a graphic representation (right), with each axis representing a factor and each corner of the cube representing a run.

If values for the response, Y, are obtained from each of the eight runs, then the analysis can provide estimates for coefficients of each term in an equation representing all of the main effects and interactions:

$$Y = b_0 \qquad \qquad \text{intercept}$$
$$+ \, b_1 x_1 + b_2 x_2 + b_3 x_3 \qquad \text{main effects}$$
$$+ \, b_{12} \, x_1 x_2 + b_{13} \, x_1 x_3 + b_{23} \, x_2 x_3 \quad \text{two-way interactions}$$
$$+ \, b_{123} \, x_1 x_2 x_3 \qquad \text{three-way interaction}$$

There are eight coefficients to estimate and eight runs; the coefficients could be obtained using matrix algebra. Without replications, no additional runs are available to estimate the variance of an error term. Because tests for statistical significance rely on a denominator representing the variance of the error term, such a full factorial design and analysis does not provide the degrees of freedom needed to obtain p-values for the statistical significances of terms.

Statisticians generally invoke the sparcity of effects principle, which indicates that, whereas main effects and two-way interactions often are statistically significant, higher order interactions such as three-way interactions ($x_1 x_2 x_3$) and four-way interactions ($x_1 x_2 x_3 x_4$) rarely prove to be significant. (The sparcity of effects principle is consistent with a Taylor series expansion for system variances, in which higher order interactions are generally negligible.)

The sparcity of effects principle is the basis for fractional factorial designs that are typically used for screening experiments. If three-way interactions can be assumed to have negligible effect on the response, or Y, then perhaps the experimenter can set another factor equal to a three-way interaction. In Figure 13.16, x_4 is set equal to the three-way interaction, $x_1x_2x_3$, as can be quickly verified using the table to the left of the figure. Factor x_4 is said to be confounded or aliased with the three-way interaction, $x_1x_2x_3$, indicating that these two terms cannot be distinguished using mathematics. Further inspection and calculations of products of combinations of columns will show that each three-way interaction is confounded with the main effect not included (i.e., x_1 is confounded with $x_2x_3x_4$ and x_2 is aliased or confounded with $x_1x_3x_4$). Moreover, two-way interactions are confounded with other two-way interactions (i.e., x_1x_2 is confounded with x_3x_4). This confounding of main effects with three-way interactions and two-way interactions with other two-way interactions is summarized by referring to this fractional factorial design as resolution IV. Resolution III would have main effects confounded with other two-way interactions and Resolution V would have main effects confounded with four-way interactions and two-way interactions confounded with three-way interactions.

Using the sparcity of effects principle, the number of runs for this fractional factorial experiment with four factors is halved: 8 runs among the possible 16, as illustrated with cubes to the right of Figure 13.16. Figure 13.17 shows a similar representation for a fractional factorial experiment with five factors, in which only a quarter of the possible combinations are run.

Figure 13.18 compares the number of runs required for a full factorial and the most highly fractionated experiment that could be useful for between 1 and 25 factors. This underscores why fractional factorial designs are generally used for designed experiments with five or more factors.

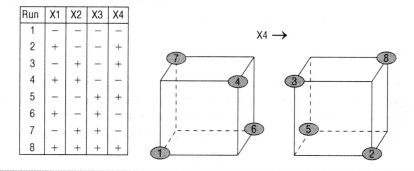

Run	X1	X2	X3	X4
1	−	−	−	−
2	+	−	−	+
3	−	+	−	+
4	+	+	−	−
5	−	−	+	+
6	+	−	+	−
7	−	+	+	−
8	+	+	+	+

Figure 13.16 Fractional factorial design with four factors at two levels (−1 and +1), 2^{4-1}_{IV}. The table (left) lists the set of levels for each of the eight runs, in which factor x_4 is set to be equal to and confounded with the three-way interaction $x_1x_2x_3$. The graphic representation of those eight runs (right) shows each of the three axes as a factor, and the cube before, to the left, and after, to the right, representing x_4. There are 2^4 or 16 possible combinations, represented as 16 corners of the before and after cubes, of which half are included as runs in the fractional factorial experiment.

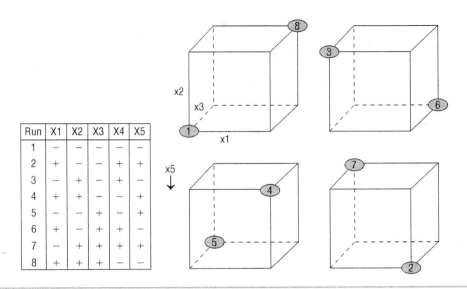

Run	X1	X2	X3	X4	X5
1	−	−	−	−	−
2	+	−	−	+	+
3	−	+	−	+	−
4	+	+	−	−	+
5	−	−	+	−	+
6	+	−	+	+	−
7	−	+	+	+	+
8	+	+	+	−	−

Figure 13.17 Fractional factorial design with five factors at two levels (-1 and $+1$), 2^{5-2}_{III}. The table (left) lists the set of levels for each of the eight runs, in which factor x_4 is set to be equal to and confounded with the two-way interaction $(-1) \times x_1 x_2$, and factor x_5 is set to be equal to and confounded with the two-way interaction $(-1) \times x_1 x_3$. The graphic representation of those eight runs (right) shows each of the three axes as a factor, and two more axes represented left to right and top to bottom for x_4 and x_5. There are 2^5 or 32 possible combinations, represented as 32 corners on the set of cubes, of which a quarter are included as runs in the fractional factorial experiment.

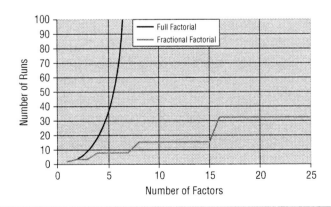

Figure 13.18 Comparison of the number of experimental runs required for a full factorial experiment ($2^{number\ of\ factors}$) and the number of runs required for a fractional factorial experiment exploring the same number of factors as main effects

The analysis of full and fractional factorial experiments can involve main effects and interaction plots. Regression analysis, as shown earlier in this chapter, is used to obtain the transfer functions. An example of similar analysis will be shown in the next section that features a more advanced version of DOE, called response surface methodology (RSM).

EMPIRICAL MODELING USING RESPONSE SURFACE METHODS

Figure 13.19 illustrates the central composite design (CCD) that can be used in response surface methods.[2] CCDs are based on full or fractional factorial designs, plus one or more center points, plus star points along the axes corresponding to the factors. The number of combinations in a CCD with n factors is 2^{n-k} runs for the full or fractional factorial (where k represents the degree of fractionation: n − 1 for a half fraction, n − 2 for a quarter fraction), plus $2 \times n$ runs for the start points, plus 1 or more runs for the center point.

With these incremental runs, the experiment now provides information for five levels for each factor (x): one value at the very low level, another at the very high level, corresponding to the two star points for that factor, several runs at an intermediate low and high level, corresponding to the −1 and +1 levels of the underlying full or fractional factorial design, and several runs at the middle level, including the center point(s) and the star points for the other factors. Proceeding from left to right for the central composite design represented in Figure 13.19, there is one run at a very low level (negative star point) of the factor running from left to right, four runs at a low level (−1), five runs at a middle value of that factor (the center point plus the four star points for the other two factors), four runs at a high level, and one run at a very high level for the factor, corresponding to that factor's other positive star point.

After the experiment has been performed and data has been collected for each of the runs in the experimental design, analysis is initiated using terms that include main effects, interactions, and curvature effects, as shown in Figure 13.20.

Initial analysis might include the full set of terms; some of these terms may not be significant. Unless the data was obtained through a simulation or similar computer approach (DACE, discussed in the next section of this chapter), a useful approach is to remove terms with p-values above a certain criteria for acceptable risk of inadvertently removing a key term. The criteria is often set to .05 or .10, such that terms with p-values much higher than the criteria are removed first; since removing a term often changes the p-values for the remaining terms, this filtering process often progresses sequentially, removing the terms with the highest p-values, analyzing the data again, then removing the

---

2. R.H. Myers and D.C. Montgomery, *Response Surface Methodology,* John Wiley and Sons, 2002.

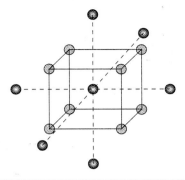

Figure 13.19 Graphical representation of a three-factor central composite design (CCD) used for response surface modeling, with each axis representing a factor. The cube represents a full or fractional factorial design, and a center point and star points are added to provide five levels of information for each factor.

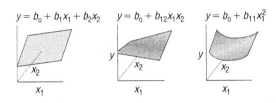

Figure 13.20 Terms in transfer function developed through response surface modeling, including main effects, interactions, and curvature terms (top), and illustrations of surface plots with only main effect terms (bottom left), interaction terms (bottom middle) and curvature terms (bottom right)

term with the next highest p-values, until only significant terms remain. Although this section provides an introduction and an overview of RSM, a "deeper dive," with examples, is provided at http://www.sigmaexperts.com/dfss/ and the book *Response Surface Methodology*[2] is recommended for a deeper understanding of this topic.

Figure 13.21 shows the completed results after the analysis of the dynamic range associated with a display device; the factors are settings in the software that affect the shape and durations for pulses sent to the pixels. In this example, most of the p-values remaining are below 0.05; the experimenter used a criteria of .10, so the two-way interaction P0 × G0 was retained with a p-value of 0.065. The main effect, G0, was retained despite a p-value of 0.214; this is a result of an application for a hierarchical rule, in which a lower order term such as a main effect cannot be removed from the equation if it shows up in a higher order term that is significant. In this instance, although G0 has too high of a p-value as a main effect, it appears in the two-way interaction P0 × G0 that was

Estimated Regression Coefficients **for dynamic range**

Term	Coef	SE Coef	T	P
Constant	7.37739	1.79668	4.106	0.000
T	0.75013	0.12278	6.110	0.000
P0	0.27409	0.01256	21.829	0.000
G0	-0.00423	0.00339	-1.249	0.214
G1	0.00828	0.00193	4.286	0.000
T*T	-0.00852	0.00213	-3.997	0.000
P0*P0	-0.00075	0.00005	-16.524	0.000
T*P0	-0.00115	0.00024	-4.851	0.000
P0*G0	0.00004	0.00002	1.861	0.065

S = 1.638 R-Sq = 93.6% R-Sq(adj) = 93.1%

Figure 13.21 Results from multiple linear regression analysis of data from a central composite design experiment for RSM, after the terms have been reduced by first eliminating higher order terms with p-values greater than .10

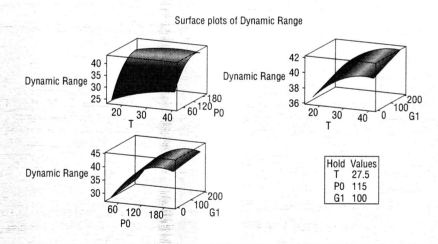

Figure 13.22 Surface plots obtained using the transfer function obtained through response surface modeling, as shown in Figure 13.21. There is noticeable curvature with respect to some of the factors.

retained, and the main effect G0 must also be retained. If the experimenter had used a criteria of .05 rather than .10, both the P0 × G0 and the main effect G0 would have been removed to simplify the equation. Using the reduced equation, contour plots and surface plots (Figure 13.22) can provide insight into the sensitivity of the response (y) to each of

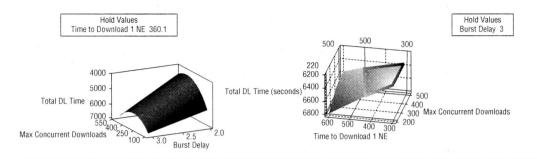

Figure 13.23 Response surface for total download time for a telecommunication system[3]; optimal region was found to require these software decisions: Time to download 1 NE: between 300 and 500, maximum concurrent downloads: between 220 & 500, Burst Delay = 3

the factors (x's) and provide a starting point for optimization, as discussed in the next chapter.

Figure 13.23 shows an example of the use of response surface modeling for the download time for a telecommunication system, depending on decisions made in software, including burst delay and the maximum number of concurrent downloads allowed.

DOE WITH SIMULATORS: DESIGN AND ANALYSIS OF COMPUTER EXPERIMENTS (DACE)

Earlier in this chapter, the topic of applying DOE with simulation programs was introduced. There are some interesting aspects of this approach, including some differences in the design and analysis for experiments run on computers rather than on actual products, devices, or prototypes. Consequently, this approach has been discussed in books and articles[4] as a powerful yet sometimes confusing way to develop simplified transfer functions that can provide insight and a means for optimization with less time and resource investment.

The key differences between DACE and DOE with actual devices include the irrelevance of replications (running the same settings a hundred times will provide the same results a hundred times) and the absence of pure error. With conventional DOE, pure

---

3. Courtesy of Vivek Vasudeva.

4. T.J. Santner, B.J. Williams, and W.I. Notz, *The Design and Analysis of Computer Experiments,* Springer, 2003.

error is introduced through uncontrolled variations in the actual values of the x's relative to the desired settings associated with the levels, variations in the environment (such as temperature and humidity), and measurement error associated with the measurements of the response (y). These sources of pure error generally do not exist with a simulation environment.

In conventional DOE, the p-values are derived from statistical tests (t-tests or F-ratio tests) that involve a ratio where the numerator relates to the effects of each term on the response and the denominator represents the error—which is assumed to be dominated by pure error.

In the absence of pure error, analysis of an experiment on a simulator only has error representing model error—the differences between the predicted values from the simplified equation and the values obtained directly from the simulator . . . or, stated another way, the error introduced by trying to extract a simplified model. Because there is no pure error, the p-values obtained do not test the hypothesis that the changes in response are due to pure error; consequently, the p-values are not dependable indicators as to whether a term is significant in experiments run on a simulator.

Because p-values are neither dependable nor useful in DACE, alternatives for determining which terms are worth keeping and which terms can be discarded in a reduced model include:

- Stepwise regression
- Best subsets regression
- Multiple linear regression, using R-squared as the key indicator: the analyst tries removing terms, determines whether the R-squared drops more than the analyst finds acceptable, and then backs off or proceeds accordingly.
- Use of the sensitivity charts with Monte Carlo simulation or YSM (discussed in the next chapter) to determine which factors' variances contribute most to the variance of the response.

The approach for DACE, the performance of designed experiments (DOE, RSM) with a simulator, includes the following steps:

- Define the control and noise factors (and possibly model parameters) for the experiment.
- If the noise factors can be varied in the simulator, then determine an experimental noise strategy.
- Design the experiment—without replications.
- Screen out parameters that have relatively little effect on the responses, using an approach that uses R-squared or sensitivity analysis rather than p-values.

Summary

The predictive engineering aspect of Design for Six Sigma requires a way to develop a model that can be used for predicting and optimizing performance. In this chapter, a fairly comprehensive set of methods for developing a mathematical model, or transfer function, was described and explored, including methods for both discrete and continuous responses. The chapter also included flowcharts that provide guidance in determining what methods are appropriate in various situations.

The next chapter builds on this chapter. Once the mathematical models or transfer functions have been developed, as described in this chapter, they can be used to predict initial performance, to optimize the performance, and predict the final performance in readiness for verification with the actual product. The methods in this chapter readily lend themselves to some very powerful methods for optimization and critical parameter flow-up.

Predictive Engineering: Optimization and Critical Parameter Flow-Up

The focus of both this chapter and the prior chapter is on predictive engineering, a major aspect of Design for Six Sigma; in fact, some experts have used these terms interchangeably. With predictive engineering, transfer functions are used to predict performance—in terms of the distribution of the critical parameters, the measurable technical requirements that are critical for the success of the product. This is referred to as critical parameter flow-up. If the predicted distributions do not provide sufficiently high confidence that the customers will be satisfied with the situation as-is, then the development team needs to optimize the performance—using the transfer functions to provide insight and guidance for the optimization.

Figure 14.1 illustrates this concept of predictive engineering, and variations on this figure will be used to illustrate some of the examples in this chapter. Y's refer to measurable technical requirements at the system or product level, whereas x's refer to control factors affecting one or more of the Y's.

For the RADIOV process, starting from the right side, the Requirements phase identifies the Y's that are important to the customers among the Y's that could be important. (This corresponds to the Concept phase of CDOV or the Define and Measure phases of DMADV.) The selection of critical parameters was discussed and described in Chapter 9.

These Y's or critical parameters are flowed down to the control factors (x's), the noises (n's) and subsystem, module, and component requirements that can be controlled indirectly (subordinate y's). These are represented by the array of line segments to the left of Figure 14.1, representing the factors (x's, n's and subordinate y's) that could be important in assuring appropriate performance and functionality. This flow-down

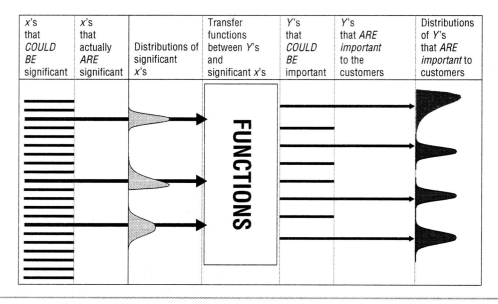

x's that COULD BE significant	x's that actually ARE significant	Distributions of significant x's	Transfer functions between Y's and significant x's	Y's that COULD BE important	Y's that *ARE important* to the customers	Distributions of Y's that *ARE important* to customers

Figure 14.1 Illustration of the flow for predicting the distributions of important Y's based on the distributions of significant x's

occurs during the Architecture or early Design phase of RADIOV, corresponding to the Design phase of CDOV and the Analyze phase of DMADV.

Various approaches can be used to identify the most significant factors, including screening design of experiments as discussed in Chapter 13. This step occurs during the Design phase of RADIOV. Within CDOV, the screening or identification of the important factors (x's) can occur in the Design or early Optimize phase; within DMADV, this can occur during the Analyze or early Design phase. These efforts can provide transfer functions, or can be followed up with more advanced methods such as response surface modeling to model the relationship between the Y's and the significant factors, also described in Chapter 13. Distributions for the significant x's can be estimated through capability studies, control charting, or similar activities with the manufacturing areas and suppliers.

Transfer functions can be used in conjunction with the estimated distributions for the significant x's to provide predictions of the distributions of the Y's, using methods such as Monte Carlo simulation and generation of system moments (including root sum of squares). This is illustrated by the distributions to the far right of Figure 14.1. This step occurs during the Optimize phase of RADIOV. Similarly, this step occurs during the Optimize phase of CDOV or for the DMADOV variation of DMADV, or in the Design phase for DMADV.

The resulting predicted distributions of the Y's that are important to the customers, shown to the right side of Figure 14.1, are verified using prototypes or early samples in the Verify phase of RADIOV, CDOV, or DMADV.

Figure 14.2 shows a detailed flowchart for critical parameter flow-up and optimization. The left side of the flowchart illustrates the initial steps of critical parameter flow-up, in which the Monte Carlo simulation or generation of system moments methods can be used to provide preliminary, predictive distributions and associated predictions for capability indices (such as Cp and Cpk) for the critical parameters (Y's) using the transfer functions combined with distributions for the significant x's. Based on the results of this initial flow-up and results from the capability indices, the critical parameters can be prioritized; if these results provide the team with sufficiently high confidence that a critical parameter will be capable, and does not warrant further efforts toward optimization, then the team can focus on other critical parameters that require such optimization efforts. For critical parameters requiring optimization, the initial flow-up results can be stored as a baseline within a Y-scorecard, for comparison with the results of flow-up

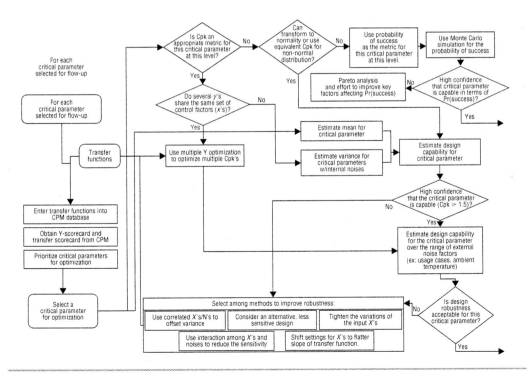

Figure 14.2 Detailed flowchart for optimization and critical parameter flow-up

capability assessments after optimization. In this sense, the Y-scorecard also provides documentation of "capability growth," in terms of the improvement achieved by the team's optimization efforts, and summarizes the team's confidence in the capability of the product successfully meeting or exceeding the customers' expectations as reflected through the selected set of critical parameters.

The next two sections will describe Monte Carlo simulation and the generation of system moments methods for critical parameter flow-up. Afterward, the critical parameter scorecard will be described, and then approaches and examples of optimization will be discussed in this and subsequent chapters. Some of the examples will involve continuations of situations for which transfer functions were obtained in Chapter 13.

CRITICAL PARAMETER FLOW-UP: MONTE CARLO SIMULATION

Monte Carlo simulation was developed as part of the highly secretive and somewhat controversial Manhattan Project during World War II. In the development of the atomic bomb, a key technical requirement—a critical parameter—was the attainment of critical mass. Dr. Stanislaw Ulam had escaped the Nazi invasion of his home country, and joined the Manhattan Project with his friend, Dr. John Von Neumann, famous for his seminal work for the design of computers. Dr. Ulam used one of the first computers, ENIAC, to perform Monte Carlo simulations for the attainment of critical mass, and for other critical problems involved in nuclear physics. He named the method "Monte Carlo simulation" for its similarity to probabilities of winning at the casinos in the city of Monte Carlo "rolling the dice," so to speak.

Figure 14.3 illustrates the process involved in the Monte Carlo simulation method, using the example of a capacitor value that is proportional to the area and inversely proportional to the oxide dielectric thickness, as derived from the antilog transform of the equation shown in Figure 13.12: Capacitance = 245 × Area/Thickness, where the constant (245) was obtained from the antilog of the intercept, 5.50. As mentioned in the previous chapter, this equation can be found in many physics and electronics textbooks, and the constant represents the product of the dielectric constant, the permittivity of free space, and unit conversions. Assuming the manufacturing engineers provided estimates of the mean (250) and standard deviation (1 square micron) for the area and dielectric thickness (mean of 3000 and standard deviation of 70 Angstroms), the Monte Carlo simulation can be run by repeatedly entering random values from the distributions for the x's (area and dielectric thickness) using this transfer function to forecast corresponding values of the y (capacitance) as described in Figure 14.3.

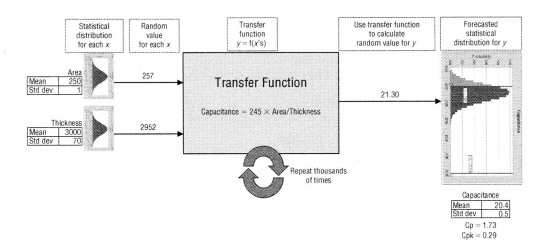

Figure 14.3 Illustration of Monte Carlo smulation for capacitance. The transfer function is entered, along with information describing the statistical distribution for each input factor or x in the transfer function. A random value is generated from the distribution for each x and entered into the transfer function to provide one forecasted value for the response or y. These steps are repeated thousands of times to generate a forecasted statistical distribution for the y. The distribution of y can be compared to specification limits to forecast the capability indices, Cp and Cpk. For this capacitance example, the Monte Carlo simulation results indicate that the Cp is acceptable at 1.73, but the Cpk is unacceptably low at 0.29.

Monte Carlo simulation similarly can be used with transfer functions obtained by most of the methods described in Chapter 13, including polynomial equations obtained empirically using DOE and RSM.

CRITICAL PARAMETER FLOW-UP: GENERATION OF SYSTEM MOMENTS (ROOT SUM OF SQUARES)

The generation of system moments method can be used to forecast values for the mean and standard deviation, which can in turn be used in calculating values for the capability indices, Cp and Cpk. A forecasted value for the mean can be obtained by simply substituting the average values for each of the factors or x's in the transfer function. For the capacitor example, substituting a value of 250 for the area and 3000 for the thickness into the transfer function, Capacitance = 245 × Area/Thickness, results in an estimate of 20.4 for the capacitance. This simple approach for the mean actually corresponds to a truncated Taylor series expansion for the critical parameter or y as a function of the x's.

$$Var(Y) = \boxed{\sum_i(\frac{\delta Y}{\delta X_i}S(X_i))^2} + \boxed{2\sum_i\sum_j r(X_i,X_j)\frac{\delta Y}{\delta X_i}\frac{\delta Y}{\delta X_j}S(X_i)S(X_j)}$$

Term for Uncorrelated *x*'s   Additional Term for Correlated *x*'s

$$\boxed{\text{Taylor Series Expansion - Terms with Correlated } x} + \boxed{\sum_i\frac{\delta Y}{\delta X_i}\frac{\delta^2 Y}{\delta X_i^2}(S(X_i))^3 * \text{Skewness}(X_i)}$$

Additional Term for Non-Linear Transfer Functions Combined with Skewed Distributions of *x*'s

Figure 14.4 Equation for predicting the variance of the critical parameter (*Y*) using partial derivatives from the transfer function, based on a truncated Taylor series expansion

Table 14.1 Generation of system moments method

X:		Partial Derivatives		Variance due to each x	
Area		dCap/dA	0.081564	0.006653	3%
Thickness		dCap/dThk	0.006797	0.226376	97%
Y:					
Capacitor Value		Variance	0.233029		
Mean	20.390994	Std Dcv	0.48273		
Cp	1.7262912	Cpk	0.269988		

Forecasting the variance of the critical parameter is considerably more complicated. Figure 14.4 breaks down the first few terms of the Taylor series expansion for the more general case, but often relevant cases exist where the factors or *x*'s might be correlated or the *x*'s might have skewed distributions that combine with a nonlinear transfer function to increase the variance of the critical parameter.[1] If we assume the latter two terms are negligible for the capacitor example (that is, the thickness and area are uncorrelated, and both the thickness and area distributions can be treated as symmetrical distributions), then the variance of the critical parameter, capacitance, can be forecasted using partial derivates of the transfer function as shown in the first term of Figure 14.4. Table 14.1

---

1. G.J. Hahn and S.S Shapiro, *Statistical Models in Engineering*, John Wiley, 1968, Appendix 7B.

shows the results of the calculations, which provides forecasts for the mean, standard deviation, Cp, and Cpk for the critical parameter, capacitance value, that are very close to the values from the Monte Carlo simulation in Figure 14.3. Forecasted Cp is calculated as the difference between the upper and lower specification limits, divided by six times the standard deviation. Forecasted Cpk is calculated as the difference between the mean and the nearer specification limit, divided by three times the standard deviation.

CRITICAL PARAMETER SCORECARD

Monte Carlo simulation and generation of system moments methods can be used with transfer functions from a variety of methods, as described in the previous chapter. The initial scorecard sets a baseline that helps prioritize optimization efforts, as discussed in the next section.

Figure 14.5 shows a critical parameter scorecard for a generic, GSM cellular phone. The critical parameters are listed in the first column as system requirements, and

System Level DFSS — Critical Parameter Management Score Card								
Project Name: Generic Cell Phone								
System Requirement	Units	Concept Phase-Spec's			Design Phase -Flow Down CGI-Design 55.9%			
		Lower Limit	Nominial	Upper Limit	Mean	Std.Dev.	CP	CPK
Display Size	mm	40.0	60.0	100.0	60.00	0.01	1000.00	666.67
Talk Time	hours	4.0	10.0	N/A	11.03	1.10		2.12
Standby Time	hours	10.0	50.0	N/A	27.30	2.73		2.11
Camera Resolution	mega pixels	4.0	5.0	6.0	5.00	0.01	33.33	33.33
Removable Memory Capacity	mega bytes	1.0	2.0	10.0	2.00	0.01	150.00	33.33
Application Processor MIPS	MIPS	12.0	16.0	N/A	15.00	3.00		0.33
Bit Error Rate	%	N/A	0.010%	0.100%	0.014%	0.030%		0.96
Total Isotrophic Sensitivity (T.I.S.)	dBm	−120.0	−100.0	−95.0	−109.89	1.04	4.01	3.24
Accurate Identification (Probability)	%	99.90%	99.99%	100%	1.00	0.01	0.02	−0.13
Baseband Processor (MIPS)	MIPS	80.0	100.0	200.0	100.00	5.00	4.00	1.33
Phone Weight	ounces	3.5	4.5	5.5	5.51	0.13	2.56	−0.03
BOM Cost	$	N/A	25.0	45.0	41.26	1.08		1.16
System Reliability	FFR at 1 year	N/A	0.1	0.3	0.32	0.01		−0.81
Drops to First Failure	# drops	2.0	5.0	N/A	5.00	2.00		0.50
Dynamic Range	dB	70.0	73.0	N/A	83.11	1.64		2.66
Battery Satisfaction Prediction	%	75%	85%	N/A	98%	3%		2.36
Total Radiated Power (T.R.P)	dBm	28.5	30.0	36.0	28.20	0.31	4.04	−0.32

Figure 14.5 Initial critical parameter scorecard for a GSM cellular phone. The critical parameters are listed in the first column as system requirements, and other columns list the corresponding units and specification limits, initial forecasts for the mean and standard deviation for each critical parameter, and initial forecasts for the capability indices, Cp and Cpk.

other columns list the corresponding units and specification limits. Monte Carlo simulation or the generation of system moments method were used to generate initial forecasts for the mean and standard deviation for each critical parameter, which were combined with the specification limits to provide initial forecasts for the capability indices, Cp and Cpk. Critical parameters having Cpk's inconsistent with goals, such as having Cpk values below the 1.5 value for Six Sigma capability, are flagged or color coded appropriately.

Later in the development process, columns would be added to the scorecard to summarize new means, standard deviations, and capability optimizations forecasted after optimization, and from values obtained with prototype or early production units through verification efforts. The scorecard can be shared with the development team and management to summarize information about the progress and technical challenges, and to clearly summarize the level of confidence that the critical parameters will meet or exceed expectations.

SELECTING CRITICAL PARAMETERS FOR OPTIMIZATION

Critical parameters are selected based on their importance to meeting customers' expectations and the level of challenge and technical risk in achieving those expectations, as described in Chapter 9. Consequently, the critical parameters have already been prioritized from a larger set of requirements that could be important, but are either less critical to satisfying the customer and/or considered to involve less technical risk.

If initial analyses indicate that the capability to meet expectations already exists for those critical parameters, as reflected in the initial forecast for the Cpk, then optimization efforts may not be required. By contrast, if the initial forecast for the Cpk is very low, and possibly even negative (indicating that the forecasted mean is outside the specification window), then optimization for the critical parameter should become a high priority.

An initial critical parameter scorecard, as shown in Figure 14.5, provides a guide for selecting and prioritizing the critical parameters for optimization efforts. The default prioritization would simply be a rank ordering of the forecasted Cpk's, but possibly excluding those critical parameters that have satisfactory capability indices from the get-go. Other considerations, such as lead times related to schedule requirements, might change the prioritization from this default rank order, and the development team should have the insight and perspective and be empowered to modify the prioritization as appropriate.

OPTIMIZATION: MEAN AND/OR VARIANCE

Once a critical parameter has been selected and prioritized for optimization, the next step is to determine what aspects of the critical parameter distribution must be optimized. Generally, these aspects relate to the mean or standard deviation of the critical parameter, or of a subordinate y that strongly impacts the capability for the critical parameter.

The ideal distribution for a critical parameter would be represented by a vertical line centered on the target value, such that the mean equals the target and the standard deviation equals zero. Deviation from this ideal distribution can be quantified by the second moment about the target, which is the average of the squared deviation from the target value. The second moment about the target is closely related to Taguchi's loss function, except that the latter includes a constant multiplied by the squared deviation to represent financial loss.

As shown in Figure 14.6, the second moment about the target can be partitioned into the amount due to miscentering (corresponding to the squared difference between the average and the target), and the amount due to variability about the average. This partitioning is also related to the difference between the Cp and the Cpk capability indices, sometimes referred to as the "k" factor.

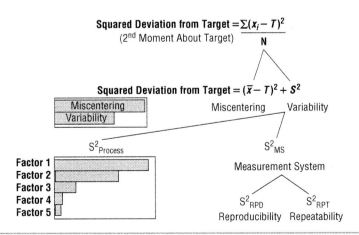

Figure 14.6 Partitioning of the squared deviation from target (which quantified the deviation from the ideal distribution) into miscentering and variability, which is further partitioned into measurement variation and process variation

Applying this analysis for the capacitance example in Figure 14.3 and Table 14.1, the target would be the center of the specification limits, $(20 + 25)/2 = 22.5$, and the difference between the mean and the target, squared, would be $(20.4 - 22.5)^2$ or 4.45. The variance would be 0.23, from Table 14.1. The second moment about the target then would be 4.68, so 95 percent of the squared deviation from the target would be attributed to the distribution's mean being off-center relative to the target.

The variance about the average can be further partitioned into the variance as a result of the measurement system and the variance as a result of manufacturing variation and other sources such as variations in environment and usage or application. The variance due to the measurement system can be partitioned through a gauge capability study or measurement system analysis (MSA) into variances due to repeatability (within operator) and reproducibility (between operators), as discussed in Chapter 16.

Variances attributable to variations in the manufacturing process, use cases, and environment can be partitioned using approaches such as source of variation (SOV), design of experiments, and the sensitivity analysis from Monte Carlo simulation or the generation of system moments, described earlier in this chapter.

Figure 14.7 illustrates a useful rule of thumb for the capability indices, Cp and Cpk. The difference between Cp and Cpk (referred to as "k") can be translated into the percent of the squared deviation from the target that can be attributed to either the variance or the difference between the mean and the target. If k, the difference between Cp and Cpk, is more than a third (0.333), then most of the squared deviation

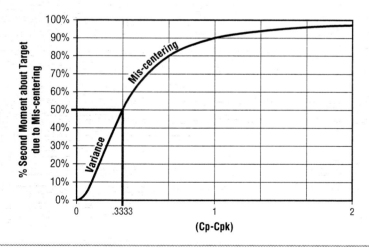

Figure 14.7 Percentage of the squared deviation from target that is a result of having the mean of the distribution off-center, as a function of k, the difference between Cp and Cpk

from target can be attributed to having the mean off-center with respect to the target. If the difference between Cp and Cpk is more than 0.333, then the key step in improving the capability is to focus on centering the mean on or near the target. By contrast, if the difference between Cp and Cpk is less than .3333, then the key step in improving the capability involves variation reduction—the topic for the rest of this section.

With the capacitor example, the difference between the Cp and Cpk is 1.73–.27, or 1.46, well more than 0.333. Consequently, the key step would be to center the average capacitance on the target by setting one of the factors or x's to an appropriate value. Both the sensitivity analysis from Monte Carlo simulation and generation of system moments method (Table 14.1) indicate that 97 percent of the variance is attributed to variation of the dielectric thickness, and the transfer function indicates that the capacitance is directly proportional to the area of the capacitor, which only contributes 3 percent to the variance of the capacitance. Consequently, it would seem reasonable to center the distribution by changing the area of the capacitor; in this instance, the distribution would be centered if the area was changed from 250 to 275, and the Cpk would improve to 1.68, achieving capability consistent with Six Sigma expectations.

Centering the mean on the target by adjusting a factor or x that affects the mean and has little impact on the variance may seem almost trivial, but it is surprisingly common and can be surprisingly effective. In the late 1980s, around the time that Motorola announced the Six Sigma program, Motorola's semiconductor product sector (now ON semiconductor) experienced low yields in the range of 60–75 percent for the ECL (emitter coupled logic) line of highly profitable integrated circuits used in high data rate applications. The low yield was found to relate to output voltage levels falling outside the specification limits. The approach shown in Figure 14.6 was applied, and showed that 98 percent of the output level deviation from ideality (squared deviation from the target) was a result of the mean being off-center relative to the middle of the specification limits. The difference between Cp and Cpk also exceeded 0.333 by a large margin (Cp's for various parts were typically greater than 2 and Cpk's were below 0.5). Subsequently, two engineers "tweaked" nearly all of the ECL products by changing resistor values to center the output voltages. The yields moved from the 60 to 75 percent range into the 95 to 100 percent range virtually overnight, and the annual profit for the ECL products improved by more than $50 million.

OPTIMIZATION: ROBUSTNESS THROUGH VARIANCE REDUCTION

If the Cpk and Cp are both well below the 1.5 level that is consistent with Six Sigma expectations, then the team will likely need to try to reduce the variance. There are five approaches that can be considered to reduce variance, and they range from simple and straightforward to surprisingly complex. In addition to these five methods, the multiple response optimization approach described later in this chapter has been used successfully in many situations to optimize Cpk's and yields directly.

The first of the five methods is straightforward and perhaps rather obvious: if the critical parameter, y, has excessive variation, and the variance of y is heavily influenced by one input factor, x_1, then the variance of y can be reduced by reducing the variance of x_1. One caveat: this approach only provides significant variance reduction if the factor focused on for variance reduction contributes more than 50 percent of the variance of y. The renowned consultant, Dorian Shainin, referred to this factor as the "red X." Reducing the variation of factors or x's that contribute less than 50 percent of the variation of y provides insignificant and perhaps undetectable improvement that can lead to frustration. The root of this red X effect is that variances add, but standard deviations do not: the standard deviation of the y is equal to the square root of the sums of the squared standard deviations due to each factor or x.

Figure 14.8 illustrates this method. Reducing the variance of x_1 is represented by the change from the initial distribution of x_1 (outline) to the improved distribution (filled). This reduction is propagated, through the transfer function, providing a distribution of Y with reduced variance. This method was used in one of the very first Six Sigma improvement efforts at Motorola.[2] A semiconductor fabrication area was having low yields that were traced to highly variable threshold voltages, the voltage to turn an MOS transistor on. Based on a suggestion from a local expert, engineers were ready to work on improving the gate oxide process. However, we had a theoretical equation for threshold voltage (found in almost any semiconductor textbook), and we could estimate the variability of the three x's that affected threshold from historical data using the equation shown in Figure 14.4. As shown in Figure 14.9, the local expert had given an unhelpful suggestion—the gate oxide thickness did not strongly affect the threshold voltage, and effort to reduce that variation would not noticeably improve the variability of threshold voltages. Oxide/interface charge density was the key factor. The engineers added an anneal that reduced the oxide interface charge density and its variability. The variability of threshold voltage dropped like a rock, and the yield increased substantially. This led to the first successful power MOS product family at Motorola, which became a large and profitable business unit.

---

2. Eric C. Maass, "System Moments Method for Reducing Fabrication Variability," *Solid State Technology*, August, 1987.

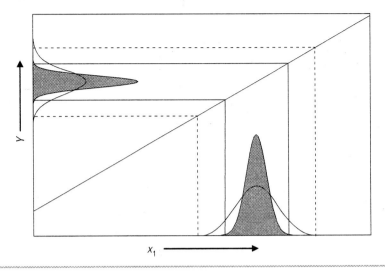

Figure 14.8 The first method to reduce the variance of parameter Y is to reduce the variation of the factor, x_1, that contributes most of the variance. The initial distribution of x_1 is shown with an outline, and the ± 3 standard deviation range is represented with dashed lines that are propagated to Y through the transfer function. The distribution of x_1 after variance reduction is shown with a filled distribution that is propagated to Y, as represented with solid lines.

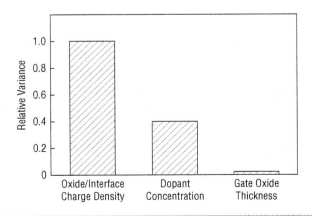

Figure 14.9 Application of the first method to reduce the variance of threshold by focusing on reducing the variation of the largest contributor to the variance, the oxide/interface charge density

In the software domain, an example of the first method might be identifying a task in an initialization sequence that contributes most of the variation in start-up time, and reducing the variability by reducing the number of memory accesses for that task.

The second method for reducing variance can be useful if the transfer function of the effect of the dominant x on the y is nonlinear. Figure 14.10 shows such a situation, with a nonlinear function representing the relationship between x_1 and Y. If the design is "tweaked" or the process re-centered such that the distribution of x_1 corresponds to a region where the response Y is less sensitive to the input factor x_1, then the same distribution of x_1 is propagated to a tighter distribution of Y. If, in reducing the variance of Y, the average has shifted too far from the target, then a different factor can be adjusted to re-center the mean for response Y back on target.

An example is provided in a paper (published when Six Sigma was first announced) that provides the first example of using Six Sigma approaches to reduce variance. In this instance, yields of integrated circuits using bipolar (NPN) transistors were low as a result of low and highly variable breakdown voltages. By switching from arsenic to antimony, a slower diffusing dopant for silicon, the mean thickness of the intrinsic epitaxy layer was shifted to a flatter portion of the transfer function between the breakdown voltage and the thickness (Figure 14.11). This process change significantly improved the yields for over a hundred types of integrated circuits, and resulted in hundreds of millions of dollars in cost savings due to reduced scrap.

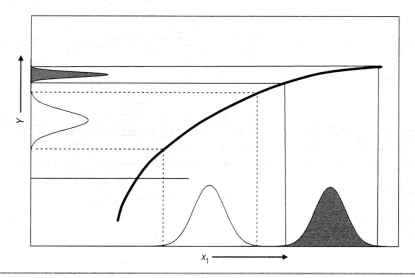

Figure 14.10 Second method to reduce the variance by shifting the distribution of factor x_1 to a region where Y is less sensitive to x_1

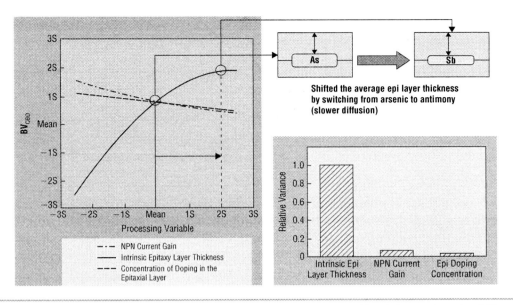

Figure 14.11 Application of the second method to reduce the variance of breakdown voltage by shifting the distribution of intrinsic epitaxy layer thickness toward a higher value (through a change in the dopant). This moved the distribution to a region where the breakdown voltage was both higher and less sensitive to the intrinisic epitaxy layer thickness.

The third method relies on an interaction between two factors or x's, one of which contributes substantially to the variation of Y. Interactions figure prominently in empirical and semiempirical transfer functions that can be obtained through DOE, as discussed in Chapter 13, and will show up as twists in a response surface, as shown in Figure 13.20. With an interaction between two factors, say x_1 and x_2, the slope of Y vs. x_1 depends on the setting for x_2—so x_2 can be set such that the slope of Y vs. x_1 is gentler, and the contribution of variation in x_1 to variation of Y can be reduced, as illustrated in Figure 14.12.

For example, an oscillator can generate a desired frequency using a tank circuit, a combination of an inductor and capacitor that sets a resonant frequency determined by the reciprocal of the product of the inductance and capacitance (an interaction). The sensitivity of the frequency to capacitance can be reduced by selecting an appropriate value for the inductance.

The fourth method involves brainstorming, to find an alternative design or approach that is less sensitive to x_1, the factor that contributes most of the variation to Y. A software-relevant example might be the initialization sequence mentioned earlier in this chapter that depends on a highly variable delay for one task in the sequence. The design team might consider a different sequence of tasks that renders the start-up

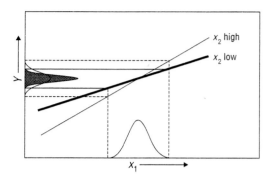

Figure 14.12 Third method to reduce the variance by setting the level of one factor in a two-way interaction, x_2, such that Y is less sensitive to x_1

time less dependent on that task. Similarly, the oscillator example mentioned earlier can be rendered less sensitive to the capacitance with an alternative oscillator design, such as replacing a single capacitor with a pair of capacitors in parallel, or using an oscillator design that does not rely on capacitors.

The fifth method is based upon the equation shown in Figure 14.4. The second term in the equation, the additional term for correlated x's, provides an intriguing method to reduce variation: if this term is negative, then it subtracts variance from the variance contributed by the x's in the first term. The second term can be negative if the correlation coefficient between two x's that contribute the majority of the variance of y is negative and the product of the two partial derivatives is positive (the slopes of Y versus each of the x's), or if the correlation coefficient is positive and one but not both of the partial derivatives is negative.

This fifth method has an enviable record in leading to patentable designs, perhaps due to the innate robustness of the resulting design combined with the complexity of the underlying concept and equation. Some examples include the difference or differential amplifier that is almost ubiquitous in analog designs, the Gilbert cell mixer used in many RF circuits, temperature compensated circuits,[3] and noise-canceling or noise feedback designs.

This method can be illustrated with a simple example for trying to set a voltage for a pair of batteries in series. Assume each battery voltage is distributed as a normal distribution with a mean of 1.5 V and standard deviation of 0.1. For the device to function properly, the combined voltage must be between 2.8 and 3.2 volts. If two batteries are

3. Phuc Pham, Lou Spangler, and Greg Davis, *Temperature Compensated Voltage Regulator having Beta Compensation*. Patent number 5,258,703; 1993.

chosen at random, the mean would be 3.0, the variance would be $2 \times (0.1)^2$ or 0.02, and the standard deviation would be 0.14. The Cpk would be $(3.2 - 1.5 \times 2)/(3 \times 0.14) =$ 0.47. About 16% of the devices would be outside the 2.8- to 3.2-volt window.

One way to apply method five would involve injecting a correlation between the two batteries by measuring the first battery, then selecting a complementary battery that has a voltage $V2 \approx (3 - V1)$. V1 and V2 would now be correlated. Because setting the voltage for the second battery involves subtraction, the product of the correlation coefficient between the x's and the two partial derivatives (from Figure 14.4) would be negative. Figure 14.13 shows the improvement in the variance of the Y, the combined voltage from the two correlated batteries in series, for various values of the correlation coefficient. If the correlation coefficient is about 0.9, the paired batteries meet the requirements with a Cpk of slightly better than 1.5.

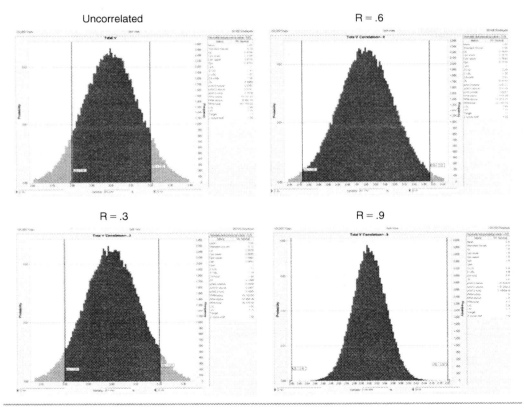

Figure 14.13 Illustration of the fifth method to reduce the variance by choosing two batteries with correlated voltages to obtain a series voltage with a tight distribution. Increased correlation between the battery voltages tightens the distribution and improves Cpk for the combined voltage.

Figure 14.14 shows a schematic for a difference amplifier. Properties of the two transistors in this circuit are highly correlated, since they are adjacent on an integrated circuit. The different amplifier amplifies the difference between inputs and noises on inputs A and B. Any variability in one input is effectively canceled out by the other; thus, this amplifier circuit is considered very robust. Variations in the properties of the two transistors are correlated, subtracted, and tend to cancel out. External noises, like variations in the ground level, affect both inputs equally and cancel out.

After tightening and centering the distribution near the target value, the critical parameter scorecard should be updated to reflect and summarize the improvement achieved, as mentioned earlier in this chapter. The five methods discussed in this section constitute the known and proven methods for reducing variation for a single parameter. More advanced methods are required if several critical parameters, Y's, need to be cooptimized. This situation is referred to as multiple response optimization.

MULTIPLE RESPONSE OPTIMIZATION

There are many instances in which optimizing one response or Y, by adjusting one factor or x, might cause problems in other Y's. In some cases, the mean and the variance or standard deviation of the Y can be treated as separate responses, involving cooptimization of two responses for each Y. This approach is generally performed using response surface modeling.

In Chapter 13, there was an example of using response surface modeling. A display device required cooptimization of the means and standard deviations for brightness,

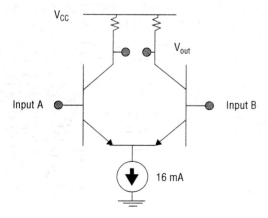

Figure 14.14 Electrical schematic for a difference amplifier. Emitters of two nearly identical bipolar transistors converge on a 16 mA current source. Differences between signals on Inputs A and B are amplified and show up at the output, V. Noises tend to cancel out, leading to a robust design.

image nonlinearity, dynamic range, and several other Y's. Optimal values for the x's, which were settings in the software that affected the shape and durations for pulses sent to the pixels, are summarized in Table 14.2. The cooptimization was accomplished using desirability functions: a desirability index was defined for each standard deviation such that the desirability would be 1 if the standard deviation was zero, and the desirability would fall linearly as the standard deviation increased, reaching and remaining zero if the standard deviation exceeded a maximum allowable value. Similarly, desirability indices were defined for each mean, following the same linear function as for standard deviations if lower was better (as in the case of image nonlinearity), or a reversed linear function if higher was better (as in the case of dynamic range), or a triangle shape peaking at the target value if an intermediate target was better (as in the case of brightness). The composite desirability was the weighted averages of the desirability index results for each mean and standard deviation; in this study, all means and standard deviations were given equal weights for composite desirability. An overlaid contour plot for the parameters is

Table 14.2 Values for several responses cooptimized for both means and standard deviations

```
Brightness-Mean = 69.841, desirability = 0.98014
Brightness-StDev = 1.191, desirability = 0.36254
Dynamic Range-Mean = 40.605, desirability = 1.00000
Dynamic Range-StDev = 1.920, desirability = 0.07992
Image Nonlinearity-Mean = 1.363, desirability = 0.31864
Image Nonlinearity-StDev = 0.325, desirability = 0.83735
Composite Desirability = 0.596
```

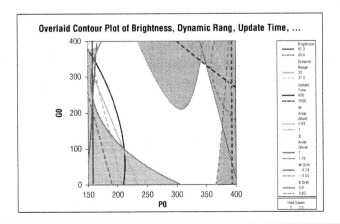

Figure 14.15 Overlaid contour plot to illustrate cooptimization for several responses associated with a display device

shown in Figure 14.15. This led to a success story at a start-up company called "e-Ink" that is a tribute to the efforts of an excellent team of engineers and scientists. They decided to "challenge the envelope" and developed an excellent, robust, reliable, cost-effective display using multiple response optimization. e-Ink's optimized ePaper displays are now used in electronic books, like Amazon's Kindle and Sony's Reader Digital Book.

COOPTIMIZATION OF CPK'S

The use of desirability functions for cooptimization for several parameters involves at least two limitations. First, this approach requires that standard deviations be obtained from measurements of units—after prototypes or early units have been built. Predictive engineering suggests that the development team should try to optimize the design before building prototypes.

The second limitation is that the use of desirability functions, and the weighted average combination of calculated desirability functions, is somewhat subjective. If the team changes the shape of the desirability function, or changes the weighting used in obtaining the weighted average combination of means and standard deviations for various parameters, a different set of recommendations might emerge. The desirability functions are one approach to the dilemma poised in trying to determine a way to combine means and standard deviations for several responses that each might be in different units; however, there is another approach.

Because each individual critical parameter to be optimized can be in different units (such as volts, seconds, kilograms, or millimeters) and have different specification limits in those units, the degree of "goodness" for each critical parameter cannot be directly combined—rather, each critical parameter can be converted to an index that is unitless.

Either Cpk or yield provides a unitless index that allows predicted performances for multiple responses to be combined. Cpk is a "higher is better" index, and a Cpk value of 1.5 is considered consistent with Six Sigma capability. Assuming normality, and assuming that the part of the distribution that would fall beyond the nearer specification limit would be rejected, the Cpk value is directly related to predicted yield in terms of meeting that requirement. Process capabilities for several critical parameters can be combined into a composite Cpk index, assuming independence, and, predicted yields for several critical parameters can similarly be combined into a composite yield index, assuming independence. Composite yield can also serve as a leading, predictive indicator for the yield of the product, assuming the product's performance is tested for this set of critical parameters, with each critical parameter tested relative to its specification limits.

Some Monte Carlo simulation programs include optimization in which the Cpk and yield (or its complement, the percent outside specification limits) can be estimated through Monte Carlo simulation, and settings for decision variables (the x's that can be adjusted through design) can be found that optimize the Cpk or yield for one critical parameter, while requiring that all other Cpks or yields meet or exceed some lower limit.

C.C. Ooi, a mechanical design engineer for Motorola in Malaysia, cooptimized several requirements for a rugged communication device for covert, secret agents (spies).[4] Ruggedness is a very critical requirement for a communication device worn by a secret agent; when covert agents enter hostile situations communication is literally their lifeline. C.C. Ooi used RSM to develop transfer functions for several responses (RWRC, LWRC, and SF1-Rightscrew), as shown at the top of Figure 14.16. She then used OptQuest with Crystal Ball to find settings for the input variables that cooptimized the Cpk's for these responses, as shown at the bottom of Figure 14.16. She completed the optimization prior to building a single physical sample, and verified it later with prototypes—eliminating at least two tool modifications and prototype builds, reducing cost and product developing time, and providing very high confidence that the communication device was sufficiently rugged to provide the lifeline needed by covert agents.

YIELD SURFACE MODELING[5]

Yield Surface Modeling (YSM) is a multiple response optimization method, developed and patented by Eric Maass and David Feldbaumer,[6] that is particularly well suited for use with simulation. Although simulation programs are usually deterministic and ill suited to handle manufacturing variability, YSM can determine settings where the critical parameters are robust against manufacturing variability. The output can be a response surface in which the response plotted as a function of the x's is either the predicted composite yield or the predicted composite Cpk, for multiple responses compared against their respective specification limits.

YSM expands on RSM, using calculations based on the set of equations developed through RSM. YSM then helps determine settings for the factors or x's that will provide acceptable composite yield and composite Cpk. YSM has been used in the development

---

4. C.C. Ooi, S.C. We, P.P. Lim, and E.C. Maass, "Developing a Robust Rugged and Reliable Mechanical Design Using Optquest," presented at the Crystal Ball 2007 Monte Carlo Awards conference.

5. Eric C. Maass, "Yield Surface Modeling," *Encyclopedia of Statistics in Quality and Reliability*, 2008.

6. David Feldbaumer and Eric C. Maass, *Yield Surface Modeling Methodology*, U.S. Patent 5 438 527, August 4, 1995.

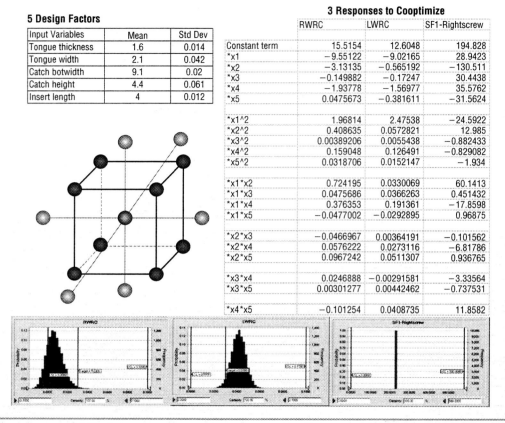

5 Design Factors

Input Variables	Mean	Std Dev
Tongue thickness	1.6	0.014
Tongue width	2.1	0.042
Catch botwidth	9.1	0.02
Catch height	4.4	0.061
Insert length	4	0.012

3 Responses to Cooptimize

	RWRC	LWRC	SF1-Rightscrew
Constant term	15.5154	12.6048	194.828
*x1	−9.55122	−9.02165	28.9423
*x2	−3.13135	−0.565192	−130.511
*x3	−0.149882	−0.17247	30.4438
*x4	−1.93778	−1.56977	35.5762
*x5	0.0475673	−0.381611	−31.5624
*x1^2	1.96814	2.47538	−24.5922
*x2^2	0.408635	0.0572821	12.985
*x3^2	0.00389206	0.0055438	−0.882433
*x4^2	0.159048	0.126491	−0.829082
*x5^2	0.0318706	0.0152147	−1.934
*x1*x2	0.724195	0.0330069	60.1413
*x1*x3	0.0475686	0.0366263	0.451432
*x1*x4	0.376353	0.191361	−17.8598
*x1*x5	−0.0477002	−0.0292895	0.96875
*x2*x3	−0.0466967	0.00364191	−0.101562
*x2*x4	0.0576222	0.0273116	−6.81786
*x2*x5	0.0967242	0.0511307	0.936765
*x3*x4	0.0246888	−0.00291581	−3.33564
*x3*x5	0.00301277	0.00442462	−0.737531
*x4*x5	−0.101254	0.0408735	11.8582

Figure 14.16 Example of cooptimization of the Cpk's for three responses, using central composite design to develop equations or transfer functions for each response (top), and then using OptQuest with Crystal Ball software to find optimal settings for the five design factors

of 60 integrated circuits, a clinical sensor, and for a mechanical assembly associated with a portable communication device. Remarkably enough, all of these products were first-pass successes with high yield.

A famous gentleman once said, "The first shall be the last";[7] in that spirit, it seems appropriate, as we near the end of this book, to discuss a case study from the very first Design for Six Sigma project, started and completed in 1989.

---

7. Quote attributed to Jesus, in Matthew 19:30.

Motorola SPS was trying to launch a new logic family of 60 integrated circuits, with expected sales of $400–$800 million. The first part suffered low yield (about 55 percent). The new logic family was about to be cancelled. Analysis showed that the low yield was a result of:

- VOH (output voltage levels—in mV) out of spec
- Leakage currents (in microamperes) out of spec
- Propagation delays (in microseconds) out of spec

Each critical parameter was in different units, and it seemed reasonable to use Cpk or yield as a unitless index that could allow us to combine these into a composite Cpk and composite yield. All three responses were primarily affected by two factors, threshold voltage (Vt) and Leff, the effective length of the distance from the source to the drain of the transistor (Figure 14.17). Each factor was set in the semiconductor manufacturing area and both factors varied around the set point. There were also two design factors that had little or no variability: output buffer design and substrate doping.

As shown in Figure 14.18, the initial design for the first product in the family was suboptimized, with considerable yield loss; all three critical parameters had low Cpk's and associated yield losses. David Feldbaumer, an expert design engineer, ran an RSM experiment on the electronic circuit simulator, SPICE. RSM provided transfer functions for each of the three responses, VOH = f(Vt, Leff), Leakage = g(Vt, Leff), and Prop Delay = h(Vt, Leff).

Because the SPICE simulator was deterministic and provided no means to estimate the Cpk for each response, the equation shown in Figure 14.4 was used with historical standard deviations of Vt and Leff and with the transfer functions to estimate the standard deviation of each response, as illustrated in Figure 14.19. As shown in Figure 14.20, these transfer functions were used to generate predictive equations for the mean and

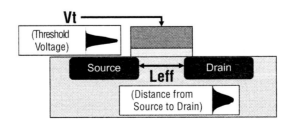

Figure 14.17 Factors involved in the three critical parameters are the threshold voltage, which is the voltage required to turn on the MOS transistor, and the Leff or effective length between the conductive source and drain

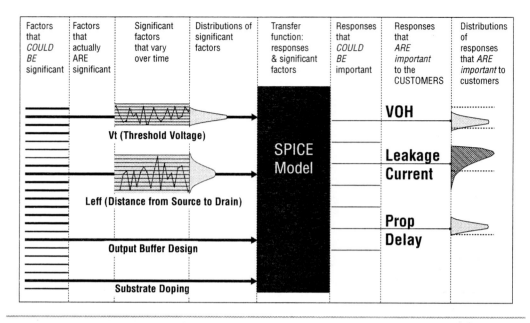

Figure 14.18 Before optimization of the Cpk's for three critical parameters, the settings for the control factors Vt, Leff, output buffer design, and substrate doping were such that the distributions of the three critical parameters involved considerable yield loss with values outside the specification limits

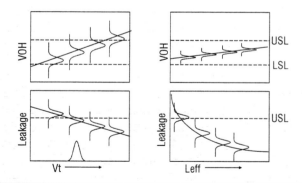

Figure 14.19 Consideration of standard deviations for the inputs in terms of their impacts on the Cpk's of the critical parameters; to the left, estimates for the critical parameters VOH and leakage are shown as functions of the threshold voltage, Vt, and can be shifted if the mean for the threshold voltage is shifted. Similar estimates are shown as functions of the effective source-to-drain length, Leff, on the right side of the figure. Specification limits for each critical parameter are also shown to illustrate the amount of yield loss depending on where Vt and Leff are located.

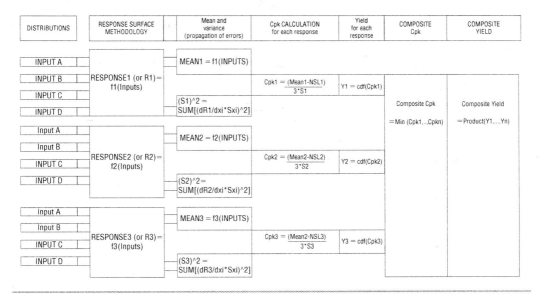

DISTRIBUTIONS	RESPONSE SURFACE METHODOLOGY	Mean and variance (propagation of errors)	Cpk CALCULATION for each response	Yield for each response	COMPOSITE Cpk	COMPOSITE YIELD
INPUT A	RESPONSE1 (or R1) = f1(Inputs)	MEAN1 = f1(INPUTS)	$Cpk1 = \frac{(Mean1-NSL1)}{3 \cdot S1}$	Y1 = cdf(Cpk1)	Composite Cpk	Composite Yield
INPUT B						
INPUT C		$(S1)^2 = SUM[(dR1/dxi \cdot Sxi)^2]$			= Min (Cpk1,...Cpkn)	= Product(Y1,...,Yn)
INPUT D						
Input A	RESPONSE2 (or R2) = f2(Inputs)	MEAN2 = f2(INPUTS)	$Cpk2 = \frac{(Mean2-NSL2)}{3 \cdot S2}$	Y2 = cdf(Cpk2)		
Input B						
INPUT C		$(S2)^2 = SUM[(dR2/dxi \cdot Sxi)^2]$				
INPUT D						
Input A	RESPONSE3 (or R3) = f3(Inputs)	MEAN3 = f3(INPUTS)	$Cpk3 = \frac{(Mean2-NSL3)}{3 \cdot S3}$	Y3 = cdf(Cpk3)		
Input B						
INPUT C		$(S3)^2 = SUM[(dR3/dxi \cdot Sxi)^2]$				
INPUT D						

Figure 14.20 Overview of method converting the transfer functions from RSM into composite Cpk and composite yield surfaces through the YSM approach

variance for each response (using the equation in Figure 14.4), and subsequently predicted Cpk's and yields for each response. These, in turn, were combined to predict the composite Cpk and the composite yield. The resulting composite yield surfaces are shown in Figure 14.21.

Using YSM with SPICE simulation, the yield for subsequent new products in the Logic IC family were each in the 90 to 100 percent range, resulting in 57 first-pass successes introduced in 28 weeks. This set a record for the company and led to hundreds of millions of dollars in sales and over $100 million in profits. The project was entered into Motorola's first Total Customer Satisfaction competition, and won the gold medal among about 5,000 teams across Motorola; the project was also featured in a training video *Team Problem Solving, the Motorola Way*.

YSM has been used in the successful development of many new products. Although most of the successful uses have been in integrated circuit development, this approach has also been used on other products including the optimization of mechanical designs for two-way radios and the development of complex medical instrumentation. To date, every new product that has been developed using YSM has been a first-pass success with high yield. This was considered so remarkable by the former president of Motorola Semiconductor Product Sector that, at his request, the approach was treated as proprietary information for several years.

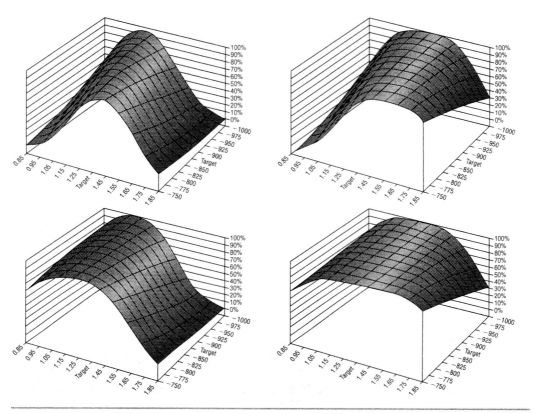

Figure 14.21 Use of YSM to review the initial design (upper left), and consider the impact of changes in the substrate doping and output buffer design (upper right and lower left). Using both (lower right) provided a robust design with consistent, high yields.

CASE STUDY: INTEGRATED ALTERNATOR REGULATOR (IAR) IC FOR AUTOMOTIVE

Another case study was presented as a Web seminar entitled **"Minitab and Crystal Ball Synergy for Multiple Response Optimization."**

The integrated alternator regulator (IAR) was a bipolar integrated circuit designed by Stephen Dow of Motorola for use in the ignition system for automobiles. Yield Surface Modeling was performed in conjunction with SPICE, a circuit simulator. A central composite design was set up to vary the levels of four factors that were monitored in the manufacturing area: current gain of bipolar transistors, capacitance of an on-chip capacitor built using a silicon nitride dielectric, and resistances for two types of resistors.

The results for four critical parameters required for acceptable performance were obtained through SPICE simulation at each combination of levels for these factors. Figure 14.22 shows the response surfaces for the mean and variance for one critical parameter, soft start time, and the combination of these into a response surface for the predicted Z or Cpk (referred to as a Cpk surface) followed by the conversion of the Cpk surface into a yield surface representing the expected percentage good parts obtained

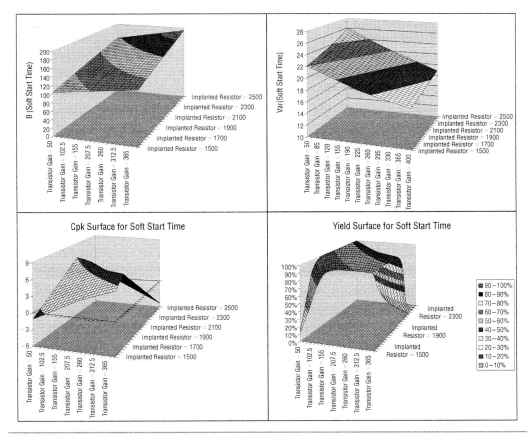

Figure 14.22 Response surface for the expected value (upper left) and the variance (upper right) for the critical parameter soft start time is translated into a response surface for Cpk (lower left) and yield (lower right)

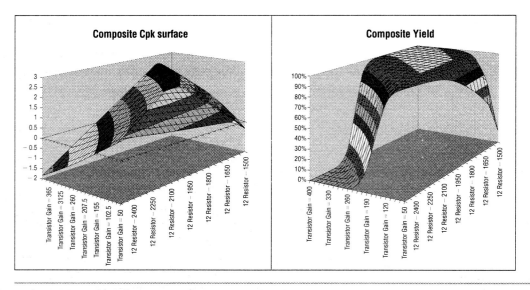

Figure 14.23 Cpk's and yields for all four critical parameters are combined into surfaces for composite Cpk (left) and composite yield (right)

with that combination of factor levels, subject to the set of assumptions. Figure 14.23 illustrates the combination of composite Cpk and yield surfaces for the four responses. Figure 14.24 puts this optimization into the format for the illustration of the flow of predictive engineering.

SUMMARY

Some experts consider the terms "predictive engineering" and "Design for Six Sigma" to be synonymous. The previous chapter began a rather comprehensive and deep exploration of predictive engineering by discussing the development of transfer functions, whereas this chapter showed how to use those transfer functions to optimize the design.

The transfer functions can first be used to predict initial performance capabilities (using Monte Carlo simulation or the generation of system moments method), which can be captured and communicated through a critical parameter scorecard. The initial results can be used to prioritize optimization efforts. Optimization of a critical parameter may require re-centering of the mean on the target or reduction of variance; five variance reduction methods were discussed, some with examples that relate to the origins of Six Sigma and the focus on reducing variability. Some of these methods were

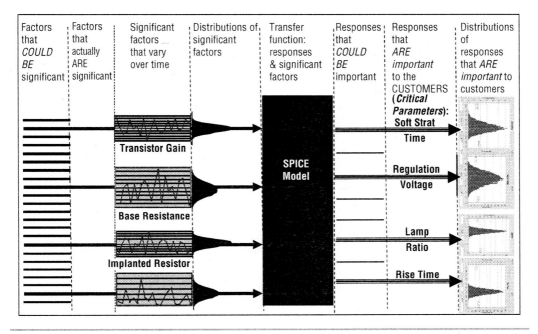

Figure 14.24 Optimization of the Cpk's for four critical parameters for the integrated alternator regulator, shown in the format for illustrating the flow associated with predictive engineering

straightforward and simple, whereas others involved advanced concepts that have often led to patentable robust designs.

Finally, some rather powerful methods to cooptimize the means and variances of several responses, or directly optimize the composite Cpk and yield, were discussed, along with examples and case studies; some of these methods are associated with an enviable record of first-pass successes in product developments.

Once the design has been optimized, the critical parameter scorecard should be revisited, to incorporate the flow-up of capabilities achieved through the transfer functions and optimization. The critical parameter scorecard then can communicate the improvement and confidence that the design team has in the optimized, robust design.

Predictive Engineering: Software Optimization

The predictive engineering approaches discussed in Chapters 13 and 14 are relevant for software as well as hardware. This chapter addresses some topics that are relevant to software in particular, including efforts to prevent or minimize software defects, and focuses on optimizing the software critical parameters that are important to meet each basic customer need and intended use of the product. These requirements must be measurable to determine the acceptability of the product in its intended application, and they may be a summary of system performance requirements. Performance requirements define *how much* or *how well* the product function or feature shall perform. It is essential that they be written in a verifiable manner and are designed in the most optimal manner. To be verifiable, the performance requirements must be measurable; they will be used to judge the outcome of the product evaluation during design verification.

MULTIPLE RESPONSE OPTIMIZATION IN SOFTWARE

The multiple response optimization of performance characteristics for software systems involves adjusting control variables to find the levels that achieve the best possible outcome or response. There are many occurrences in software design where finding optimal settings for a control variable x to optimize a response, or Y, can impact another response; in some incidences, the settings may cause the software code to break. In software systems, there usually are many response variables to optimize simultaneously. In Figure 15.1, the "connector speed" is a control factor, or x, that affects two responses, or Y's: transmit data to server and receive data from server.

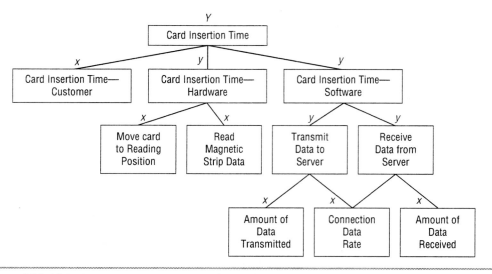

Figure 15.1 Multiple response optimization software example

Multiple response optimization is a systematic, quantitative approach to constructing software systems to ensure that you meet multiple and sometimes competing performance objectives for the software project.

USE CASE MODELING IN OPTIMIZATION

The next major decisions in optimization are to select the overall key performance indicators (metrics). Software key performance indicators (KPIs) are critical parameters (functional and nonfunctional) metrics used to help an organization define and evaluate how successful the software release is, typically in terms of making progress toward meeting its customer requirements and needs. KPIs are typically tied to an organization's strategy and customer scorecards. See Figure 15.2 for an example of a KPI scorecard.

Each critical parameter listed in the KPI scorecard is modeled to understand the behavior, and then optimized. Use case modeling is a popular method used to understand the intended behavior of a system, and contains three major elements (Figure 15.3).

"A use case diagram describes relationships between a set of use cases and the actor(s) involved." This statement is intended to focus on relationships between use cases and the actor or actors involved with those use cases, but not relationships between use cases themselves. *Note that actors can have a relationship with multiple use cases.* Although use cases can technically have relationships to one another, this is an

Key Performance Indicators Scorecard

Continuous Critical Parameters	Units	Lower Spec Limit	Target (optional)	Upper Spec Limit
Measure				
Call Performance Rate—Termination	%		98	
Call Performance Rate—Origination	%		98	
Call Performance Rate—Long Call	%		98	
FTP Performance	%			
CDG3 Feature Test Pass Rate	%	97		100
Time to Power Up	**seconds**			10
TL9000 (FFR < .5%)	%			
Field Quality—NPI Cycle Time	**days**			60

Discrete Critical Parameters
Measure

	DPMO Target
Number of Software Releases	1
Carrier Lab Submissions (Stability of Product)	2
Meeting Program Ship Dates	

Figure 15.2 Key performance indicator scorecard

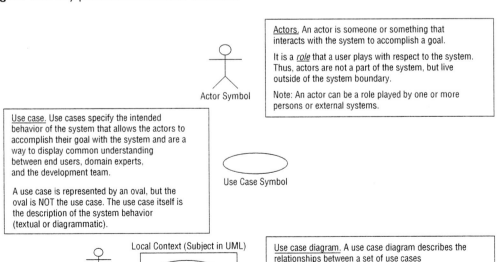

Actors. An actor is someone or something that interacts with the system to accomplish a goal.

It is a _role_ that a user plays with respect to the system. Thus, actors are not a part of the system, but live outside of the system boundary.

Note: An actor can be a role played by one or more persons or external systems.

Actor Symbol

Use case. Use cases specify the intended behavior of the system that allows the actors to accomplish their goal with the system and are a way to display common understanding between end users, domain experts, and the development team.

A use case is represented by an oval, but the oval is NOT the use case. The use case itself is the description of the system behavior (textual or diagrammatic).

Use Case Symbol

Local Context (Subject in UML)

Actor   Use Case

Use case diagram. A use case diagram describes the relationships between a set of use cases and the actor(s) involved, within some particular context or subject.

Figure 15.3 Use case models symbols

advanced use case topic that will not be covered in this book. More information about use cases can be found in a book dedicated to use cases, software requirements management, or UML.[1]

Early involvement in the use case identification process is better to help avoid specifying unreasonable requirements. The critical use cases are those that are important to responsiveness as seen by users, or those for which there is a performance risk. Generally, critical use cases are those use cases for which the system will fail, or be at a level less than the required level of success, if performance goals are not met. Not every use case will be critical to performance; a relatively small subset of the use cases (<20 percent) account for most of the uses (>80 percent) of the system. The performance of the system is dominated by these heavily used functions. Thus, these should be the first concern when assessing performance; focus on use cases that are important to responsiveness as seen by users, or on those for which there is a performance risk. Critical use cases can be modeled, and set up for simulation or emulation, in preparation for optimization.

Next, identify critical scenarios. For each critical use case:

a. Focus on the scenarios that are executed frequently, and on those that are critical to the user's perception of performance.
b. Identify scenarios that are likely to utilize maximum computer resources.
c. For some systems, it may also be important to include scenarios that are not executed frequently, but whose performance is critical when they are executed. For example, recovery from a crash may not occur often, but it may be critical that it be done quickly.

Describe the use case scenario in the following manner:

- Problem statement: typically asks for a performance measure
- Scenario: state the scenario
- Condition: For platform performance, include the hardware and software and system conditions that are not available. For functional performance, include the conditions of the system that do not vary during performance analysis. State one condition per line.
- Constraints: List the constraints for analysis
  - Variables: Include the parameters that you want varied during the analysis. For platform performance, these are typically workload parameters. For each of them, give the varying range (or discrete values)

---

1. Alistair Cockburn, *Writing Effective Use Cases,* Addison-Wesley, 2001.

- Type: Derived/not derived
- Level: Suggested level of modeling
- Output: Format of the desired output

The next step is to determine what key performance metrics will be collected in the benchmarking process for the use case. For the example your use case could collect the following:

- CPU usage
- Memory use
- Object creation time
- Reaction time
- Event handling time
- Image loading and rendering time
- Frame update rate of animation

Each critical scenario should have at least one associated performance objective. Performance objectives specify quantitative criteria for evaluating the performance characteristics of each scenario. Performance objectives should be established early in the development process. Initially, these will be end-to-end requirements. Later, particularly for real-time systems, it may be desirable to break an end-to-end performance objective into sub-objectives that are assigned as performance budgets to each part of the processing for an event.

Performance objectives can be expressed in several different ways, including response time, throughput, or constraints on resource usage like MIPS, ROM, and RAM. For information systems, or Web applications, the response time for a performance scenario will typically be described from a user perspective, that is, the number of seconds required to respond to a user action or request, or the end-to-end time needed to accomplish a business task. For embedded real-time systems, response time is the amount of time required to respond to a particular external event. Throughput requirements specify the number of transactions or events to be processed per unit time. Constraints on resource usage may be limitations on the overall utilization of a resource, for example, "Total CPU utilization must be less than 65 percent." Constraints may also be limits on the amounts of various resources used by a given scenario or a portion of it.

For each combination of scenario and performance objective, specify the conditions under which the required performance is to be achieved. These conditions include the workload mix and intensity. The workload mix specifies the kinds of requests made for the scenario.

By building and analyzing models of proposed software architectures and designs, we can evaluate their suitability for meeting the performance objectives for a given set of conditions.

The construction and evaluation of performance models follows the simple-model strategy, i.e., use the simplest model that identifies problems with the system architecture, design, or implementation plans. The simplest model is the software execution model or a Monte Carlo simulation (see Chapter 14 for more information on Monte Carlo simulations). Performance models are derived from the critical scenarios. Each model should specify the processing steps for the selected scenario. Model specifications also include the number of times that the scenario executes in a given interval (e.g., 10/sec) or the arrival rate of requests, and the amount of service needed for each of the software resources for each processing step in the model.

EVALUATE THE MODEL

The principal model results are the elapsed time for each scenario and the overall computer resource utilization. Other information from solving the models, such as the maximum queue length, may help in establishing design parameters (e.g., buffer size).

Models should be constructed and solved for each combination of design alternative and execution environment being considered. A design change in the x factor may improve some responses and make others worse. By quantifying these changes, trade-offs are able to be evaluated and the best alternative chosen.

Sensitivity studies should then be performed by varying a model parameter and observing the effect on the response. A large change in a response variable, such as elapsed time, for a small change in a model parameter (x or "little y") indicates that the model is sensitive to that quantity. Early in the development process, sensitivity studies are valuable for pinpointing areas where more data is needed. If a model is sensitive to the estimated value of a computer resource requirement, a precise estimate of that requirement will need to be obtained. Later, sensitivity studies will tell what to expect when resource requirements or processing loads vary from those expected.

If optimizing the software indicates that there are problems, there are two alternatives:

1. Modify the product concept: Modifying the product concept means re-architecting or redesigning the software, or refactoring the existing design to improve its performance. If a feasible, cost-effective alternative is found, modify the corresponding scenario(s) or create new ones to represent the new software plans. Solve the model(s) again to evaluate the effect of the changes on performance. This step is

repeated until the performance is acceptable, or until no additional cost-effective improvements can be found.

2. Revise performance objectives: If no feasible, cost-effective alternative exists, then it is necessary to modify the performance objectives to reflect this new reality. While this is not the most desirable outcome, it is better to know it early and plan accordingly than it is to wait until the end of the project and discover that there is a problem.

In an extreme case, it may not be worthwhile to precede with revised performance optimization objectives. Again, it is better to cancel the project early than to incur the cost of development, only to learn that the result is not usable and the software is not stable.

SOFTWARE MISTAKE PROOFING

Optimization will be easier if we have less defects in our software code. A design engineer individually creating a defect is rarely a problem in software, however collectively as a group it is a major problem. The defect discovery rate decreases as the number of defects increases.

Why is proofing for mistakes and errors so important?

- ~ 60 percent of manufacturing losses are attributed to operator or inspector error—half have the design as a contributing factor
- ~ 50 percent of field failures are related to errors—primarily software and virtually all have design as a contributor
- Motorola concluded that 300 to 3,000 DPMO were caused by errors
- Similar data exists for the nuclear power industry, NASA accidents, and many other industries

With software mistake proofing techniques most of the errors can be eliminated by:

1. *natural mappings*
2. *affordances*
3. *visibility*
4. *feedback*
5. *constraints*

1. *Natural mapping* means building logical one-to-one correspondences that allow the operation or task to become more obvious.

If a proper phone user interface operation requires a label, it could be designed better. In this case, the phone user interface has a natural mapping that makes the "green" start button you click on the left obvious (see Figure 15.4). It is a better design than where no natural mapping exists, as in the phone on the right. Phone A relies on labels to instruct the user on how to start an application or make a call. It is a better design than Phone B where no natural mapping exists. However, the use of labels or instruction often means it could have been designed better.

2. *Affordances:* The next method of putting knowledge in the world is called affordances. Affordances are properties that suggest how an object could be used. *Which API would you prefer to use (and remember) to resize a window?*

1. **setSize(w, h);**
2. **configure(x, y, sx, sy);**
3. **setBoundingRect(x0, x1, y0, y1);**

The first one is easier, because it "affords" you an instant understanding of the parameters.

3. *Visibility* means making relevant components visible, and effectively displaying system status.

- Which compilation log would you want to wade through?
    - 1,500 warnings, 2 errors
    - 0 warnings, 4 errors

Phone A   Phone B

Figure 15.4 Natural mapping of phone user interface

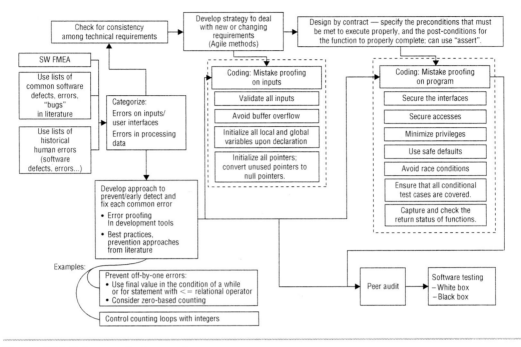

Figure 15.5 Software mistake proofing process

4. *Feedback* means providing an immediate and obvious effect for each action taken.
- Automatically updated GUIs
- Buttons "press" even if action is delayed

5. *Constraints:* Guidance on what to do is encoded as constraints. For example, a simple constraint for feature interaction could be:
- Given factor A = "attachment" with levels {picture, video} and factor B = "picture type" with levels {.jpg, .bmp, .png}.
- The constraints might be:
    If "attachment" = video, "picture type" cannot be .jpg
    If "attachment" = video, "picture type" cannot be .bmp

The solution is to consider what design changes would reduce the chances of the most prevalent errors. Six Sigma methods often consider transferring approaches developed in one field of knowledge, where a similar problem has already been solved, to another field where such an approach may prove valuable. Figure 15.5 shows a high-level process flow for software mistake proofing that combines an anticipate-and-prevent approach from FMEA with approaches developed for the field of secure programming[2] to guard against software vulnerabilities and attacks.

The following are key techniques in the software mistake proofing process, primarily based on approaches from secure programming:

- *Validate all inputs:* Only allow acceptable matches to input criteria. This is consistent with the Poka Yoke mistake-proofing approach of source inspections. Make sure you don't check for just illegal values; there's always another illegal value. Use known illegal values to test validators and limit maximum character length.
- *Strings and numbers:* Watch out for special characters; control characters, including linefeed, ASCII NuLL, metacharacters for shell, SQL, etc. (e.g., *, ?, \, .".), and internal storage delimiters (e.g., tab, comma, <, ☺). Make sure encoding and decoded results are legal. For numbers, check against the acceptable range (minimum and maximum), and be cognizant of and avoid common errors like the off-by-one coding error.
- *Avoid buffer overflow:* Avoid (or carefully use) risky functions, which include gets(), strcpy(), strcat(), *sprintf(), and *scanf(%s). The alternative involves choosing a consistent approach such as Standard fixed length, by using strncpy(), strncat(), snprintf(), or using a standard dynamic length like malloc().
- *Secure the interfaces:* Simple, narrow, nonbypassable; avoid macro languages.
- *Minimize privilege granted:* permanently give up privilege as soon as possible; minimize the time the privilege is active; minimize the modules given the privilege; break the program up to do so. Consider resource limiting.
- *Use safe defaults:* Install defaults as secure, then let users weaken security if necessary after initial installation. Don't install a working "default" password; install programs owned by root and nonwriteable by others (inhibits viruses); Load initialization values safely; fail safe: stop processing the request if surprising errors or input problems occur.

---

2. Jason Grembi, *Secure Software Development: A Security Programmer's Guide,* Delmar Cengage Learning, 2008.

Asoke K. Talukder and Manish Chaitanya, *Architecting Secure Software Systems,* Auerbach Publications, 2008.

David A. Wheeler, *Secure Programming for Linux and Unix,* http://www.dwheeler.com/secure-programs/Secure-Programs-HOWTO/, 2003.

Mark G. Graff, Kenneth R. Van Wyk, *Secure Coding: Principles and Practices,* O'Reilly Media, 2003.

Gene Spafford, Simson Garfinkel, and Alan Schwartz, *Practical UNIX and Internet Security, Third Edition,* O'Reilly Media, 2003.

Applying a mistake proofing process like this allows the team to design components to be self tuning and to minimize the need for reconfiguration during integration.

SOFTWARE STABILITY

After we optimize the software we might want to ensure that our software release is stable. One way to determine stability is to develop *a defect discovery rate collection plan.* Defect discovery rate (defects per hour of test, defects per hour of cycling). It is not based on total defect count that is skewed by bursts of activity. One would use the rate of change in the defect discovery rate to determine software stability. Upward changes in the rate indicate software is not ready for release and when the rate change is near zero, the software is ready. We maintain a defect discovery rate plot to measure this (Figure 15.6).

- First, define the hazard function $h(t)$—the rate of defect discovery per unit time:

$$h(t) = \frac{dH(t)}{dt} \quad e.g., \text{"} h = 14 \ CRs \,/\, day \quad during \ t = TA - 6\text{"}$$

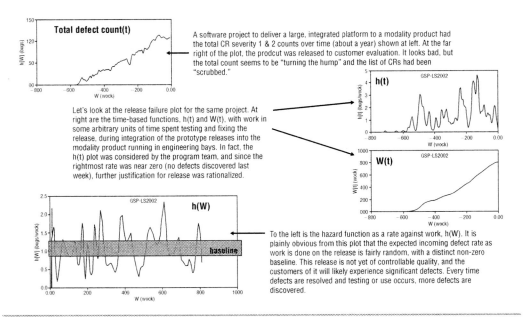

Figure 15.6 Defect discovery example

This software project also delivered a large, integrated platform to a modality product, and actually a large part of the source code is the same as the previous example. However, the intended use of the software was different (different use cases), and the requirements were more well-defined up front. The h(t) function shows a sharp dropoff near the end of the project:

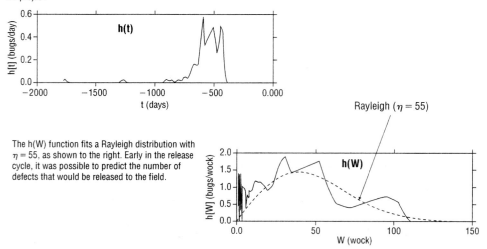

The h(W) function fits a Rayleigh distribution with $\eta = 55$, as shown to the right. Early in the release cycle, it was possible to predict the number of defects that would be released to the field.

Figure 15.7 Example of Rayleigh model use for prediction of software release capability

- Also define the work function $W(t)$, the cumulative test & fix work done by time t, and its first derivative, $W'(t)$. The hazard rate can be changed to a rate against work $h(W)$ as follows:

$$h(W) = \frac{\partial H(t(W))}{(\partial)W} = h(t)(W)\left(\frac{\partial W(t)}{\partial t}\right)^{-1} = \frac{h(t(W))}{W'(t(w))}$$

- Now plot $h(W)$ against W to obtain the release defect discovery rate plot (see Figure 15.7).
- Note that for constant effort applied, $W'(t)$ is constant and the variable change is not needed for the analysis.

The Rayleigh distribution is derived from the case where two normally distributed, independent variables are combined in a 2D distance equation. The distance distribution is then the Rayleigh distribution. Physical problems this applies to include circuit boards (if vertical and horizontal distributions are independent), the distribution of wind speed over time, and, of course, software defect find and fix rates![3]

---

3. Stephen H. Kan, *Metrics and Models in Software Quality Engineering*, Addison-Wesley, 1995.

Here is an example of how the software team will use the Rayleigh model to predict if the software release is stable.

Once the critical parameters capability analysis and defect discovery rate plots are available, an action plan can be developed and executed to evaluate significant design change proposals and optimization to bring the system critical parameters into specification limit. The team prioritize the defects to be fixed according to cost of failure analysis and generate a prioritized backlog.

If large design changes are called for by an action plan in the Optimize phase, it is best to consider the impacts of the various design elements on the overall system critical parameters, to make sure that major critical parameters will not be compromised by the action plan. For example, if the requirement to be portable is a high-level critical parameter for this system, then an action plan that calls for changing the OS abstraction and the architecture (e.g., Google Android architecture) design elements is called into question—it may resolve some problems but create much larger critical parameter misses. If the reaction plan has the high-level critical parameters "locked in" some key design elements then the action plan must accommodate this. Use the design attributes and functional requirements to crosscheck the action plan to begin performance optimization on your critical parameters.

SUMMARY

This chapter addressed multiple response optimization, creating use cases to model critical parameters to optimize, evaluating the model, and preventing software mistakes from occurring to facilitate optimization efforts. The main goal of this chapter was to show the reader a unique set of options to allow the reader to interactively optimize single or multiple response variables, given key performance indicators and use case scenarios. Note that for software design, the appropriateness profiler options are not based on simple re-parameterization of the software use case model parameters; that could lead to erroneous results. Chapter 18 will provide more insight into optimizing software for testing, including discussions of pair-wise testing. This approach is not based on a simple re-parameterization of the use case model to an unconstrained model, such as optimizing factor settings that are not valid feature integrations; instead, all computations are performed based on constraints. Thus, when searching for the optimum factor settings given the appropriate function for one or more response variables, it is assured that only the constrained software features are tested.

Specifically, for multiple response Y variables, an appropriate function can be specified that reflects the most desirable value for each response variable, and the importance of each variable for the overall desirability.

Verification of Design Capability: Hardware

16

POSITION WITHIN DFSS FLOW

Verification of design capability is aligned with the Verify phase of the DFSS process, whether RADIOV, CDOV, or DMADV. Figure 16.1 is the detailed flowchart for this chapter, showing the steps and methods involved in verifying the capability of the hardware. Verification of the software aspects will be discussed in Chapter 18, which focuses on verification through software testing. With DFSS, "verification" covers both verification and validation; requirements are traceable to the VOC, and verification through capability analyses and testing the answers to the questions "Are you building the right thing?" and "Are you building the thing right?"

MEASUREMENT SYSTEM ANALYSIS (MSA)

Once the critical parameters have been selected and specification limits have been set, it seems reasonable to monitor progress towards achieving expectations. As discussed in Chapters 7 and 9, the critical parameters have been defined in measurable terms—but it is uncertain whether the measurement systems are capable of measuring the critical parameters. This discussion of measurement system analysis applies to critical parameters, and also to measurable parameters identified through the flow-down of the critical parameters.

For continuous parameters, measurement system analysis (MSA) involves determining accuracy and precision. Accuracy refers to comparison of the mean of measured values of a calibration standard or "golden unit" to the assumed value provided for that calibration

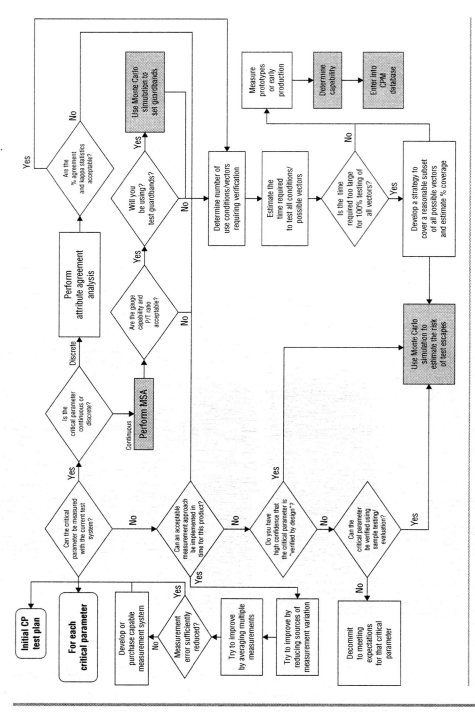

Figure 16.1 DFSS flowchart, drilled down to detailed flowchart for verification of the capability of the design

standard. The calibration standard used in the measurement system analysis has an assumed value that can be traced to a value provided by a standards agency, such as the National Institute of Standards and Technology (NIST) in the United States (formerly known as the National Bureau of Standards). The difference between the mean of measured values and the provided "true value" for a calibration standard may be referred to as "bias." Precision refers to the variation in the measured values due to the measurement system itself.

Observed variance can come from two sources of variation—variation in the actual value, and variation as a result of the measurement system. If the variation caused by the measurement system is too large, it impacts the confidence in the values obtained and any decisions based on evaluation of those values. If the variation due to the measurement system dominates the variation in the values (and there have been many instances of this)—the measurement system could pretty much be replaced by a random number generator—with virtually the same low confidence and the same general uselessness.

Measurement error is considered to arise from two sources. Reproducibility (like reproducing results in scientific literature) refers to the ability of one operator to reproduce similar results on either the same or a similar gauge on the same part. Repeatability refers to the variability among measurements made by a single operator, using a consistent setup on the same gauge on the same part.

Many firms have adopted two indices for determining whether a measurement system is considered capable for making useful measurements: the gauge repeatability and reproducibility (GR&R) index and the precision-to-tolerance ratio (P/T ratio). A widely used rule of thumb is that the GR&R and P/T ratio must both be less than .30. For the P/T ratio, this corresponds to having the ±3 standard deviations of the measurement system consume less than 30 percent of the tolerance window (the range between the upper and lower specification limits). Figure 16.2 shows an example of a measurement systems analysis for a continuous variable. A deeper dive into measurement system

Gage R&R Study - XBar/R Method

Source	StdDev (SD)	Study Var (6 * SD)	%Study Var (%SV)	%Tolerance (SV/Toler)
Total Gage R&R	0.118133	0.70880	64.09	11.83
Repeatability	0.118133	0.70880	64.09	11.83
Reproducibility	0.000000	0.00000	0.00	0.00
Part-To-Part	0.141509	0.84906	76.77	14.17
Total Variation	0.184338	1.10603	100.00	18.46

Figure 16.2 Gage capability study. In this example, the GR&R is 64 percent and the P/T ratio is about 12 percent.

analysis, along with examples for use with Minitab software, can be found at http://www.sigmaexperts.com/dfss/.

GR&R is a ratio of the standard deviation associated with the measurement error (a root-sum-of-squares for repeatability and reproducibility standard deviations) divided by the total variation of measurements made on a set of parts. The 30 percent criteria for GR&R assumes that the total variation observed in the MSA study is similar to and consistent with the expected level of total variation that would be observed when the product is in production. By contrast, if the parts selected for the MSA study were all produced at the same time, in the same batch, then the denominator for the GR&R index would be artificially low, and inflate the GR&R index to indicate that the gauge is not capable.

Measurement error is generally considered to be an aspect of "pure error," as in design of experiments (DOE). If there is no significant bias in the measurements, measurement error will have a mean of zero and a standard deviation equal to the square root of the sum of the variance components due to repeatability and reproducibility—after subtracting the contribution of the variance from repeatability from the variance from reproducibility. In other words, if there is substantial measurement error as a result of repeatability with the same operator, and if one were to "fool the system" by pretending to use two different operators and simply have the same operator change names—some level of false "reproducibility" variance would be observed simply as a result of the variance within the operator, or the repeatability variance. This false reproducibility variance due to repeatability variance is therefore subtracted, although not allowed to take the reproducibility variance to a negative value.

For critical parameters that are discrete or attribute, measurement error can be analyzed through attribute agreement analysis, in which defective and non-defective product is inspected or tested multiple times by multiple inspectors or testers, and the degree of agreement or disagreement is assessed to determine the confidence that can be placed in the measurement system. A result from attribute agreement analysis is shown in Figure 16.3.

IMPROVEMENTS FOR INADEQUATE MEASUREMENT SYSTEMS

If the results of the measurement system analysis fail to meet acceptable guidelines, then there are a series of actions for improving the measurement system. If the problem is primarily related to an offset of the mean relative to the true value, referred to as bias or an accuracy issue, then the solution may involve simple recalibration—that is, shifting the mean of the measurement system to correspond to the true mean.

If the problem is primarily related to measurement variability, referred to as precision, then approaches such as those used to reduce variability for processes could be relevant—that is, the measurement system is considered a process with excessive variation. The first

Attribute Agreement Analysis for Results

Within Appraisers

Assessment Agreement

Appraiser	# Inspected	# Matched	Percent	95 % CI
Fred	20	20	100.00	(86.09, 100.00)
Lee	20	18	90.00	(68.30, 98.77)

Matched: Appraiser agrees with him/herself across trials.

Fleiss' Kappa Statistics

Appraiser	Response	Kappa	SE Kappa	Z	P(vs > 0)
Fred	G	1.0000	0.223607	4.47214	0.0000
	NG	1.0000	0.223607	4.47214	0.0000
Lee	G	0.6875	0.223607	3.07459	0.0011
	NG	0.6875	0.223607	3.07459	0.0011

Figure 16.3 Attribute agreement analysis, comparing two appraisers

step is to try to determine the source of the measurement system; if most of the problem is a result of reproducibility between operators, then the source of measurement variation may be related to the inexperience of one or more of the operators—improved or additional training or improved procedures may be an effective solution. If the reproducibility problem is between measurement systems, then analysis of the performance of the systems may be warranted; a simple solution may be to use only the more trustworthy measurement system. If the issue is primarily within operator, then it may be appropriate to take one or a few units to test varying aspects of the test system, perhaps as a designed experiment, in order to determine whether the problem is with the contacts or interface to the unit or device under test, with the link from the contacts to the sensors, or from the sensors to the information processor that determines the measured value. For example, the measurement error might be due to variations in the contact resistance or integrity of the contact made to the device or unit under test, lengths of the wiring or cabling between the sensor and the contact (especially when the test involves frequencies on the same order of magnitude as the cable length, leading to impedance matching issues), or variations of timing measurements within the system.

Often, the measurement variation can be traced to one of these issues, and the measurement system can be improved to the point where it is considered capable. If all else fails, there are two alternatives that remain:

- Average multiple measurements of the same unit
- Obtain another, more capable measurement system

An approach using the average multiple measurements makes use of two facts. First, the numerator of the P/T ratio and the GR&R are affected by measurement error, whereas the denominator of the P/T ratio is not; the denominator of the GR&R is partly affected by the measurement error, since it is the total standard deviation—combined part-to-part variation and measurement variation. Second, the central limit theorem states that the standard deviation of averages of independent measurements is lower than the standard deviation of the individual measurements themselves. This approach will not work if the P/T ratio or GR&R is very close to 1—that is, if the measurement system is effectively acting like a random number generator, with the measured values bearing no relationship to the actual values.

If the GR&R is adequate but the P/T ratio misses the mark, then the P/T ratio can be improved by using the average of a set of independent measurements for each unit or device under test. The number of independent measurements for each unit that should be averaged is the rounded result of the square of the P/T ratio obtained divided by the square of the P/T ratio desired. For example, if the P/T ratio obtained is .4, and the desired P/T ratio is less than .3, then the P/T ratio can be improved if the recorded result from the MSA is no longer an individual measured value but rather the average of two (the rounded value of $(0.4/.3)^2$) independent measurements of each unit.

If the GR&R is inadequate, then it can be similarly improved by averaging n independent measurements of each unit, where n is determined from the equation:

$$n = \left(\frac{G_C^2}{G_D^2} \right) \left(\frac{1 - G_D^2}{1 - G_C^2} \right) \tag{16.1}$$

where G_C represents the current value of the GR&R, and G_D represents the desired or maximum acceptable value for the GR&R.

Another topic that bears mentioning is the situation where the measurement analysis changes or destroys the unit under test—often referred to as destructive testing. One approach is to find sets of pairs of units, where each pair of units are considered to be nearly identical but there is a substantial difference pair-to-pair. The nearly identical pair may be parts that were manufactured around the same time, in the same batch. The units are destructively tested/measured in a way similar to the MSA approach for non-destructive testing, treating the nearly identical units as if they are replicates. The measurement system analysis for destructive testing uses a nested rather than crossed study approach, to account for the nested variance aspect of using similar but not identical units.

The approaches discussed in the section should provide the development team with confidence in the measurement system. If the measurement system cannot be rendered capable, then perhaps the capability can be "assured by design." This approach was used for a very fast family of logic integrated circuits—the chips were so fast that there was no test or evaluation system available that could test the speed or logic gate delay of the products. Instead, extensive simulations, combined with correlation of simulation to actual results, were used to build confidence that the parts would meet the speed requirements without the use of testing to verify the high-speed performance.

In some cases, the capability can be verified, but on a test or evaluation system that does not have sufficient capacity to test all units. In this case, the capability can be verified by results from an appropriate (random or stratified) sample of the product.

If all these approaches fall short, and the performance promised to customers cannot be verified through design, evaluation, measurement, or testing, then integrity might require that the enterprise share this information with the customers and decommit from providing verification and assurance that the requirements/expectations will be consistently met.

THE RISK OF FAILURES DESPITE VERIFICATION: TEST ESCAPES

The purpose of verification is to assure the enterprise and its customers that the products will consistently meet or exceed expectations. In the spirit of anticipating problems, it is worthwhile to consider possible causes that could lead to failures at the customer despite efforts at verification and containment.

There are a small number of potential reasons why the customer could receive or experience failures despite efforts at verification:

1. The measurement system could measure something that has inadequate correlation to the requirements.
2. The measurement system could experience measurement error that falsely passes bad parts.
3. The measurement system could malfunction, allowing bad parts to pass.
4. The testing and measurement could involve sample testing: testing on some but not all units, so that untested bad units could be delivered to the customer.
5. A subset of the possible test cases (representing the full set of use cases or conditions) are performed, but the condition leading to failure at the customer is not included, allowing test escapes.
6. The parts could go bad after being measured.

The potential issue of inadequate correlation to requirements should be addressed twice in the new product development process. When the critical parameter (or parameter flowed down from the critical parameter) is identified and rendered measurable, the evaluation and test engineers should be engaged so that they fully understand why this parameter was selected and how it ties to customer satisfaction. Later, when early prototypes or mockups or emulations of the product become available, marketing should become engaged and use these with selected customers as part of lead user analysis or concept testing, as described in Chapter 7. With the Agile development approach (Chapter 11), representatives for the customers are engaged and the ongoing customer feedback should ameliorate this potential problem.

One approach for handling measurement error is "guard-banding"—setting test limits inside the specification limits. The difference between the sets of limits can be set to balance the risk to the consumer of receiving parts that are actually outside the specification limits (but pass through measurement error) with the risk to the producer of rejecting parts that meet requirements. This balance can be achieved through appropriate use of Monte Carlo simulation, as shown in Figure 16.4.

The risk of measurement system malfunction can be handled in several ways—through frequent calibrations, ongoing MSA approaches, control charting of the measurement system, and applying Poka Yoke (mistake-proofing; see Chapter 15) approaches to the measurement system.

The risks associated with sample testing—testing less than 100 percent of the units and/or less than 100 percent of the use cases—can be assessed through the use of Monte

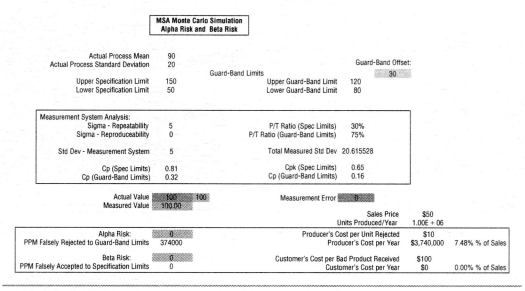

Figure 16.4 Monte Carlo simulation for optimizing the selection of guard-bands for test limits

Carlo simulation with a statistical model of the sample testing method to assess the magnitude of the risk and consider the risk and time required for alternative approaches.

The risks for parts initially passing because they are initially "good," and then failing, involve reliability assessments, as covered in Chapter 17.

DETERMINE THE CAPABILITY

Once the measurement system has been characterized and verified for usefulness, and the risks have been assessed and managed, the measurement systems can be used to determine process capability based on measurements from prototypes or early production samples.

As discussed in Chapter 10, there are two key indices used to assess design capability: the Cp (also known as Pp) and the Cpk (also known as Ppk). Equations for these two indices are given below.

$$Cp = \frac{USL - LSL}{6s} \qquad (16.2)$$

$$Cpk = min\left[\frac{USL - xbar}{3s}, \ \frac{xbar - LSL}{3s} \right] \qquad (16.3)$$

Before performing capability analysis, the assumption for independence or stability of the data should be checked through an approach such as the non-parametric runs test or the appropriate use of a control chart (applying a subset of the Westinghouse rules). The assumption of normality should be checked through normality testing methods such as a normal probability plot or the Anderson-Darling test. If the data is non-normal, then it might be possible to transform the data (using an approach such as the Box-Cox transformation), or a non-normal distribution that fits the data can be identified and an approach for non-normal capability assessment using that non-normal distribution can be used. An example of the results from a capability analysis study is shown in Figure 16.5; this "capability six-pack" includes checks on the assumptions of independence and normality.

Six Sigma performance is defined as having a Cp greater than or equal to 2 and a Cpk greater than or equal to 1.5. The capability results can be entered into the critical parameter management database. If the capability results indicate that the capability expectations are met, consistent with Six Sigma performance expectations, the team has cause to celebrate. If they fall short, the first step is to determine whether the problem is a result of the average being far from the target or center of the specification limits, or due to excessive

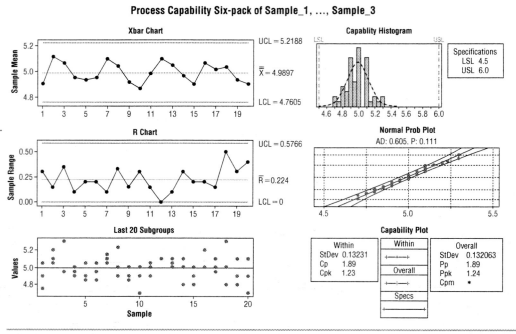

Figure 16.5 Capability six-pack, including checks on the assumptions for normality and independence and an assessment of the long-term capability in terms of Pp and Ppk (also referred to as long-term Cp and Cpk). Some examples of capability analyses using Minitab software can be found at http://www.sigmaexperts.com/dfss/chapter16msa_capability.

variation in the data. If the Cp is more than 1.5 and the difference between the Cp and Cpk is more than a third (0.333), then the problem is primarily due to mis-centering; in this case, the system can usually be recentered with a slight adjustment to one of the control factors or the value of a component. If the Cp is too low and/or the difference between Cp and Cpk is less than a third, then the situation can be improved by methods for reducing the variance, as discussed in Chapter 14. The flow-up represented in the critical parameter management database can be used to study the causes and help the team determine directions to improve the capability through recentering or tightening distributions among the components or at the interfaces.

SUMMARY

The steps involved in verification of capability can and should be initiated shortly after the critical parameters are identified, in terms of analyzing and assessing the measurement system to be used for evaluating and testing the critical parameters and, later, for some flowed-down requirements. If the measurement system is not capable, then a series of steps can be considered to improve the measurement system, ranging from determining the causes to replacing the measurement system or instituting the replacement of individual measurements with the average of repeated measurements.

The capability of meeting requirements can then be assessed using capable measurement systems; hopefully, this step simply verifies the predicted capability that has been estimated and forecasted through the capability flow-up (Chapter 14). If capabilities fall short of expectations, the methods discussed in Chapter 14 can be applied, and after improvement, the critical parameter flow-up can be repeated.

The risks of delivering products that do not meet customer expectations, despite the verification of capability, can be considered and managed in terms of six categories of test-escape risks.

Verification of Reliability and Availability

Eric C. Maass and Vivek Vasudeva

CUSTOMER PERSPECTIVE

Verification of reliability is aligned with the Verify phase of the RADIOV, CDOV, or DMADV process of DFSS. Reliability can be considered to represent performance and functionality over an extended time period, such as over the expected useful life of the product. Customers expect the products they purchase to be reliable—to function adequately over an appropriate period of time—whether this expectation is explicitly stated and captured in the voice of the customer (VOC) or is tacit. Reliability perceptions are a major factor in purchase decisions,[1] and unfavorable experiences undercut customer satisfaction. Since it is considerably more difficult to gain a new customer as it is to retain an existing customer,[2] reliability has a clear and direct impact on future sales to existing and potential customers.

---

1. David A. Garvin, "Competing on the Eight Dimensions of Quality," *Harvard Business Review,* Nov.–Dec. 1987.

2. "All research points to the fact that the cost of selling to an existing customer is about 1/10 to 1/5 the cost of selling to a new customer." Hornbill Systems LTD;

   "65% of a company's business comes from existing customers, and it costs five times as much to attract a new customer than to keep an existing one satisfied." Gartner;

   "A mere 5% reduction in customer defections increases company profits by 25% to 85%." Fredereich F. Reichheld and W. Earl Sasser, Harvard University;

   "The cost of acquiring customers is 10 times the cost of keeping them. Rescuing defected customers costs 100 times more than keeping existing customers." 2003 McKinsey report.

A common engineering definition for reliability is derived from MIL-STD-721C, "the probability than an item can perform its intended function for a specified interval under stated conditions." Key aspects of reliability are that it represents an assessment of probability, that measurements related to reliability are often destructive (and therefore nonrepeatable) tests, and that measurement of reliability must estimate the reliability over a long period of time from an assessment over a considerably shorter period of time; hence, acceleration is required.

It is probably worth noting that customer's definitions of reliability may differ somewhat from the engineering definition. The latter relate to metrics such as mean time to failure (MTTF), where failure is defined as a loss of functionality or unacceptable degradation of performance. By contrast, customers' definitions of reliability are more similar to what is referred to as availability in engineering parlance. The key measure in both is the time associated with inconvenience—the duration of the time that the customer experiences frustration. Imagine two scenarios, each starting with the same reliability issue: the customer has dropped his cellular phone one time too many, and it no longer functions.

- In the first scenario, he tries to get his cellular phone repaired or replaced, and finds that—because he has a special service through his company—he will need to wait 10 days to receive a new cellular phone that is being shipped to him.
- In the second scenario, as soon as the network detects his non-functional cellular phone, a new phone is sent to him through a courier, such that—within five minutes—his doorbell rings, and the courier hands him his new, replacement cellular phone.

Both scenarios have the same level of reliability—the phone stopped functioning after the same number of drops—but the first scenario dissatisfied the customer much more. The second scenario provided much higher availability, which would favorably impress the customer and could lead to improved customer satisfaction.

Availability can be measured in terms of the minutes per year that the product functionality is unavailable to the customer, or as the percent of the time that the system is unavailable. For example, a goal of "five nines" availability for a telecommunication system (or a base station as part of that system) would require that system to be available 99.999 percent of the time, or available for all but at most five minutes per year. Availability as a percentage can be obtained as the ratio of the mean time between failures (MTBF) divided by the sum of the MTBF and the MTTR (mean time to repair or replace). Reliability is often measured in terms of the MTTF (mean time to failure) or MTBF (mean time between failures), depending on whether or not the system is repaired. Clearly, substantial improvement in reliability (as reflected in improved MTTF or MTBF) would improve both reliability and availability. A case study in modeling availability is provided in the Appendix at the end of this chapter.

AVAILABILITY AND RELIABILITY FLOW-DOWN

Availability can be flowed down by allocating the maximum acceptable downtime (five minutes per year for "five nines availability") to the various subsystems or functions. This can lead to a simple model for availability, in which the total downtime per year is the sum of the planned and projected unplanned downtimes per year. More complex analyses may involve state diagrams that illustrate transitions from failure to recovery (repair or replacement) and associated transition matrix representations of Markov chain models[3] (Figure 17.1).

Reliability flow-down usually requires an underlying model for the reliability; the initial model would generally be a series model for the reliability of the product, which can be described by analogy to light bulbs connected in series (Figure 17.2). If a single

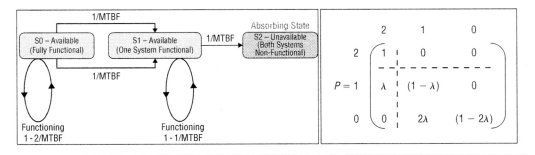

Figure 17.1 Availability analysis: State diagram for hot standby without repair (left) and associated transition matrix representation of Markov chain mode (right)

3. Jan Pukite and Paul Pukite, *Modeling for Reliability Analysis: Markov Modeling for Reliability, Maintainability, Safety and Supportability Analyses of Complex Systems*, Wiley-IEEE Press, 1998.

Wallace R. Blischke and D.N. Prabhakar Murthy, *Reliability: Modeling, Prediction, and Optimization*, Wiley Interscience, 2000.

Figure 17.2 Light bulbs in series, as an analogy for series model for reliability of an electronic system

light bulb in the series fails, the "system" (consisting of the set of light bulbs) fails. Similarly, if the hard drive of a computer, or the microprocessor, or the RAM of a computer system fails, the computer system fails. If the cellular phone transmitter, or receiver, or audio system fails, then the cellular phone is no longer usable as a cell phone—but might be somewhat useful as a practice hockey puck.

Because reliability is defined as a probability (the probability that the system can provide its intended function for a specified interval), then the series reliability model for the system can be translated into a mathematical model that is useful for flow-down and flow-up:

$$\text{Reliability(System)} = \prod_{i=1}^{N} \text{Reliability}_i \qquad (17.1)$$

where i can represent the ith subsystem, function, or component.

Brief tutorials and an associated Excel workbook that includes some simple examples and an "availability game" can be accessed at http://www.sigmaexperts.com/dfss/. This "availability game" inspired the availability modeling approach discussed in the case study for availability in the Appendix to this chapter.

Fault tree analysis provides a tool for flowing down and flowing up reliability, although it tends to deal with the converse, the probability of failure. The probability of failure for a system is logically and mathematically related to an "OR" combination of the various possible causes of failure, as displayed in a logic diagram-like format in Figure 17.3.

BATHTUB CURVE AND WEIBULL MODEL

The definition of reliability, "the probability than an item can perform its intended function for a specified interval . . ." includes a probability aspect, as described in the last

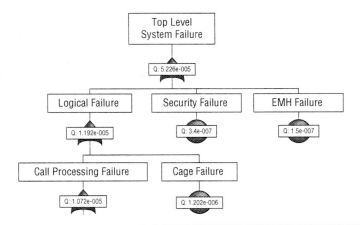

Figure 17.3 Portion of a fault tree analysis model for a network failure, using software from Relex

section, and also a specified interval aspect. The interval may be measured in terms of continuous measures such as elapsed time and number of miles driven, or in terms of ordinal discrete measures such as the number of activations, the number of drops, or the number of uses.

The interval aspect of reliability is often described in terms of a bathtub curve, which relates remarkably well to the Weibull distribution often used to describe reliability distributions such as times-to-failure. The Weibull distribution has two parameters, a shape parameter and a scale parameter (the characteristic life, or the time at which 63.2 percent of the components can be expected to fail). The Weibull distribution can be described in terms of

- $f(t)$, instantaneous probability of failure—the probability that a unit will fail at time t
- $F(t)$, the cumulative probability of failure—the probability that a unit will fail by time t
- $R(t)$, the probability of survival—the probability that a unit will survive until time t
- $h(t)$, the hazard function or failure rate—the probability that a unit will fail at time t given that it has already survived until time t. This is equivalent to the ratio $f(t)/R(t)$

The top part of Figure 17.4 shows a bathtub curve, in which the vertical (y) axis is the hazard function, $h(t)$, also known as the failure rate, and the horizontal (x) axis is time or number of drops or another index representing the interval aspect of reliability. The bathtub curve is divided into three regions:

- Early life failures (also known as infant mortality), for which the Weibull shape parameter is less than 1.

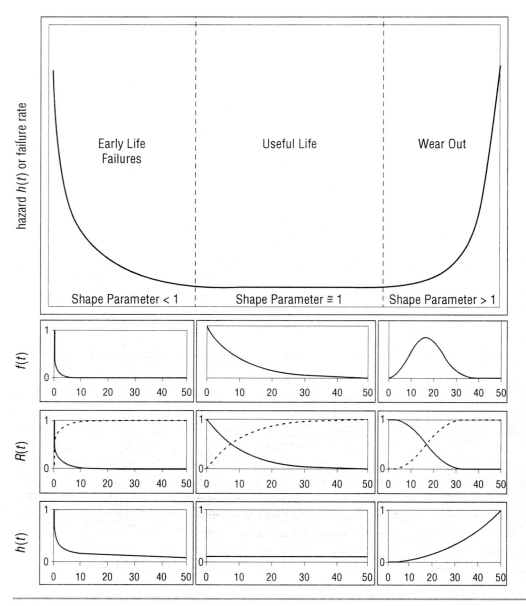

Figure 17.4 (Top) Bathtub curve with three regions representing three sets of reliability degradation overlaid on a graph of the hazard or failure rate versus time to failure. (Bottom) Weibull distributions for each region shown as $f(t)$, cumulative distributions: reliability or survival rate shown as $R(t)$ (solid line) overlaid with the cumulative probability of failure, $F(t) = [1 - R(t)]$ (dashed line), and hazard or failure rate shown as $h(t)$, corresponding to each of the three regions, with relevant Weibull shape parameters.

- Useful life (also referred to as constant failure rate), for which the shape parameter is approximately equal to 1.
- Wear out, for which the shape parameter exceeds 1.

The bottom part of Figure 17.4 shows $f(t)$, $F(t)$, $R(t)$, and $h(t)$ graphs corresponding to each region of the bathtub curve. The hazard function is the ratio of the $f(t)$ and the $R(t)$ and relates to the hazard function vertical axis for the corresponding region of the bathtub curve.

The three regions usually involve different failure mechanisms, as will be described shortly.

SOFTWARE RELIABILITY

By contrast, software reliability is sometimes represented by a different version of a "bathtub curve," as shown in Figure 17.5. Some might note that the initial region of the software bathtub curve bears some resemblance to the early life failures region of the standard bathtub curve of Figure 17.4, and that the subsequent peaks and declining failure rates corresponding to each software upgrade seems to behave similarly.

The early life failures region of the standard bathtub curve for hardware includes test escapes, which provides an interesting analogy to the corresponding region for the bathtub curve for software. From that perspective, software reliability relates to software defects that can be considered to behave as test escapes from a hypothetical "ideal software test." The hypothetical ideal software test would provide comprehensive, 100 percent coverage of all conditions, and would prevent software defects from escaping and impacting the customer—thereby providing higher software reliability.

The appendix to this chapter includes a case study focused on software reliability, along with the case study for system availability, which includes both hardware and software availability.

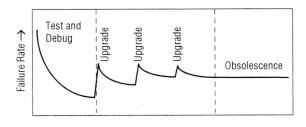

Figure 17.5 "Bathtub Curve" analogy for software reliability

EARLY LIFE FAILURES/INFANT MORTALITY

Early life failure relates to the leftmost region of the bathtub curve in Figure 17.4, and generally involve mechanisms relating to either latent defects, test escapes, or high susceptibility to harmful events. Electronics systems involve metal lines connecting circuits and devices within integrated circuits and on printed circuit boards; latent defects can occur on thin metal lines that have weak points, such as notches in the lines or thinning as the lines go over steps in the topology of the surface. These are examples of latent defects.

Components that are bad, but do not fail tests, can show up in early life failures as test escapes. Reasons for test escapes include sample testing (i.e., not all parts are tested), incomplete testing (i.e., not all use cases are tested or not all test vectors or paths are exercised), errors in testing, inadequate testing, and measurement error (in which a part that is marginally bad shows up as good because of variations in the measurement system).

Historically, high failure rates have been experienced with electronic components such as integrated circuits that have inadequate ESD (electrostatic discharge) protection. Such a component will be initially good, and will pass tests by the supplier, but subsequent handling and environmental changes can provide situations where the component is "zapped" by static electricity or a pulse of relatively high voltage that damages the component.

"Burn in" is a method that is often used to remove the weak parts with latent defects and parts that are especially susceptible to static charges. The parts are stressed for a period of time such that weak components fail, and the components are tested to remove the damaged parts from the set of parts shipped to customers. For software defects that behave similarly to test escapes, extensive test/debug efforts, accelerated software testing (with higher clock speeds or higher usage rates) and approaches such as fault injection testing, can provide benefits analogous to the benefits of burn-in screening for defects.

It is worth noting that early life failures, and their apparent correlation to latent defects and parts near the spec limits such that measurement error can pass a marginally bad part, prompted a quality and reliability engineer named Bill Smith to approach the CEO of Motorola, Bob Galvin, with the Six Sigma concept, as described in Chapter 1.

USEFUL LIFE/CONSTANT FAILURE RATE

The useful life portion of the bathtub curve involves a constant failure rate, or a shape factor of about 1 for a Weibull distribution model, in which case the Weibull model behaves as an exponential distribution-based reliability model. The exponential model is considered memoryless, in that the failure rate a year later is the same as a year earlier within this region of the bathtub curve. By analogy to an actuarial bathtub curve, in which the early life failure region corresponds to infant mortality and wear out corresponds to old age and eventual death, the constant failure rate corresponds roughly to the adult years, where the major causes of death are random events such as automobile accidents that are about as likely to impact a person at age 45 as at age 35.

For complex electronic systems (as well as complex nonelectronic systems) with a host of potential failure modes for each of several subsystems, the net effect of system reliability is often assumed to behave as random events that can be modeled using a constant failure rate, such as with an exponential distribution or a Weibull distribution with a shape factor of about 1.

WEAR OUT

Wear out mechanisms are fairly common for mechanical components; in fact, Weibull's original paper discussed some mechanical wear out mechanisms as applications and examples.[4] Fans, hinges, and connectors all experience wear out, with shape factors greater than 1.

Some electronic components experience notable wear out mechanisms; packages of light bulbs often quote parameters related to the scale parameter, the characteristic life of the bulb. Naturally, vacuum tubes experience similar wear out mechanisms, as the filaments degrade through chemical and electrochemical processes in a partial, imperfect vacuum. Flash memories also experience wear out as they degrade with extensive rewriting to the same locations.

DETAILED FLOWCHART FOR RELIABILITY OPTIMIZATION AND VERIFICATION

Figure 17.6 is the detailed flowchart aligned with this chapter, showing the steps and methods involved in verifying the reliability for the system, subsystem, or component.

The first few steps of the flowchart involve preparation and gathering appropriate information, which may already be available from prior steps and efforts; these include the test plan, and a P-diagram that might provide insight into noise factors and environmental factors that can suggest potential approaches for accelerated testing. The need for accelerated testing for reliability verification might be obvious, but let's state

---

4. W. Weibull, *A Statistical Distribution Function of Widespread Applicability,* ASME, 1951.

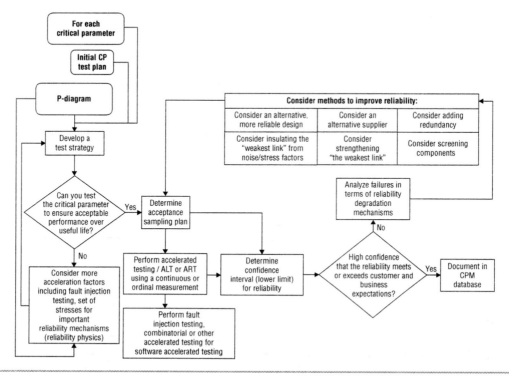

Figure 17.6 DFSS flowchart, drilled down to detailed flowchart for verification of the reliability

the obvious in the interest of clarity: reliability is the probability of successful continued operation into the future, generally far beyond the testing time. It is highly risky to assume that a short duration test in normal operating conditions assures continued operation over a longer period of time under a range of operating conditions that can include harsh conditions.

ACCELERATED LIFE TESTING

For many situations, reliability physics provides useful guidelines for accelerated life testing. The goal of accelerated life testing is to appropriately accelerate failure mechanisms that could substantially degrade performance, but not introduce unreasonable failure mechanisms unlikely to be experienced in realistic situations. For example, elevated temperature is a common variable that accelerates many reliability degradation mechanisms that involve chemical reactions; as such, the associated acceleration factor (the ratio of projected lifetime under normal operating conditions to the measured lifetime under stress or accelerated testing) may be estimated through the Arrhenius equation that reflects the impact of temperature on chemical reaction rates. However, although appropriate and useful acceleration might be achieved through testing at temperatures of about 150 degrees Celsius to accelerate mechanisms at normal temperatures of about 25 degrees Celsius, an excessive elevation temperature of 1000 degrees Celsius might introduce mechanisms unlikely to be experienced under normal operation—such as melting of materials or a tendency for some materials to burst into flames.

Common acceleration factors include the aforementioned elevated temperature, but also elevation of voltage, current through conductors, humidity, current, or clock rates. Table 17.1 summarizes some of the failure mechanisms and associated acceleration factors commonly encountered in electronic systems consisting of electronic components such as resistors, capacitors, inductors, transistors, diodes, and integrated circuits.

Figure 17.7 shows a plot for the estimation of time to failure for failures of insulation used in an electronic system, where the system is tested at various elevated temperatures and the results are extrapolated to provide a confidence interval for the projected lifetime at normal operating temperature. The data set and screen shots for performing this analysis are available at http://www.sigmaexperts.com/dfss/ For mechanical systems or properties, acceleration might be achieved through elevated temperatures, pressures, or stress conditions such as higher drop heights, deeper submersion under water, or mechanical loading.

Table 17.1 Some failure mechanisms and acceleration variables for electronic components

Mechanism	Acceleration Factors
ESD	Voltage Spike Amplitude
Dielectric Breakdown	Voltage, Temperature
Electromigration	Current, Temperature
Corrosion	Humidity, Voltage, Temperature
Mobile Ions, Surface Charges	Voltage, Temperature
Hot Carrier Injection	Voltage, Low Temperature

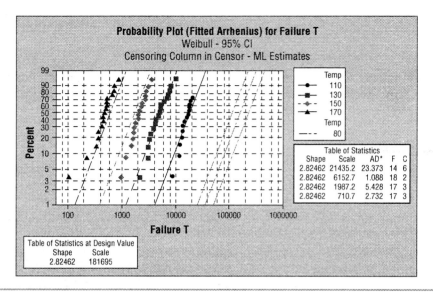

Figure 17.7 Probability plot for fitting failures at elevated temperatures to a Weibull model combined with an Arrhenius acceleration model for insulators. The resulting model is extrapolated to normal operating temperature (80 degrees Celsius) to project a confidence interval for lifetime.

As suggested in Figure 17.6, acceleration variables can also be considered for software reliability concerns, including more extensive and intensive testing, fault injection testing, greater system loading, and appropriately higher clock rates or data rates.

WEIBAYES: ZERO FAILURES OBTAINED FROM ALT

If no failures are obtained through accelerated testing, then with a few assumptions, an upper confidence limit can be estimated for the reliability and lifetime under normal operating conditions. With the WeiBayes approach, the data is assumed to come from a Weibull distribution, and an estimate must be provided for the shape parameter for that distribution. The maximum likelihood method is generally used with this estimation, and Bayes' theorem for conditional probabilities may be invoked to estimate the worst case reliability that could result in the observed results of zero failures, with the given confidence levels. Table 17.2 shows some results from the application of the WeiBayes approach using Minitab's reliability tools.

Table 17.2 Results from applying the WeiBayes approach for 20 samples, no failures, tested for 100 hours and assuming a Weibull shape parameter of 2

Variable: Life   Frequency: Freq

Right censored value: 20   Type 1 (Time) censored at 100

Estimation Method: Maximum likelihood

Bayes Analysis Distribution: Weibull

Parameter Estimates

Parameter	Estimates	Standard Error	95% CI Lower	95% CI Upper
Shape	2			
Scale	*	*	258.383	*

Characteristics of Distribution

	Estimates	Standard Error	95% CI Lower	95% CI Upper
Mean (MTTF)	*	*	228.986	*
Standard deviation	*	*	119.696	*
Median	*	*	215.118	*

RISK OF FAILURES DESPITE VERIFICATION: RELIABILITY TEST ESCAPES

The prior chapter discussed the risk of performance failures despite verification; the same discussion is relevant for the verification of reliability, and the same causes apply. As mentioned previously in this chapter, these same test escape risks relate to software reliability as well. At the risk of redundancy but in the hopes of providing some convenience, here again are the set of possible causes of test escapes, as described in the prior chapter on verification of capability/performance:

1. The measurement system could measure something that has inadequate correlation to the requirements.
2. The measurement system could experience measurement error that falsely passes bad parts.

3. The measurement system could malfunction, allowing bad parts to pass.
4. The testing and measurement could involve sample testing: testing on some but not all units, so that untested bad units could be delivered to the customer.
5. A subset of the possible test cases (representing the full set of use cases or conditions) are performed, but the condition leading to failure at the customer is not included, allowing test escapes.
6. The parts could go bad after being measured.

METHODS TO IMPROVE RELIABILITY AND AVAILABILITY

In general, improving reliability or availability involves a trade-off between the level of reliability achieved and the associated cost and size considerations. One can readily envision ways to dramatically improve the reliability of a handheld cellular phone: encapsulate the phone in light, foamy materials to protect against clumsy phone droppers such as your humble author, provide three redundant systems for each component such that all three components must fail for the phone to fail, place those redundant systems far enough apart to minimize the chance that the same event would wipe them out simultaneously—and, for availability, provide the instant detection and replacement service for a dead phone as described near the beginning of this chapter.

Improvement of reliability and availability often involves a practical trade-off: improving the reliability within reasonable cost constraints and subject to other constraints (such as the customers' lack of enthusiasm for a handheld device with the size and weight of a brick). Possible methods for improving reliability are provided in Figure 17.6, but reproduced and enlarged in Figure 17.8 for convenience and ready reference. These methods can be brainstormed with the team as they endeavor to improve reliability, and handle a trade-off of improvement of reliability with cost and size constraints.

Consider methods to improve reliability:		
Consider an alternative, more reliable design	Consider an alternative supplier	Consider adding redundancy
Consider insulating the "weakest link" from noise/stress factors	Consider strengthening "the weakest link"	Consider screening components

Figure 17.8 Methods to improve reliability

SUMMARY

Whether or not it is explicitly stated as a VOC, reliability and availability are key expectations for customers that affect their purchase decisions and their retention as customers. Reliability and availability can be readily flowed down to the subsystems and functions that are vital for the operation of the product, using a series model for reliability.

A bathtub curve model for the failure rate of a system, subsystem, component, or function can have three regions: early life failures, useful life, and wear out; software reliability follows a different model that bears some similarity to the early life failure region for hardware reliability. Each region involves different mechanisms, and these mechanisms can be accelerated or—in the case of early life failures—"burned in" to render that region less risky for the customers. Accelerated life testing involves using acceleration factors based on reliability physics, and can be used to extrapolate projected life under normal operating conditions. Methods to improve reliability include adding redundancy and replacing or isolating the "weakest links"; however, these methods must be evaluated in the context of pragmatic considerations such as cost and size.

APPENDIX: CASE STUDIES—SOFTWARE RELIABILITY, AND SYSTEM AVAILABILITY (HARDWARE AND SOFTWARE AVAILABILITY)
(Courtesy of Vivek Vasudeva, Software DFSS Black Belt, Motorola Networks)

SOFTWARE RELIABILITY: A CASE STUDY IN A ZERO DEFECT INITIATIVE

Software reliability is a daunting challenge. Developing reliable software helps an organization meet many challenges. We have all heard the phrases from management like, "I do not have resources for this," or "I do not have the time in my schedule for this." The facts are that reliable software will actually save an organization a lot more money than it costs, improve cycle time, and increase productivity of the organization. Table 17.3 shows a cost model used to describe the costs to find and fix software defects, based on phase injected and phase found.

Figure 17.9 graphically describes the same cost model as in Table 17.3. Between 70 percent and 80 percent of the cost to develop highly reliable software is in the testing phases (development test, system test, and post-release test), and 20 percent to 30 percent in the development phases (requirements, design, and coding). This indicates that leaving

Table 17.3 Cost model for finding and fixing software defects based on phase injected and phase found

Cost at Phase Found	Phase Injected		
Phase Detected	**Requirements**	**Design**	**Code**
Requirements	$240		
Design	$1,450	$240	
Code	$2,500	$1,150	$200
Development Test	$5,800	$4,450	$3,400
System Test	$8,000	$7,000	$6,000
Post-Release Test	$30,000	$20,000	$16,000
Customer	$70,000	$68,000	$66,000

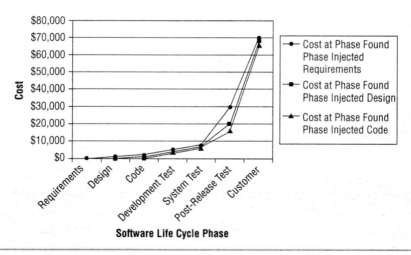

Figure 17.9 Graphical representation of cost versus software life cycle phases

reliability to develop during testing phases—as is the case in many organizations—is the most expensive and the least efficient option. Defects found later in the life cycle cost exponentially more to fix and cause a bigger delay to the schedule, and can severely impact customer satisfaction.

The zero defect initiative was focused on developing reliable software for the common platform software team. The goals of the zero defect initiative were:

- Target zero known high-severity software defects, a maximum of 10 low-severity defects at entry into system test, including third-party vendor issues.
- Target zero known high-severity software defects, a maximum of 10 low-severity software defects for first deployment at customer site.

These goals were extremely challenging for the software team. Generally, there is an arrival rate of software defects from previous releases, and an arrival rate from the current software release, which may include new technology challenges such as new operating systems, new hardware, and new third-party vendor software. A software team has a closure rate—an average rate at which they resolve their defects. The management has to study carefully all four factors that change dynamically on a regular basis and plan to have a software backlog of 10 minor defects regardless of the size of the software. The four factors are

1. Defect arrival from previous software release
2. Defect arrival from current software release (this involves new technology, hardware, operating systems, etc.)
3. Traditional defect closure rate versus required defect closure rate
4. Third-party vendor software issues, their arrival and closure rates

Please see Figure 17.10 for the analysis spreadsheet—this sheet was monitored closely every week, for arrivals, closures, and the "net" number of defects. For a month or so, the team was monitoring these results daily!

Every organization usually has some "backlog" defects (known software defects) to contend with. This "living backlog" often defines the productivity of an organization: the larger the living backlog, the less productive the testing phases, as they have to block the "known scenarios." They may be blocked from running many tests, or else they will obtain duplicate defects if they do run tests of known issues or scenarios that are awaiting software fixes. By having larger defect backlogs, test organization cannot meet their schedule, and will deliver poor-quality software as a result.

The zero defect initiative was intended to find a steady state of software defect quantity: a number of open defects that the test organization could tolerate and still be effective and productive. The zero defect initiative team consisted of:

1. A Six Sigma black belt
2. A project manager, who monitored incoming arrivals and drove the closure efforts

Initial Backlog	Week Start Monday	Week End Sunday	Start Number	Pred Arrival SR15 + SR16	Predicted Arrival SR17	Pred Arrival SR17 factoring out D&T	Pred SR17.0 Vendor Arrival	Pred Arrival Total	Pred Arrival Cumulative	Actual Arrival	Pred Closure	Pred Closure Cum	Actual Closure	Pred Net (Cumul)	Actual Net
Load 1 delivery 4/23	4/30/2007	5/6/2007	62	1	7	5	1	9	71		9	9		62	
	5/7/2007	5/13/2007		0	10	6	1	11	82		8	19		65	
	5/14/2007	5/20/2007		1	17	10	1	19	101		8	28		76	
Load 1 impact from NEs	5/21/2007	5/27/2007		0	15	9	1	16	117		8	36		84	
	5/28/2007	6/3/2007		1	20	12	1	22	139		10	50		96	
	6/4/2007	6/10/2007		0	26	16	1	27	166		65	128		58	
Development Test Start Load 2	6/11/2007	6/17/2007		0	50	30	1	51	217		15	146		94	
Earliest NE BT start 5/15 iSG	6/18/2007	6/24/2007		1	46	28	1	48	265		45	200		97	
	6/25/2007	7/1/2007		0	41	25	1	42	307		15	218		124	
	7/2/2007	7/8/2007		0	23	14	1	24	331		45	272		103	
	7/9/2007	7/15/2007		0	20	12	1	21	352		15	290		109	
	7/16/2007	7/22/2007		0	18	11	1	19	371		45	344		83	
	7/23/2007	7/29/2007		0	17	10	1	18	389		15	362		86	
Last discovery for SI	7/30/2007	8/5/2007		0	13	8	1	14	403		40	410		60	
	8/6/2007	8/12/2007		0	7	4	1	8	411		10	422		58	
System Integration Build	8/13/2007	8/19/2007		0	5				417		30	458		34	
Pre-SI Start 8/21	8/20/2007	8/26/2007		0	7				425		8	467		34	
	8/27/2007	9/2/2007		0	5				431		20	491		20	
	9/3/2007	9/9/2007		0	3				435		5	497		19	
	9/10/2007	9/16/2007		0	2				438		10	509		12	
SI Start 9/19 ZDI	9/17/2007	9/23/2007		0	3	2	1	4	442		4	514		12	

ZDI – Initial Date Backlog estimate

Figure 17.10 Analysis spreadsheet for monitoring arrivals, closures, and net number of defects

3. Software development middle managers, who drove defect resolutions by the engineers to meet the weekly backlog goals
4. Director of engineering (who made this happen by committing to this project)

The zero defect initiative team engaged in the following activities:

1. Predict incoming arrivals from previous software releases
2. Predict incoming arrivals from current software release
3. Determine a closure rate (plan staffing based on peak arrivals)
4. Arrival and closure of third-party vendor defects
5. Present results weekly at the organization level

Predicting incoming arrivals from previous releases and the current software release was facilitated by available significant historical data. A constant arrival rate of about

120 percent of the actual from the previous release was assumed, which helped the team plan for peaks. The current release entailed new hardware, with a new operating system, and limited reuse of existing code. For this project, the team supported the following applications:

1. Base station controller—BSC (two platforms, the main network element and the I/O card rack)
2. Application controller—DAP (two platforms)
3. Home location register—HLR
4. Surveillance gateway—SG
5. SSC—A new controller for the new platform

The platform had a high-availability functional area (HA), a middleware layer functional area (MW), an input/output layer functional area (IO), and an operating system functional area (OS). The team analyzed historical reuse for each failure analysis along with the defect densities of each of the platforms. Figure 17.11 shows the capability analysis for the defects for historical releases and the current software release.

CCP Release	Defect Density	Fault Density	Percentage of Defect Density Over Fault Density (Capability)	Defects in Release	Features in Release	KLOC	Vendor Defects	New Product Impact
SR12	0.77	1.24	37.90%	379	Platform and OS	100	73	iCP
SR13.4	2.58	5	48.40%	119	Neither	8.5		MDG4
SR14.0	0.79	1.62	51.20%	414	Platform and OS	115	98	DAP
SR15.0	0.5	0.94	46.80%	97	Capacity	34	4	None
SR16.0	0.62	0.69	10.20%	56	OS	15.7	37	DAP/iCP
SSR6.2	0.52	0.58	10.30%	21	Platform	6		HLR-HSC
SR13.0				252		1.2	120	
R20						2.74		
R19				35			14	
SR17.0	?	?	?	?	?	?		DAP/iCP/ iSG/IHLR /SSC

Previous Software Releases

Current Software Release

Figure 17.11 Capability analysis for defects by software release

Figure 17.12 estimates unique FA content per product. Ideally, there would be total reuse, with no unique content per product, but the platform was not completely common and there were some unique requirements/code. The expected reuse factor was somewhere between 2 and 2.5; the team chose a reuse factor of 2.5 due to a new platform and a new operating system. Figure 17.13 shows the defect arrival prediction based on this logic and historical data. The new source lines of code were estimated at 38.4 KLOC, and were multiplied by the reuse factor to estimate defect density.

From the defect arrival prediction model summarized in Figure 17.13, 374 defects were predicted, but the defect prediction was adjusted for about 30 percent estimated duplicates, which increased the prediction to about 500. The team also developed a discrete event simulation model for defect injection (Figure 17.14), which proved to be a valuable tool and was also used as a "sanity check" for the manual defect predictions.

The next question was where and when most of these defects would be discovered. Historical profiles of arrivals were used to predict defects discovery, using the defect prediction and control spreadsheet from Figure 17.13. The arrival peaks were consistently about two to three weeks after start of a testing phase. The arrivals were strictly monitored each week and corrected based on actual defects. Closure rate for the team was relatively consistent at one defect per engineer per week. Management allocated more staff as the defect arrivals peaked.

Third-party arrivals were consistent with historical data, and the team set fixed response times to one month within the vendor contracts. A clause was inserted for

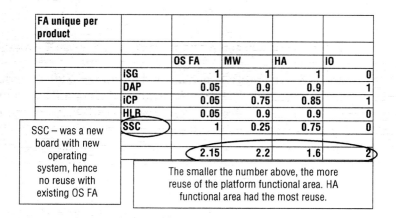

Figure 17.12 Analysis of unique content indicated which functional areas had the most reuse

Release	SR12	SR14	SR15	SR16	Current release SR17	SR17 Adj	SR17 Baseline Capability	Average of ALL the releases	SR15 and SR16 Average	SR17Adj3
HA	75	138	24	38						
MW	78	147	47	22						
IO	128	114	15	9						
OS	37	49	21	10						
General	62	1	1							
MMII	5									
BUTI		7	1							
Total	385	456	109	79	149.6727273	297.4745455	449.0181818	1029	188	374.1818182
KLOC	100	115	34	15	38.4	76.32	115.2	264	49	96
def/KLOC		3.85	3.965217393	3.20588235	5.266666667	4(Expected), 153 defects total			3.897727273	3.836734694	3.897727273
										7.795454545
										0.779545455
										1.559090909

Annotations on the table:

- *After reuse factor of 3 applied, number of defects and KLOC*
- *The team decided to use a reuse factor of 2.5*
- *The release that used FMEA had 20% lower defect density (3.2 def/KLOC) **vs** average defect density of 4.0*
- *FMEA Release*
- *Source lines of code estimated for current release*

Figure 17.13 Defect arrival prediction model

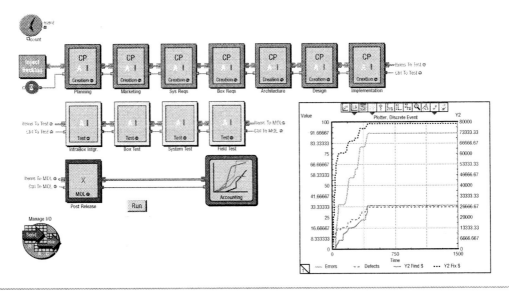

Figure 17.14 Discrete event simulation model for defect injection and defect discovery

the zero defect initiative that allowed workarounds or quick fixes from the third-party vendors until a formal fix was received. Below is a sample chart presented weekly to the senior director of engineering; Figure 17.15 shows total arrival prediction, but does not include the closures; Figure 17.16 shows the net (arrivals—closures) zero defect initiative chart.

Some dramatic results are shown in Figures 17.17 and 17.18. Three months into the effort, the results were so close to predictions that the team needed to convince senior management no cheating was involved; then, the team received recognition. This program was highly efficient in terms of cost of testing and meeting schedule deadlines, and the customers were extremely satisfied. Four releases later, this group continues to receive the best customer satisfaction ratings across Motorola. According to the team, "Software can be designed for a level of reliability; it is worth building the reliability into the software instead of testing for reliability!" The ROI for this effort was estimated at several million dollars, and the team's efforts were recognized for favorable impact on customer satisfaction.

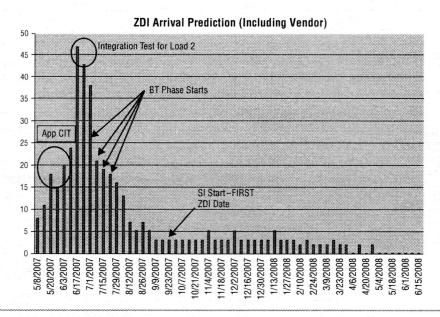

Figure 17.15 Total arrival prediction, not including the closures

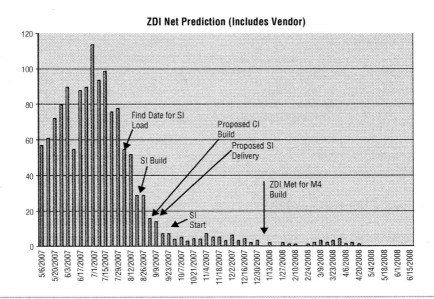

Figure 17.16 Net (arrivals—closures) zero defect initiative chart

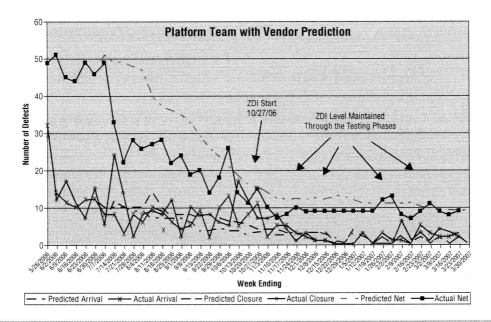

Figure 17.17 Improvement in defects in alignment with predictions, including third-party vendor results

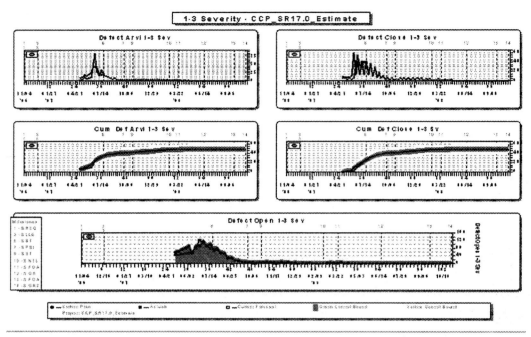

Figure 17.18 Three months into the software release, the prediction was according to plan

Case Study: Modeling Availability for a Cellular Base Station

This case study, courtesy of Kevin Doss, Vivek Vasudeva, and Eric Maass focuses on a critical parameter, availability, of a base station developed by the networks organization at Motorola. Availability measures the percentage of the time that the system is up, operational and able to perform its functions; it is related to the mean time to failure (MTTF), a reliability metric, and the mean time to repair (MTTR): Availability = MTTF/(MTTF + MTTR). Availability for a cellular base station involves software and hardware components—either failure would render the system unavailable.

A critical parameter flow-down (qualitative and quantitative) for the availability metric, annual downtime in minutes, is shown in Figure 17.19. Annual downtime allowed had an initial upper limit of 40 minutes of downtime per year, representing an availability of 99.992 percent, and a long-term goal of 99.885 percent availability (less than 26 minutes per year of downtime).

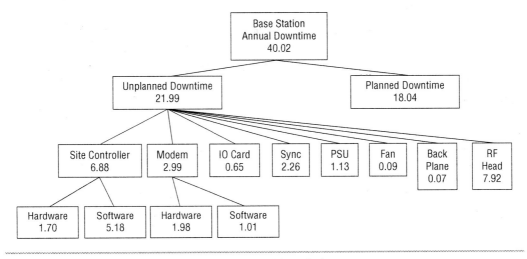

Figure 17.19 Critical parameter flow-down for availability of a base station

In this case study, modeling tools were developed in order to apply DFSS methods to optimize the base station design to meet this level of availability. This was to be accomplished by a software refactoring effort that reduces the application software failure rate by reducing the reset times of the site controller and the modem.

A new availability modeling approach was developed by Kevin Doss, partly inspired by the system availability game at http://www.sigmaexperts.com/dfss/ In this availability game, each player tries to find a way to improve availability to the 5 nines level (99.999 percent) at the lowest possible cost—handling the trade-off between cost and availability. The game showed that Excel could be used for availability modeling, and a more sophisticated version using Excel and Monte Carlo simulation could conceivably replace another approach that involved special and expensive availability modeling software based on setting up and solving Markov chains. Figure 17.20 shows the Markov chain approach for the site controller (one of the highest sources of downtime issues in the quantitative flow-down of Figure 17.19), and Figure 17.21 shows a Monte Carlo simulation-based approach using Excel with Crystal Ball, modeling recoveries involving application reset; similar availability models were incorporated that addressed failures that could be recovered with a board reset and the more severe situations that required a site visit to manually assist with recovery.

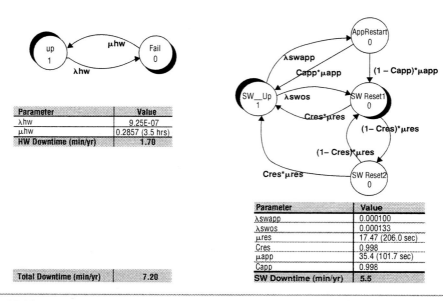

Parameter	Value
λhw	9.25E-07
μhw	0.2857 (3.5 hrs)
HW Downtime (min/yr)	1.70

Parameter	Value
λswapp	0.000100
λswos	0.000133
μres	17.47 (206.0 sec)
Cres	0.998
μapp	35.4 (101.7 sec)
Capp	0.998
SW Downtime (min/yr)	5.5

Total Downtime (min/yr)	7.20

Figure 17.20 Availability modeling of the site controller subsystem using the Markov chain approach

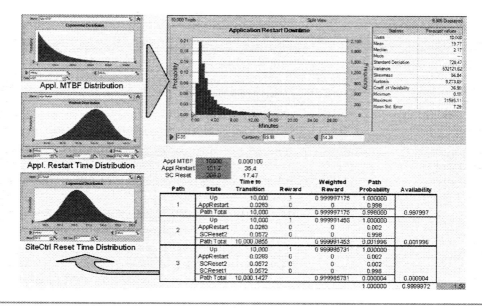

Figure 17.21 Availability modeling for the site controller subsystem of the base station, using the Monte Carlo simulation approach for application restart as a recovery mechanism

The prior Markov modeling-based approach was limited to assuming exponential distributions for time-to-failure and time-to-repair distributions, whereas the Monte Carlo-based approach allowed flexibility in the selection of distributions, allowing distributions based on historical data. The Monte Carlo simulation-based approach using Excel and Crystal Ball (an Excel add-in that enables Monte Carlo simulation) provided comparable results to the Markov-based approach when similar assumptions were used. Using more realistic assumptions based on historical data with the Monte Carlo simulation-based approach was found to provide 17 percent better accuracy in predicting field performance, a reduction in prediction error from 23 percent to 6 percent, and an improvement in R-squared from less than 70 percent to 92 percent in changing from the Markov-based model to the more flexible Monte Carlo simulation-based model using Excel with Crystal Ball. Furthermore, the Monte Carlo-based approach involved considerably lower costs than the software for the Markov-based approach for availability simulation. The Monte Carlo-based approach also allowed for combination of the Excel models for the subsystems, allowing both flow-down and flow-up for the availability as a critical parameter (Figure 17.22) as part of predictive engineering.

Because the Monte Carlo simulation approach improved accuracy and allowed predictive engineering, the team used the model to predict the impact software refactoring would have on availability of the base station, as shown in Figure 17.23.

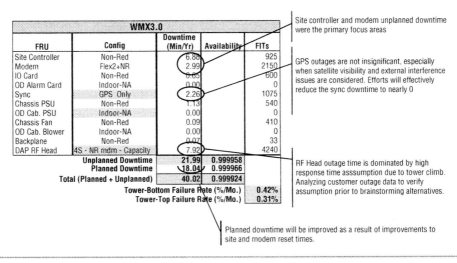

Figure 17.22 Use of Excel spreadsheet model with Monte Carlo simulation to flow-down and flow-up predictive model for availability, and to align efforts for achieving availability goals

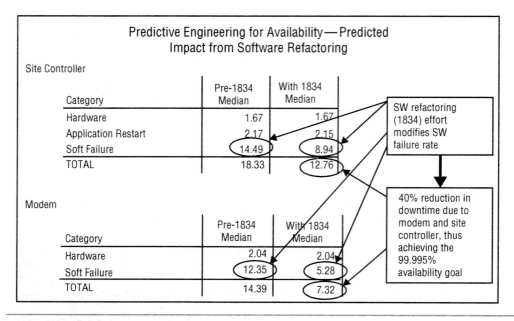

Figure 17.23 Application of predictive engineering for availability, predicting the impact software refactoring could have on availability of the base station

The team wanted to verify that their software and hardware failure rate assumptions were "in the ball park." These failure rates needed to be measured in the customer environment with their configurations; the team used a two prong approach:

1. System test—longevity testing, simulating, and emulating the devices and other hardware using the customer load profiles, and increasing the customer loads in terms of capacity to account for a range of field conditions. Longevity testing for a base station typically takes about two weeks on actual customer hardware.
2. Gathering availability data at customer deployment, using their configuration and hardware. This is the ultimate test of software and hardware failure rates in the customer environment.

Data from these two approaches can be compared and used to improve the effectiveness of longevity testing with emulation and to develop optimal test cases to obtain customer desired availability in our labs. Comparative methods can be used to implement effective longevity testing in system test, within the company, to have high confidence that the customer deployment will involve acceptable availability results.

Verification: Software Testing Combined with DFSS Techniques

Software testing is aimed at verifying features, behaviors, or functions to determine whether the customer critical requirements (critical parameters) and required results are satisfactorily met. Some common types and terms used in software testing are included in the **glossary of common software testing terms** at the end of this chapter.

The purposes of software testing can include reducing defects before moving to the next lifecycle phase, validation, soak, or reliability testing. Although software testing is sometimes considered something of "an art," and oftentimes considered very complex, test plans and test cases can be methodically developed to reduce complexity and increase confidence in the functionality and performance of the software. Software testing can be a dynamic assessment, in which sample input is run through software programs, and the actual outcome is compared with the expected outcome.

Software testing typically includes:

- White box or unit testing for particular functions or code modules that involves knowledge of the internal logic of the code.
- Black box or functional testing for requirements and functionality that does not require knowledge of the internal logic of the code.
- Performance, stress, or load testing of an application under heavy loads, such as testing a Web site under a range of loads to determine at what point the system's response time degrades or fails.

- Regression testing after fixes or modifications of the software or its environment.
- Security testing to determine how well the system protects against unauthorized internal or external access, or willful damage.

End-to-end testing is based on meeting overall requirements specification in a situation that mimics real-world use, such as interacting with a database, using network communications, or interacting with other hardware, applications, or systems in a complete application environment. End-to-end testing can be represented by phases or combinations of these approaches, such as a front-end phase of development testing that includes unit testing, feature testing, and build/integration testing, and a back-end phase of testing that includes inter-operability testing (IOT, which checks functionality, appropriate communication, and compatibility with other systems with which it interfaces), system testing (black box testing based on overall requirements specifications for all parts of a system), field testing, and customer validation testing, as illustrated in Figure 18.1.

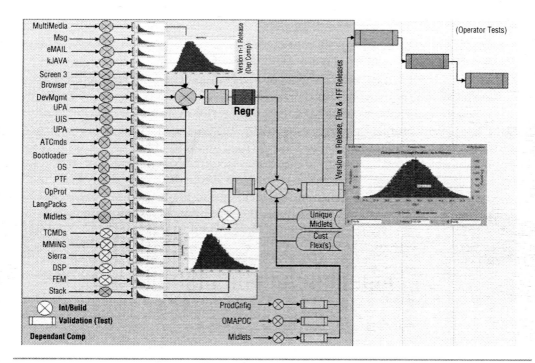

Figure 18.1 Example of end-to-end software testing for software development process from a group within Motorola, along with modeling of the associated cycle times through various stages

This end-to-end testing approach incorporates software testing methods that are compatible with DFSS principles.

Traditionally, the degree to which a program is exercised by a test suite can be assessed through coverage criteria:

- Function coverage
  - Has each function in the program been executed?

- Statement coverage
  - Has each line of the source code been covered? A set of test cases is generated to ensure every source language statement in the program is executed at least once, forcing execution of every statement. However, it often is infeasible to execute every statement, and there might be issues regarding the testing of execution sequences and branching conditions in the program.

- Branch or condition coverage
  - Has each branch or condition (such as true/false decisions) been executed and covered? This coverage focuses on the number of equivalent branches or decisions.

- Path coverage
  - Has every possible route through a given part of the code been executed?

- Entry/exit coverage
  - Has every possible call and return of the function been executed?

There are five key principals of testing that are found useful when developing test cases:

1. **Define the expected output or result.**
   More often than not, the tester approaches a test case without a set of predefined and expected results. The danger lies in getting what you expected and the tester seeing what he/she wants to see. Without knowing the expected result, erroneous output can easily be overlooked. This problem can be avoided by carefully predefining all expected results for each of the test cases.

2. **Don't test your own programs.**
   Programming is a constructive activity. To suddenly reverse constructive thinking and begin the destructive process of testing is a difficult task. It is likely that the programmer will be "without sight" to his/her own mistakes.

3. **Inspect the results of each test completely.**
   As obvious as it sounds, this simple principle is often overlooked. In many test cases, an after-the-fact review of earlier test results shows that errors were present but overlooked because no one took the time to study the results.

4. **Include test cases for invalid or unexpected conditions.**
   Programs already in production often cause errors when used in some new or novel fashion. This stems from the natural tendency to concentrate on valid and expected input conditions during a testing cycle.

5. **Test the program to see if it does what it is supposed to do, and doesn't do what it is not supposed to do.**
   A thorough examination of data structures, entry and exit criteria, and other output can often show that a program is doing what is not supposed to do and therefore still contains errors.

SOFTWARE VERIFICATION TEST STRATEGY USING SIX SIGMA

Verification of software functionality and robustness to use cases is aligned with the Verify phase of the RADIOV process, corresponding to the Verify phase of the DMADV and CDOV processes for DFSS. With the software testing strategy, the team removes defects through reviews or inspections before verification and validation. Removal of defects in software through reviews or debugging is closely associated with testing, but is a correction process, not an assessment process.[1] Testing is used to find defects, rather than to prove that the program works.

Example:
The requirement: the monitor software must accept input in the range of [1–10].
Test to prove it works: try the value 5.
Test to find bugs: try the values 0, 1, 3, 15, 60.

With DFSS, verification links design input to design output and validation links the user or customer needs to the system (see Figure 18.2).

Think of the software module to be tested as a transfer function and design the tests accordingly. The software module has one or more inputs and one or more outputs; the transfer function converts the inputs to the outputs. The software development team must develop a system test plan for verification that the functional and non-functional requirements are met. The process of creating a plan for executing test cases helps ensure that the team will:

- Thoroughly test the critical parameters and transfer functions under appropriate operating conditions and with an appropriate number of samples
- Look for and address problems and opportunities that surface

---

1. Marciniak, *Encyclopedia of Software Engineering,* John Wiley & Sons, 1994.

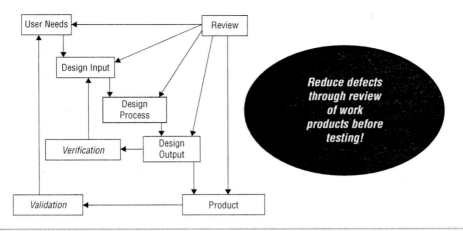

Figure 18.2 Reducing defects through reviews

- Use resources efficiently
- Verify the validity and completeness of test cases using an inspection process for correctness and thoroughness
- Consider the risks associated with the design

The following questions should be asked during system verification and validation planning:

1. Are the critical parameters adequately specified?
2. Has the system verification plan been developed? (Early in the project)
3. Have program checkpoints been created for testing your critical parameters during the project life cycle? What if the plan is not on track?
4. Have the appropriate tools been purchased?
5. Has the team been trained to use the tools?
6. Is a configuration and build plan in place?
7. Has a regression test plan been put in place for the system?
8. What testing will be automated?
9. Do you have a resource plan to match the testing/test tools maintenance needs?

Risk management planning considers the risks associated with the design; risk identification and abatement activities occur throughout the project timeline to:

- Help document and objectively prioritize critical parameters
- Help clarify and manage what is known
- Minimize the impact of what is not yet known.

Risk assessment consists of three basic steps:

1. Classify risks into related groups.
2. Evaluate attributes for each risk: probability, impact, and time frame.
3. Prioritize the risks and identify those that are the most critical.

Roles and responsibilities of risk assessment:

Project manager (PM)—The PM has overall responsibility for ensuring risk management occurs on the project, and directs or approves risk management planning efforts and all substantive changes to the risk control plans. The SPM assumes overall responsibility for all risk management actions associated with risks assigned to the software project.

Development manager (DM)—The DM coordinates risk management planning to ensure that risks are identified, risk assessment data is prepared, and control plans are complete.

In addition, the development manager:

- Coordinates or performs risk status and report updates
- Prepares and maintains the risk plan
- Evaluates risk control action effectiveness and recommends control action changes
- Coordinates the continuous review of the risk plan to keep the set of risk issues and associated risk assessment data and control plans current with project conditions

Risk owner/green belt/black belt—The risk owner/GB/BB is the individual who was identified by the DM/PM as the responsible individual for overseeing the risk.

- Identifies and assesses risks to produce probability and impact information
- Develops risk avoidance (mitigation) and contingency plans
- Provides status data for respective risk issues and assists in evaluating risk control action/effectiveness
- Documents threshold criteria for high and medium risks
- Supports identification of new risks

PM and risk owner presession preparation:

- Identify participants
- Clarify/define management expectations for project
  - Project mission/objectives and schedule/milestones
- Gather existing data and record issues on Post-it

- Discuss major risk categories
- Schedule time and location
- Communicate to participants (time, location, agenda/purpose)
- Risk owner gives an overview of the FMEA process (see FMEA example in Chapter 17) that will be used for identification of potential risk, causes, and effect.

FMEAs reduce the risk of potential failures of the system (hardware or software) by identifying failure severity, likelihood of occurrence, and ease of detection. The objective of a FMEA is to look at all of the ways a product, process, or software can fail, analyze risks, and take action where warranted (see Figure 18.3).

Task Failure Modes	Fails to execute
	Executes incompletely
	Output incorrect
	Incorrect timing—too early, too late, slow, etc.
	Shared memory region corruption
Thread Failure Modes	Fails to execute
	Not scheduled to run
	Executes incompletely
	Output incorrect
	Incorrect timing—too early, too late, slow, etc.
	Blocked on semaphore
Message Failure Modes	Queue full
	Data corruption in queue
	Failure to write to queue
	Queue disappears
Mutual Exclusion Failure Modes	Wait lock (contention)
	Priority inversion on lock
	Incorrect process/thread synchronization
	Deadlock
	Starvation
	Wait while hold (holding one lock and waiting for another)
System Failure Modes	Input value incorrect (logically complete set)
	Output value corrupted (logically complete set)
	Blocked interrupt
	Incorrect interrupt return (priority, failure to return)
	Priority errors
	Resource conflict (logically complete set)
	OS failure

Figure 18.3 List of common software failures

The team can use a P-diagram to find potential error states, and then use the P-diagram as input to the FMEA. The P-diagram also summarizes noises, some of which might be considered for stress conditions for testing purposes. Examples of P-diagrams were provided in Chapters 8, 10, and 11. After completing the FMEA, the team should then develop the software test cases using the failure modes found in the FMEA and common software failure modes.

CONTROLLING SOFTWARE TEST CASE DEVELOPMENT THROUGH DESIGN PATTERNS

Software design patterns may be leveraged to expedite the testing process in a high maturity organization's development process. A software design pattern systematically names, motivates, and explains a general design that addresses a recurring design problem in object-oriented systems. It describes the problem, the solution, when to apply the solution, and its consequences. It also gives implementation hints and examples. The solution is a general arrangement of objects and classes that solve the problem. The solution is customized and implemented to solve the problem in a particular context.

Each pattern describes a problem that occurs over and over again in the environment and then describes the core of the solution to that problem so that the solution can be reused many times. The 23 patterns defined in *Design Patterns: Elements of Reusable Object-Oriented Software*[2] are categorized as creational, structural, and behavioral design patterns, and are known as the Gang of Four, or GoF (Figure 18.4).

Creational design patterns abstract the instantiation process. They help make a system independent of how its objects are created, composed, and represented. A class creational pattern uses inheritance to vary the class that's instantiated, whereas an object creational pattern will delegate instantiation to another object.

Structural patterns are concerned with how classes and objects are composed to form larger structures. Structural class patterns use inheritance to compose interfaces or implementations.

Behavioral patterns are concerned with algorithms and the assignment of responsibilities between objects. Behavioral patterns describe not just patterns of objects or classes but also the patterns of communication between them. These patterns characterize a complex control flow that's difficult to follow at run-time.

---

2. Erich Gamma et al., *Design Patterns: Elements of Reusable Object-Oriented Software*, Addison-Wesley Professional, 1994.

		Purpose		
		Creational	**Structural**	**Behavioral**
Scope	**Class**	Factory Method (107)	Adapter (139)	Interpreter (243) Template Method (325)
	Object	Abstract Factory (87) Builder (97) Prototype (117) Singleton (127)	Adapter (139) Bridge (151) Composite (163) Decorator (175) Façade (185) Proxy (207)	Chain of Responsibility (223) Command (233) Iterator (257) Mediator (273) Memento (283) Flyweight (195) Observer (293) State (305) Strategy (315) Visitor (331)

Figure 18.4 Gang of Four Design Patterns

Several design patterns provide ways for new functionality to more easily and more feasibly interact with existing or legacy software.

For example, new features may require access to capabilities spread throughout the existing and/or legacy code; the façade pattern can be used to provide a simpler, unified interface (hiding the complexities of where these supporting capabilities appear in the legacy base) to a set of classes or some subset of a system that increases testability.

A test strategy designed to access subsystem interface through the façade pattern can be even more useful than more traditional testing strategy. Intuitively, we can see that the application of the façade pattern will reduce the number of test cases needed to achieve a similar degree of method coverage. In most cases we will be able to invoke a collection of methods through a single call to the façade.

As an example, Figure 18.5 shows a façade procedure *calculator* that contains two subprocedures *add* and *sub.* Depending on the value of *mode,* the procedure *calculator* will call either procedure *add* or procedure *sub* to perform the requested calculation.

Another example of using design pattern to improve test cases is an adapter pattern. CAL (CORBA abstraction layer) is an adapter pattern applied to the problem of making an application independent of the CORBA implementation. However, the adapter pattern will probably add at least a level of indirection to the design. The trade-off between future flexibility and present complexity has to be considered carefully.

Design patterns can be valuable aids in understanding the very complex relationships that exist within a software system, and can make it easier to generate test cases for detecting failures.

Procedure Calculator

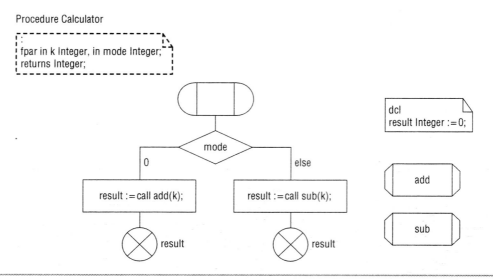

Figure 18.5 A façade procedure: providing a unified interface to two subprocedures *add* and *sub*

IMPROVING SOFTWARE VERIFICATION TESTING USING COMBINATORIAL DESIGN METHODS

The combinatorial design method, a relatively new software testing approach, can reduce the number of tests needed to verify the system functions. Combinatorial designs[3] are mathematical constructions. At Motorola, we have developed a system tool called MotoOATSGen, (orthogonal-array based testing), which uses combinatorial design theory to generate tests. Several of the Motorola Mobile Device teams are using the OATS system for system and interoperability testing.

All-pairs testing or **pairwise testing** is a combinatorial software testing method that, for each pair of input parameters to a system (typically, a software algorithm), tests all possible discrete combinations of those parameters. Using carefully chosen test vectors, this can be done much faster than an exhaustive search of all combinations of all parameters, by "parallelizing" the tests of parameter pairs. The number of tests is typically $O(nm)$, where n and m are the number of possibilities for each of the two parameters with the most choices.[4]

Combinatorial optimization algorithms solve instances of problems that are believed to be hard in general, by exploring the usually-large solution space of these instances.

---

3. M. Hall Jr., *Combinatorial Theory,* Wiley Interscience, New York, 1986.

4. http://en.wikipedia.org/wiki/All-pairs_testing.

Table 18.1

User Interfaces	Carrier Network	Browser	Display Type
User Interface 1	Verizon	Opera	WVXGA
User Interface 2	AT & T	Motorola	VGA
User Interface 3	T-Mobile	Netscape	VSGA

Combinatorial optimization algorithms achieve this by reducing the effective size of the space, and by exploring the space efficiently.

For example, Table 18.1 helps organize how to test three new user interfaces for cellular phones.

With all combinations the total would be $3^4 = 81$; but, the team considered 81 configurations to be way too many test cases! Pairwise testing would include all pairs of interactions between test factors and provides a small set of test cases, which is a practical alternative to testing all combinations.

There are now four test configurations factors with three values. Replacing numbers in the orthogonal array $(9;3^4)$ reduces the set to needing just nine configurations to cover all 54 pairs.

There are several academic and industry papers on this subject, including some texts,[5] which can be used for an in-depth study and understanding of combinatorial testing.

Table 18.2

User Interfaces	Carrier Network	Browser	Display Type
0 User Interface 1	0 Verizon	0 Opera	0 WVXGA
1 User Interface 2	1 AT & T	1 Motorola	1 VGA
2 User Interface 3	2 T-Mobile	2 Netscape	2 VSGA

---

5. Some books that discuss combinatorial testing include:

Alan Page, Ken Johnston, and Bi Rollison, *How We Test Software at Microsoft,* Microsoft Press, 2008.

Alexandre Petrenko et al. (eds.), Testing of Software and Communicating Systems: 19th IFIP TC 6/WG 6.1 International Conference, TestCom 2007, 7th International, Springer, 2007.

Table 18.3

User Interfaces	Carrier Network	Browser	Display Type
User Interface 1	Verizon	Opera	WVXGA
User Interface 2	AT & T	Motorola	WVXGA
User Interface 3	T-Mobile	Netscape	WVXGA
User Interface 3	Verizon	Netscape	VGA
User Interface 1	AT & T	Motorola	VGA
User Interface 2	T-Mobile	Opera	VGA
User Interface 2	Verizon	Opera	VSGA
User Interface 3	AT & T	Netscape	VSGA
User Interface 1	T-Mobile	Motorola	VSGA

For situations in which the number of input parameters or use cases exceed the capabilities of standard design of experiments-based methodologies to provide satisfactory coverage with a reasonable number of test vectors, a relatively obscure field involving supersaturated design approaches might provide alternative approaches that could be considered.[6]

SUMMARY

Software verification testing includes planning, risk assessment, strategy, and execution. Testing is essential in the development of any software system. Systems today are more

---

6. Some key articles on supersaturated designs:

D.R. Holcomb and W.M. Carlyle, "Some Notes on the Construction Methods and Evaluation of Supersaturated Designs," *Quality and Reliability Engineering International,* 18, 299–304, 2000.

W.W. Li and C.F.J. Wu, Column-Pairwise Algorithms with Applications to the Construction of Supersaturated Designs, *Technometrics,* 39(2), 171–179, 1997.

N.K. Nguyen, An Algorithmic Approach to Constructing Supersaturated Designs, *Technometrics,* 38(1), 69–73, 1996.

complex and the possibility of producing thousands of test cases is more common today then ever before, requiring a more efficient and effective way to complete software verification. Using DFSS tools such as FMEA and combinatorial testing coupled with methodologies such as designed patterns aids in the reduction of test cases and increases the effectiveness of software verification.

BIBLIOGRAPHY

William C. Hetzel, *The Complete Guide to Software Testing, 2nd ed.* Wellesley, MA: QED Information Sciences, 1988.

Gamma, Helm, Johnson, and Vlissides, *Design Patterns: Elements of Reusable Object-Oriented Software*, Addison-Wesley, 1995.

Encyclopedia of Software Engineering, Volume 2, Wiley Inter-science, 1994

GLOSSARY OF COMMON SOFTWARE TESTING TERMS

- **Acceptance testing**—final testing based on specifications of the end-user or customer, or based on use by end-users/customers over some limited period of time.
- **Ad-hoc testing**—similar to exploratory testing, but often taken to mean that the testers have significant understanding of the software before testing it.
- **Alpha testing**—testing of an application when development is nearing completion; minor design changes may still be made as a result of such testing. Typically done by end-users or others, not by programmers or testers.
- **Beta testing**—testing when development and testing are essentially completed and final bugs and problems need to be found before final release. Typically done by end-users or others, not by programmers or testers.
- **Black box testing**—testing based on requirements and functionality (rather than knowledge of the internal design of code).
- **Comparison testing**—comparing software weaknesses and strengths to competing products.
- **Compatibility testing**—testing how well software performs in a particular hardware/software/operating system/network/etc. environment.
- **Context-driven testing**—testing driven by an understanding of the environment, culture, and intended use of software. For example, the testing approach for life-critical medical equipment software would be completely different than that for a low-cost computer game.
- **End-to-end testing**—similar to system testing; the "macro" end of the test scale; involves testing of a complete application environment in a situation that mimics

real-world use, such as interacting with a database, using network communications, or interacting with other hardware, applications, or systems if appropriate.

- **Exploratory testing**—often taken to mean a creative, informal software test that is not based on formal test plans or test cases; testers may be learning the software as they test it.
- **Failover testing**—typically used interchangeably with "recovery testing."
- **Field testing**—testing the complete software in situations similar to actual customer use.
- **Functional testing**—black box type testing geared to functional requirements of an application; this type of testing should be done by testers. This doesn't mean that the programmers shouldn't check that their code works before releasing it (which of course applies to any stage of testing).
- **Incremental integration testing**—continuous testing of an application as new functionality is added; requires that various aspects of an application's functionality be independent enough to work separately before all parts of the program are completed, or that test drivers be developed as needed; done by programmers or by testers.
- **Install/uninstall testing**—testing of full, partial, or upgrade install/uninstall processes.
- **Integration testing**—testing of combined parts of an application to determine if they function together correctly. The "parts" can be code modules, individual applications, client and server applications on a network, and so on. This type of testing is especially relevant to client/server and distributed systems.
- **Load testing**—testing an application under heavy loads, such as testing of a Web site under a range of loads to determine at what point the system's response time degrades or fails.
- **Mutation testing**—a method for determining if a set of test data or test cases is useful, by deliberately introducing various code changes ("bugs") and retesting with the original test data/cases to determine if the "bugs" are detected. Proper implementation requires large computational resources.
- **Orthogonal array testing**—tests pairwise combinations of object interactions (instead of all possible combinations). Allows estimation of the average effects of a control factor (feature) without the results being distorted by the effects of other factors.
- **Performance testing**—term often used interchangeably with "stress" and "load" testing. Ideally, "performance" testing (and any other "type" of testing) is defined in requirements documentation or QA or test plans.
- **Recovery testing**—testing how well a system recovers from crashes, hardware failures, or other catastrophic problems.
- **Regression testing**—retesting after fixes or modifications of the software or its environment. It can be difficult to determine how much retesting is needed, especially near the end of the development cycle. Automated testing tools can be especially useful for this type of testing.

- **Sanity testing or smoke testing**—typically, an initial testing effort to determine if a new software version is performing well enough to accept it for a major testing effort. For example, if the new software is crashing systems every five minutes, bogging down systems to a crawl, or corrupting databases, the software may not be in a "sane" enough condition to warrant further testing in its current state.
- **Security testing**—testing how well the system protects against unauthorized internal or external access, willful damage, and so on; may require sophisticated testing techniques.
- **Stress testing**—term often used interchangeably with "load" and "performance" testing. Also used to describe such tests as system functional testing while under unusually heavy loads, heavy repetition of certain actions or inputs, input of large numerical values, large complex queries to a database system, and so on.
- **System testing**—black box type testing that is based on overall requirements specifications; covers all combined parts of a system.
- **Thread-based testing**—integrate and test minimal sets of classes that are necessary to respond to a single event.
- **Unit testing**—the "micro" scale of testing, intended to test particular functions or code modules. Unit testing is typically performed by the programmer (rather than by testers), because it requires detailed knowledge of the internal program design and code. A well-designed architecture with tight code facilitates unit testing; otherwise, it can be very difficult. Unit testing may require developing test driver modules or test harnesses.
- **Usability testing**—testing for "user-friendliness." Clearly, this is subjective, and will depend on the targeted end-user or customer. User interviews, surveys, video recording of user sessions, and other techniques can be used. Programmers and testers are usually not appropriate as usability testers.
- **User acceptance testing**—determining if software is satisfactory to an end-user or customer.
- **Use-based (cluster) testing**—incrementally test sets of classes at successive levels of dependency (clusters).
- **Validation testing**—testing the complete software and system at the customer site.
- **Verification testing**—a short set of tests, which exercises the main functionality of the software or system.
- **White box testing**—testing based on knowledge of the internal logic of an application's code, considering coverage of code statements, branches, paths, and conditions.

Verification of Supply Chain Readiness

POSITION WITHIN DFSS FLOW

The last stage of the DFSS flow is verification of the readiness of the supply chain to deliver the product dependably—that is, to start the product launch on time to the updated schedule, to meet customer expectations in terms of quantities and timing of deliveries (allowing for some anticipated flexibility in terms of potential up side), and to have high confidence that the supply chain will be able to consistently provide products and respond quickly and acceptably in the face of perturbations and problems.

This step and set of deliverables is aligned with the latter part of the Verify phase of the DFSS process. The DFSS flowchart, available for download at http://www.sigmaexperts .com/dfss/, provides a high-level overview (Figure 19.1). Figure 19.2 is the detailed flow-chart for this chapter, showing the steps and methods involved in verifying supply chain readiness. At this stage of DFSS, the system's hardware and software requirements have converged, so the steps shown in Figure 19.2 would be appropriate for a wide variety of products, including but not limited to complex electronic systems.

The key deliverables for this phase are:

1. Verification and associated confidence that the manufacturers in the supply chain will meet their tolerance expectations as they launch and continue production.
2. Verification and confidence that the components will be correctly assembled.
3. Verification that the interfaces will meet expectations in terms of information flow, energy flow, material flow, and thermal matching.

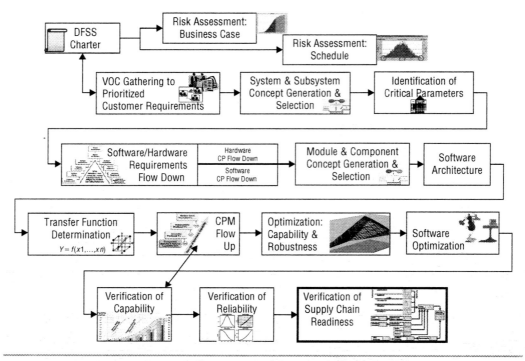

Figure 19.1 DFSS flowchart overview highlighting step for verification of supply chain readiness

4. Verification and confidence in the launch schedule, including meeting qualification, availability of equipment and resources, and meeting logistical requirements to manufacture and deliver according to the project schedule for the launch.
5. Verification and confidence in the ongoing delivery capabilities, allowing for variability in yields and cycle times and providing a reasonable flexibility to respond to potential upsides in orders—lest the supply chain constrain sales and profits!

Although verification of supply chain readiness is the last part of the Verify phase in terms of sequence in the DFSS process, the steps involved can and should be initiated much earlier—almost immediately after the critical parameters have been flowed down to the modules and components and the associated tolerances allocated.

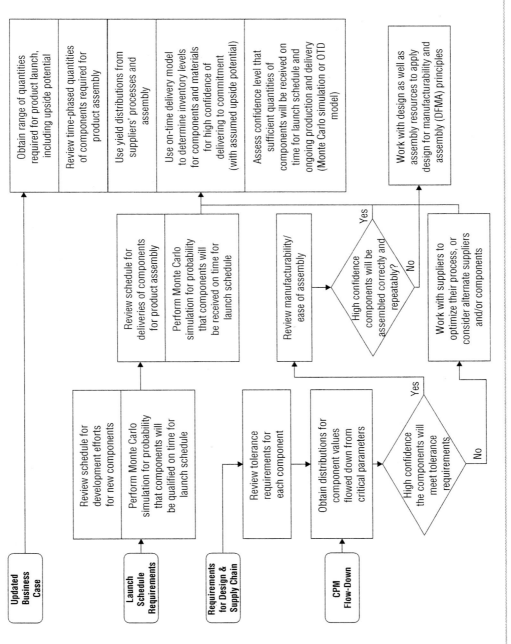

Figure 19.2 DFSS flowchart, drilled down to detailed flowchart for verification of supply chain readiness

VERIFICATION THAT TOLERANCE EXPECTATIONS WILL BE MET

Electronic systems require both electronic components and mechanical components to meet expectations for form, fit, and function. In the hardware requirements flow-down (Chapter 10), tolerance expectations initially were allocated to each component involved in the critical parameters, and they were adjusted based on ongoing feedback with the supplier and information on similar components (Chapter 14). Finally, measurements from early prototypes or early production output were used to verify the capability (Chapter 16). In theory, this should be sufficient to ensure that the tolerance expectations will be met; however, it is prudent to return to the suppliers, review again the tolerance expectations for their respective components, and ensure that their recent experience and information confirms that they are confident that they can continue to meet tolerance expectations as the product ramps into full-scale production. The authors have run into occasions where this follow-up was essential for a successful product launch.

CONFIDENCE IN ROBUST PRODUCT ASSEMBLY (DFMA)

Verification that the components meet tolerance expectations is necessary but not sufficient to ensure successful form and fit associated with the product. The components must be assembled to make the product—and the assembly process introduces more ways for things to go wrong.

Design for manufacturability and assembly (DFMA) is much like robust design for the product assembly process—but understanding that the assembly process does not exist in a vacuum. Robust design for product assembly should be integrated with the product design, starting with the selection of concept for the product and iterated with the selection of concepts for subsystems, subassemblies, and modules. Particular attention should be paid to the mechanical and material aspects of the product design, as optimization for robustness of product assembly should strongly influence mechanical concept and material selection.

The intent of DFMA is to reduce the sources of variation and the opportunities for errors during the assembly process. DFMA can be considered both a philosophy and a set of general best practices for easy of assembly and best practices specific for the types of materials, connectors, and assembly processes. Some key best practices associated with DFMA are summarized in Table 19.1; a version of this table with descriptive images for each aspect can be found at http://www.sigmaexperts.com/dfss/. The last suggestion in Table 19.1 would be an example of a best practice specific to injection molding. The philosophy and general best practices include efforts toward minimizing the number of parts that must be assembled—considering that more parts represent more

Table 19.1 Some key aspects and practices associated with design for manufacturability and assembly

Aspect	DFMA Suggestion	Comments
	Keep holes away from corners, edges, or bends	Improve reliability
Holes	Use a simple drilling pattern	Holes parallel or perpendicular to reference
	Avoid long, narrow holes	. . .unless drilling for oil
	Avoid driling at an angle	. . .unless your neighbor struck oil
Alignment	Use Alignment features, self-aligning approaches	Ease of alignment → ease of assembly
	minimize resistance to insertion	Ease of insertion → ease of assembly
Insertion	Use axial and rotational symmetry if possible	Eliminate impact of wrong side insertion
	If symmetry impossible, make assymmetry painfully obvious	Reduce opportunity for error
	Reduce number of parts and number of varieties of parts	Reduced part count → higher reliability, lower opportunity for error, faster assembly, lower cost
	Use a single piece that can meet several functional requirements	Eliminate need for multi-part assemblies
Simplification	Use a single part unless multiple parts really necessary	Reasons for separate parts include: requiring motion between parts, requiring different materials, requiring isolation, or requiring separate parts to avoid impact to the assembly of other parts.
	Color code if different parts shaped similarly	Facilitate Poka Yoke approaches
	Standard Components	Lowest cost, lower risk
	Standard Dimensions	Lower opportunity for errors

continued

Table 19.1 Some key aspects and practices associated with design for manufacturability and assembly

Aspect	DFMA Suggestion	Comments
Simplification	Minimize machining by pre-shaping (ex: casting, extruding)	Lower cost, lower risk
Storage	Shape to prevent tangling, jamming when stored	Lower opportunity for errors
	Restrict sticks and grooves to one surface	Lower opportunity for errors
Surfaces and sides	Consistent thickness	Lower opportunity for errors
	Avoid thin walls that can break	Improve reliability
	Avoid long thin sections that can vibrate	. . .unless an integrated tuning fork is a key "Feature"
	Avoid sharp corners	improve reliability
	Allow access for tools during assembly, machining, etc.	Don't bury important components
Injection Molding	Minimize section thicknesses	Cooling time is proportional to square of thickness

opportunities for error and impacts cost, assembly time, and reliability. These general best practices also include efforts toward simplification of orientation and alignment of parts that are to be attached or connected. These and other DFMA approaches are described in much greater detail in books focused on DFMA, DFA, and DFM.[1]

---

1. James Bralla, *Design for Manufacturability Handbook,* McGraw-Hill Professional, 1998.

 Geoffrey Boothroyd, Winston Knight, and Peter Dewhurst, *Product Design for Manufacture & Assembly, Revised and Expanded,* CRC, 2001.

 Corrado Poli, *Design for Manufacturing: A Structured Approach,* Butterworth-Heinemann, 2001.

 Daniel Whitney, *Mechanical Assemblies: Their Design, Manufacture, and Role in Product Development,* Oxford University Press, 2004.

Some best practices for DFMA apply specifically for certain types of material and assembly processes. Some may be specific for the selected assembly site; these best practices might be documented in a design manual or design guide developed and provided by the assembly site. Ideally, these best practices could be incorporated into the CAD environment used by the design engineers, avoiding violations of design rules and subtly and automatically encouraging best practices.

The approaches of DFMA set the stage for successful assembly. Mistake-proofing approaches (Poka Yoke) and statistical process control approaches in the assembly facility supplement the up-front DFMA approaches to help ensure robust and successful assembly.

VERIFICATION OF APPROPRIATE AND ACCEPTABLE INTERFACE FLOWS

As mentioned in earlier chapters, things often go wrong at the interfaces between subsystems and modules as well as at the man-machine interfaces. The approaches described in this book should mitigate those risks but, as the saying goes, "Trust, but verify."

The key aspects of interfaces are risks to the flows between the subsystems or modules or between systems and the user. These interfaces and flows should have been included in the DFMEAs, or perhaps a DFMEA session focused on interfaces would have been held to identify, prioritize, and mitigate these risks. The information, material, and energy flows at the interfaces can be analyzed on prototypes and early production units, to verify that the flows are adequate, consistent, reliable, and dependable. The key risks might involve the information flow; for electronic systems, these can be the flows of information in analog or digital form. Stress testing of the interface functionality in transferring information (in which the stress corresponds to the presence of excessive noise to simulate extremes of use cases and use environments) can verify that the flows are reliable and will be dependable in actual usage. Similar stress testing can verify electrical energy flow and thermal matching between the subsystems or modules that are interfaced. Different stress conditions may be required to verify dependable material flow, such as air flow in a computer system or base station or paper and ink flows in a printer. Testing under stress conditions was discussed in Chapter 17.

CONFIDENCE IN THE PRODUCT LAUNCH SCHEDULE

Product launches often seem to be the "happy proving ground" for Murphy's Law: Anything that can go wrong, will go wrong. Consequently, the best counteraction is to anticipate what can go wrong, consider and implement preventative actions where feasible, and develop an array of contingencies.

The launch of a new product in the form of an electronic system might be dependent on the development of new software, new integrated circuits, or a new technology. As discussed in Chapter 6, each of these involve high risks for the project schedule—but, beyond the project schedule, these also involve risks to the launch schedule as a result of uncertainties as to what might have been missed for something that is new and therefore unproven in the field.

Successful product launches generally use some variation on this approach:

- Develop a clear set of criteria defining a "successful launch." Motorola adopted the term "flawless launch" for its set of criteria. These criteria should be leading indicators, since lagging indicators such as field returns or warranty returns occur too late.
- Starting from the success criteria, work backward, defining what actions, criteria, and events are necessary to achieve the successful launch. At each stage, probe with the team whether each new action, criteria, and event truly is necessary, and if the set of actions, criteria, and events, together, are sufficient to ensure that the corresponding success criterion will be met.
- Use the resulting project plan for the launch as the roadmap to ensure success.
- Supplement this project plan with a Product Launch FMEA to brainstorm any other risks that might have been missed.

This approach is appropriate for the systems team, the test team, the hardware team, the software team, the supply chain team, and the marketing team. Business success for a new product is often at least as dependent on the success of the marketing efforts as the success of the engineering efforts.

CONFIDENCE IN MEETING ON-TIME DELIVERY AND LEAD-TIME COMMITMENTS[2]

Product launch is not the end of the story—there truly is life after launch. Verification of supply chain readiness involves preparation for an effective handoff to the supply chain for product delivery. This preparation entails special challenges for electronic systems that naturally require electronic components. This obvious point leads to a key insight: when you embed electronic components such as integrated circuits in your product, you also embed the highly complex, long cycle time semiconductor manufacturing process into your supply chain.

---

2. Eric C. Maass, "Modeling the On Time Delivery and Inventory for Semiconductor Supply Chains" (Ph.D. Dissertation, Arizona State University, 2008).

Considering the supply chain of an electronic system as a whole and intentionally ignoring the lines between companies and organizations, the semiconductor manufacturing processes dominate the supply chain in terms of cycle time and inventory. From this perspective, managing the supply chain involves decisions and approaches that comprehend the impact of the complexity of semiconductor manufacturing, without getting enmeshed in the details.

Semiconductor manufacturing includes wafer fabrication, wafer probe, shipment to an assembly site, assembly, final test, and shipment to the customer (Figure 19.3). Wafer fabrication is a complex flow involving hundreds of operations, complex and high-maintenance equipment, reentrant flows, and highly variable yields. It is typically the longest part of the manufacturing process, with variable cycle times on the order of

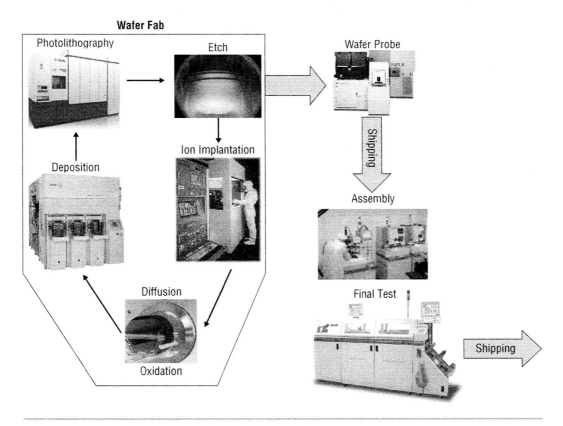

Figure 19.3 Semiconductor manufacturing: wafer fab, wafer probe, assembly, final test, and shipment

weeks or months. Without a strategic buffering approach, these cycle times would inflate lead times accordingly and add considerable uncertainty regarding the ability to meet delivery commitments.

Supply chain decisions involve trade-offs among on-time delivery, lead time, and inventory levels and associated costs. A cohesive set of models implemented in a spreadsheet has been used for "what if" analyses and as a preprocessor for providing inputs into the production and inventory control systems. This model includes a statistical model for semiconductor yields and cycle time that are incorporated into a model for on-time delivery.

Aspects of uncertainty that must be comprehended to confidently plan for delivery commitments are the uncertainties of demand, yield, and cycle time. In some cases, the enterprise providing electronic products might have a clear idea of the range of unit demand over the given period of time. In other cases, there might be considerable uncertainty, with no more than a "most likely" guess. Similarly, supply chain information might consist of "most likely" estimates for the yields and cycle times of the components. This subsection provides approaches to estimate distributions for these aspects using generalizations based on analyses of historical data.

Regarding demand uncertainty, the risks associated with overestimation of demand involves building excess inventory that hopefully can be sold later but might need to be discarded in the worst-case scenario. The risks associated with underestimating demand involves missed opportunities, perhaps delaying shipments to customers or losing those potential orders in the worst-case scenario. Unit demands over time periods can be assumed to be characterized with two parameters, such as the most likely and a standard deviation or a CV (coefficient of variation, the ratio of the standard deviation to the most likely). Although it is very possible that marketing can and will offer a best guess for the most likely unit sales for the product, it is a very rare marketer who can estimate the standard deviation or coefficient of variation for unit sales.

Assuming that unit sales data from electronic component suppliers reflects the demand for the electronic products that use these electronic components, historical data for unit sales of electronic components might provide useful rules-of-thumb for the CV for a business producing electronic systems. Historical data indicates that the coefficient of variation for monthly semiconductor demand is 25 percent or more;[3,4] Figure 19.4

3. R.O Roundy and M. Cakanyildirim, "SeDFAM: Semiconductor Demand Forecast Accuracy Model, *IEE Transactions on Scheduling and Logistics,* Vol. 34, No. 5, pp. 449–465, 2002.

4. V. Tardif and M. M. Nielsen, "Understanding Forecast Variability in Contract Electronics Manufacturing," Journal of Electronics Manufacturing, Vol. 11, No. 2, 2002.

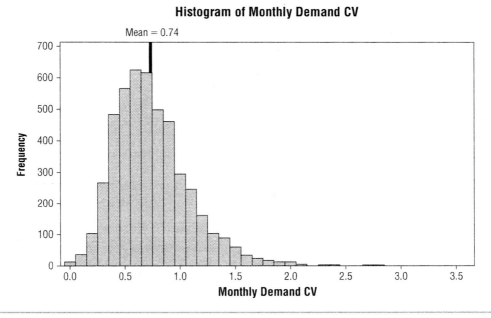

Figure 19.4 Distribution of coefficients of variation for month-to-month demand for 5,000 semiconductor products

shows the distribution of the coefficient variation for monthly demands for about 5,000 semiconductor products, indicating that CV for semiconductor components typically ranges from 25 to 100 percent.

If the underlying distribution for unit demand is lognormal, than the range of +/– one standard deviation for a normal distribution, which represents about two-thirds of the area under the normal curve, would correspond to multiplying or dividing by one plus the coefficient of variation. In other words, a coefficient of variation of 100 percent would indicate that about two-thirds of the distribution of unit demands would lie within the range of half to double the most likely unit demand. Similarly, a coefficient of variation of 50 percent would correspond to two-thirds of the distribution lying between .67 and 1.5 times the most likely unit demand, and a coefficient of variation of 30 percent would relate to the range of .77 to 1.3 times the most likely unit demand.

Yields of electronic components tend to follow a beta distribution, which is relevant for parameters confined to the range of 0 to 100 percent. Semiempirical studies have shown that the standard deviations of the various types of yields—manufacturing mechanical yield prior to assembly (as a result of mechanisms like breakage), electrical yield, and mechanical yield after assembly—tend to have predictable relationships between the mean and standard deviation of the yield. This is convenient, as suppliers and manufacturing and assembly sites typically have estimates for each of these yields but don't have estimates of the standard deviation of these yields. Relationships between the standard deviations and yields for each of these types of yield are shown in Figure 19.5.

Cycle times of electronic components tend to follow a three-parameter gamma distribution,[5] which is relevant for parameters with a lower boundary—corresponding to the theoretical minimum or value-added cycle time—but no upper bound (as many may attest, cycle times will occasionally flirt with infinity). Analysis of historical data provides rules of thumb: the theoretical cycle time is about a third of the most likely cycle

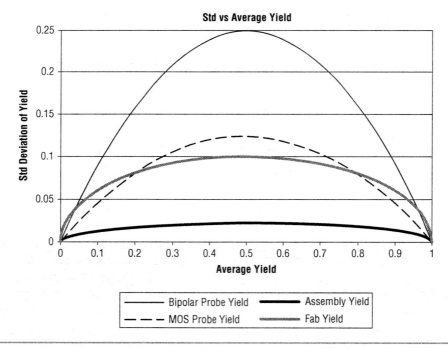

Figure 19.5 Relationships between the standard deviation and average yield for bipolar or MOS probe yields, assembly yield, and fab yield

---

5. O. Rose, "Why Do Simple Wafer Fab Models Fail in Certain Scenarios?," *Proceedings of the 2000 Winter Simulation Conference*

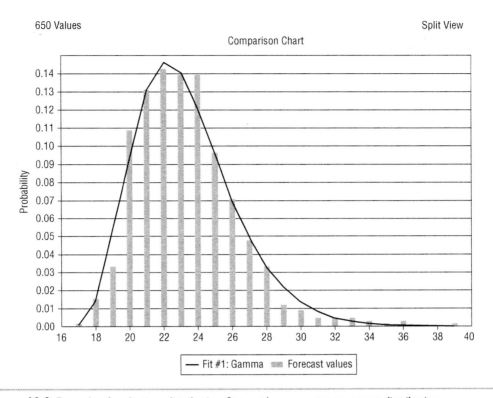

Figure 19.6 Example of cycle time distribution fit to a three-parameter gamma distribution

time, and the coefficient of variation tends to be about .133, corresponding to a value for beta of 25 (Figure 19.6).

These models for the demand and yield distributions can be used to model the inventory and associated inventory cost impact for inventory-related decisions. The models for the cycle time distributions can be used to assess the associated lead-time impact. Combined, these approaches can be used to select an inventory strategy that will support a level of confidence in consistently achieving on-time delivery at an acceptable cost. This is illustrated by Figure 19.7, which evaluated the cost of excessive inventory and the risk-related cost of insufficient delivery to find an optimal on-time delivery goal and associated strategy; the optimal goal of about 95 percent on-time delivery in Figure 19.7 is from a real application for an integrated circuit used in a cellular phone.

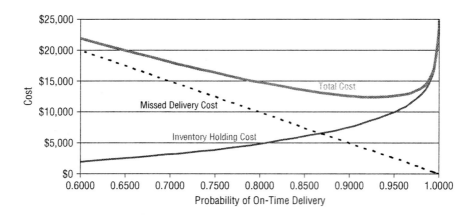

Figure 19.7 Inventory holding cost and the risk-related cost of missed delivery versus the probability of on-time delivery. An on-time delivery goal of 95 percent was found to be optimal for this integrated circuit used in CDMA cellular phones.

The classic objective of logistics suggests several key response variables: the right quantities at the right time are reflected in delivery dependability, as measured by the lead time achievable with a given probability of on-time delivery. On-time delivery contributes to profitability by improving repeat business with existing customers. Customer satisfaction improves long-term profitability by improving the probability of repeat business, and the perceived dependability of the supplier. Poor on-time delivery can significantly impact sales of the final product, and the unavailability of an inexpensive semiconductor can impact sales of products selling for hundreds or thousands of dollars and idle a factory. These risks can be ameliorated through safety stock, but with a trade-off of inventory holding costs and the risk that a redesign can render the inventory obsolete.

To assess these risks, the supply chain for a product using semiconductors can be modeled as a network of operations. The cycle time of a supply chain for a multichip module or populated printed circuit board includes the cycle time of final assembly and the cycle times of the components. The supply chain cycle time can be treated as a complex combination of the cycle times of the component processes (Figure 19.8), analogous to the series and parallel combinations of resistors (Figure 19.9), or system reliability models composed of subsystems in series and parallel. In Figure 19.8, each of the cycle time distributions have the same horizontal axis. Proceeding from left to right in the manufacturing flow, the cycle time distributions are shown as histograms representing the cycle time experienced before and during that step.

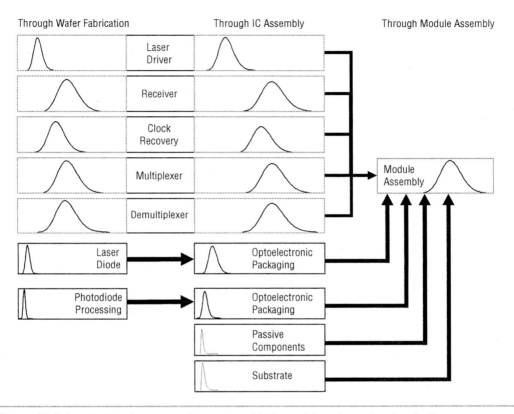

Figure 19.8 Supply chain for an optoelectronic multichip module and associated cycle times

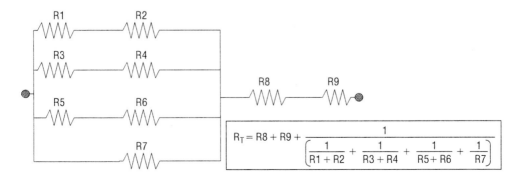

Figure 19.9 Example of a resistor network as series and parallel combination of resistors, and its associated mathematical model, as an analogy to a supply chain network

The supply chain is assumed to consist of sequential stages of manufacturing flows for each of several components, subsequently assembled into a module (Figure 19.10). Stages where semiconductor component inventory could be held include:

- metallized wafer inventory bank partway through semiconductor processing
- probed wafer inventory bank
- die inventory before assembly bank
- finished goods inventory bank
- component inventory bank before module assembly

The case with no strategic inventory for any components at any of the inventory banks is set as the baseline for combined cycle times and costs. In an ongoing manufacturing situation, there will be incidental inventory in work in progress (WIP) and stored at the various banks, as illustrated in Figure 19.10. Relatively high volume products may have sufficient incidental inventory levels to meet delivery commitments with short lead times. For lower volume products, strategic inventory levels can be set to align the inventory levels at these banks with on-time delivery and lead-time goals. The decisions are whether to hold strategic inventory for any or all components at one or more of the various banks, and what quantity is required to achieve the desired confidence in meeting commitments. These strategic inventory decisions involve a trade-off of the probability of on-time delivery, quoted lead times, and inventory costs.

Strategic inventory decisions range from minimal cost (no strategic inventory) to minimal lead time (sufficient quantities held at finished goods inventory), and are illustrated by a case study. Between these extremes, the impact on the lead time from holding strategic inventory at one or more inventory banks at a given confidence level is described by the following reasoning: if a link in the supply chain is depleted of material, the gap in the supply chain can interrupt delivery of final product to the final customer. The delay can be estimated as the cycle time between the prior inventory stage and the depleted inventory stage required to replenish the depleted inventory point. A "rush" can reduce the cycle time for this product, but will increase variability in manufacturing and impact cycle time and on-time delivery for other products. Prevention of interruption in the supply chain requires sufficient days of inventory to provide confidence that the supply chain will not perturbed by yield- or cycle-time variation.

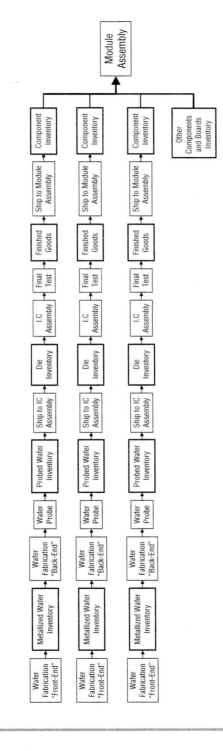

Figure 19.10 Diagram for supply chain with semiconductors assembled into a module. Stages where inventories can be managed include metallized wafer, probed wafer, die, finished goods, and component inventory banks.

CASE STUDY: OPTOELECTRONIC MULTICHIP MODULE

An idealized supply chain would behave as if all of the semiconductor manufacturing, assembly, testing, and module assembly were completed within one company. This was the case for the Optobus module, an optoelectronic multichip module (MCM) requiring several integrated circuits and optoelectronic. The supply chain involved semiconductor processes ranging from CMOS and BiCMOS to compound semiconductors, each involving a series of manufacturing steps and costs. The components were manufactured in parallel until module assembly.

Strategic inventory decisions need not be the same for all components, but the lead time is largely determined by the component with the longest effective lead time, so some sets of strategic inventory decisions would be ineffective. For example, holding all but one component in finished goods inventory could keep the lead time dependent on that component such that the cost of strategic finished goods inventory for other components would be wasted. Different strategic inventory decisions for different components in the same module could make sense if some components have much shorter cycle times. Decisions regarding strategic inventory can reduce the lead time, without resorting to rush lots, but with added cost for holding the inventory. Starting from the baseline, strategic inventory could be progressively added as safety stock for the longest cycle time component until each constraint is resolved, like providing buffer for the bottleneck (also referred to as the "Herbie" in the theory of constraints,[6] after the name of a Boy Scout who provided a key insight regarding bottlenecks, and referred to as the "time trap" in Lean Six Sigma[7]). The efficient frontier in Figure 19.11 was developed using a model based on the flowchart of Figure 19.8. The efficient frontier represents the trade-off between strategic inventory cost and lead time that can be committed with 90 percent confidence, as determined by decisions regarding strategic inventories for a set of ICs assembled in the optoelectronics module.

Although the case study applied this approach to a supply chain for an optoelectronic multichip module assembled using several semiconductor components, the models developed and applied for electronic supply chains can be generalized to supply chains involving other types of semiconductor components as well as to other types of components in general. The supply chain model consists of three sets of models, each represented by a set of equations with coefficients and exponents that

6. E. M. Goldratt and J. Cox, *The Goal, 2nd revised edition,* North River Press, 1992.

7. Michael L. George, *Lean Six Sigma,* McGraw-Hill, 2002.

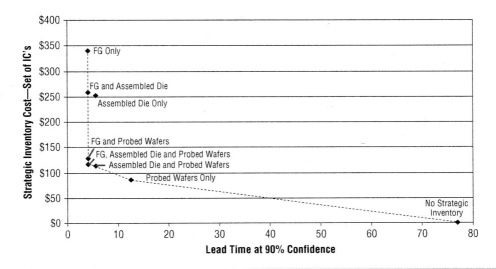

Figure 19.11 Efficient frontier representing the trade-off between strategic inventory cost and lead time that can be committed with 90 percent confidence, as determined by decisions regarding strategic inventories for a set of ICs assembled in the optoelectronics module

can be estimated through regression analyses: models for the yield distributions, models for the cycle-time distributions, and models for the combination of cycle-time distributions.

The spreadsheet involves inputs and results on one worksheet (Figure 19.12), while coefficients and exponents for the models, as described in this section, are organized on a separate worksheet (Figure 19.13), where they can be set to default values or modified based on knowledge of aspects of the supply chain. The spreadsheet facilitates customization of the model for applications, including supply chains with combinations of other semiconductors, combinations of components some of which are not semiconductors, or a supply chain consisting entirely of components that are not semiconductors. The key output from the model is a set of recommendations regarding inventories that would support the organizations' goals for on-time delivery and for expectations and commitments for lead times that can be communicated to the customers with confidence.

| Component | CYCLE TIME INFORMATION | | | | YIELD INFORMATION | | | | | |
	Wafer Fab and Probe Cycle Time	Assembly Cycle Time	Final Test Cycle Time	Admin & Transit Cycle Time	Wafer Fab Yield	PDPW (Potential Die Per Wafer)	Probe Yield	Assembly Yield	Final Test Yield	Module Assembly Yield
Multiplexer	50	2	1	3	90%	1000	90%	97%	98%	99%
Receiver	49	2	1	3	90%	1000	90%	98%	95%	99%
Clock Recovery	40	2	1	3	90%	1000	80%	97%	90%	99%
Laser Driver	20	2	1	3	90%	1000	95%	98%	97%	99%
Laser Diode	10	2	2	1	85%	10000	60%	80%	65%	99%
Demultiplexer	50	2	1	3	90%	1000	90%	97%	98%	99%
Photo Diode	5	1	1	1	85%	10000	98%	95%	97%	99%

Figure 19.12 Excel worksheet with inputs and results to faciliate the use of the model

| Component | | Cycle Time Coefficients | | | |
| | | Wafer Fab and Probe | | | |
		Mean/Min Ratio	CV (S/Mean)	% Utilizn for Basis	% Utilizn for Timeframe
Multiplexer		3	0.13	80%	80%
Receiver		3	0.13	80%	80%
Clock Recovery		3	0.13	80%	80%
Laser Driver		3	0.13	80%	80%
Laser Diode		3	0.13	80%	80%
Demultiplexer		3	0.13	80%	80%
Photo Diode		3	0.13	80%	80%

Figure 19.13 Excel worksheet, with coefficients for the models, to facilitate the use of the model

SUMMARY

Several expectations should be quantified and assured during the Verify phase to anticipate and try to mitigate risks associated before, during, and after the product launch:

- That the manufacturing and assembly processes will meet their tolerance expectations dependably
- That the assembly process has been set up for success, since the product has been designed for ease of assembly
- That risks at the interfaces have been anticipated and mitigated
- That risks associated with the launch schedule have also been mitigated

- That the supply chain is prepared for success in meeting the demand by supplying the quantity of products on time (allowing for variations in yield and cycle time), and has prepared contingencies to respond to potential upsides if the product is more successful in the market than expected

Verification of supply chain readiness provides a gentle hand-off to the people involved in sustaining and supporting the product while it is in production. It also concludes the Verify phase of the RADIOV, CDOV or DMAD(O)V process—and thereby concludes the DFSS process.

Summary and Future Directions

The goal for this book was to provide a clear roadmap and guidance for developing products, particularly high-tech products and systems. Some of these products are more complex than others, involving both software and hardware. Many chapters provided a rather detailed flowchart that linked into an overall flowchart for DFSS, and many chapters provided real case studies and examples, to try to provide the reader with exactly what he or she needs to successfully apply the appropriate tools and improve how they develop new products.

This book grew from a desire to help people help their companies effectively develop robust and compelling new products in a competitive and dynamic environment. Developing new products is high risk but, when successful, it literally is "the gift that keeps on giving." A successful product makes a big difference for the organization financially, lifts the energy and optimism among the people, and, sometimes, perhaps more often than most people may realize, can make a difference in the world.

Does that perhaps sound over the top, grandiose even? Perhaps, but consider how new products have permeated our lives over the past 30 years. Cellular phones are ubiquitous, and taken for granted—but they were first developed at Motorola in the 1980s; how many lives have been affected, how many people have felt safer and even saved lives through communication in emergency situations? How much do our police and firemen depend on communication to be effective in protecting us?

Can you imagine handling the people, the flights, the throngs of people at a busy airport without computer systems? Can you imagine your business working without computers? Would you even consider trading in your new computer and going back to the computer that you used just a few years ago, which seemed so powerful at the time?

Can you imagine a hospital without the medical systems that we have today? The nuclear magnetic resonance systems that were breakthrough chemical analysis tools a few decades ago have led to magnetic resonance imaging systems that allow precise diagnoses that were inconceivable just a few years ago.

New products drive the business, drive the economy, and affect society, often in unexpected ways.

Our goal in writing this Design for Six Sigma book, then, was to help people involved in new product development step back and consider what they might be able to do, what tools they can use, to handle the risks, to reduce the variation or impact of variation, minimize defects, and develop robust, compelling new products.

The early chapters discussed the new product development environment, and how to handle the macro-perspective: deployment, change management, risk management, recognition of people and teams, and certification based on accomplishments. The issues of business case or financial risk and schedule slippage risks were discussed along with approaches.

From there, the chapters proceeded through the DFSS flowchart, from gathering the VOC, through translation into measurable requirements that are prioritized, some of which receive special attention as critical parameters, through innovating and considering alternative architectures and concepts, through flowing down the requirements to subsystems, through optimizing and flowing up—to build a robust, reliable, and compelling new product. The flowchart in Figure 20.1 summarizes the steps and the concepts covered in this book.

FUTURE DIRECTIONS

This section provides an opportunity for the authors to make some recommendations for Design for Six Sigma leaders to consider to enhance their programs and tools, and also to look at some trends and opportunities to forecast what directions might be forthcoming. Forecasting the future is fun, fleeting, and perhaps fatally flawed. Nonetheless, in the spirit of sharing, let us suggest some possible paths toward promising product developments, in approximately the same sequence as the DFSS flowchart provides.

1. At a strategic level for the business, the CEO sets the vision, the mission statement, and the strategies. The strategy then must then be flowed down into a set of tactics, metrics, and actions that will later be assessed in terms of progress in achieving the vision. This is very analogous to critical parameter management's flow-down and flow-up, except that business metrics have replaced technical metrics; some

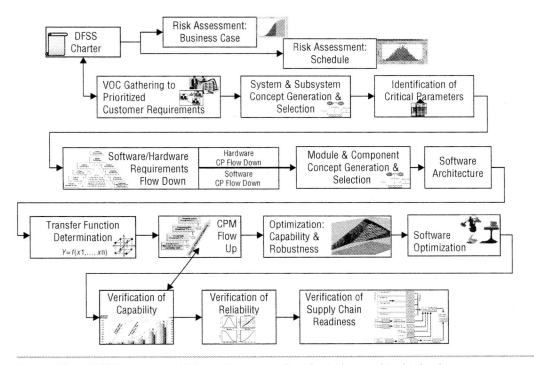

Figure 20.1 DFSS flowchart, highlighting the steps and methods discussed in this book

variation on this approach apparently is being used at some successful companies, including Toyota.[1]

2. At a product portfolio level for a product roadmap, there is an interesting integrated risk management process proposed by Dr. Johnathan Mun[2], which builds further on the methodology discussed in Chapter 5 and includes these steps:
   - Generate lists of projects and strategies to evaluate
   - Develop base case projections for each project using time-series forecasting
   - Develop static financial models using discounted cash flow models
   - Use Monte Carlo simulation for dynamic analysis

---

1. Thomas L. Jackson, Sandi Claudell, and Dave Hallowell, *Crystal Ball for Strategy Deployment,* 2007 Crystal Ball User Conference Denver, Colorado, May 22, 2007.

2. Dr. Johnathan Mun, "A Quick Primer on Risk and Decision Analysis for Everyone: Applying Monte Carlo Simulation, Real Options, Forecasting, and Portfolio Optimization," http://www.realoptionsvaluation.com/download.html.

- Frame and analyze the portfolio and individual projects with real options analysis
- Perform stochastic portfolio optimization and resource allocation

3. The portfolio of projects should not be limited to new products, but also should include some services. In some cases, the services might be linked to new products, such as the on-site installation and ongoing support of wireless base stations. Many of the DFSS methods described in this book apply well to services, especially the early stages such as gathering the voice of the customer (VOC). The optimization stage should include dynamic modeling for the timeliness of providing the services and responsiveness in general, which can be achieved with Monte Carlo simulation or discrete event simulation, each of which can be combined with appropriate lean principles.

4. Project and program management methods are powerful but may need further improvement—and, in our experience, many project and program managers are eager to consider ways to make their efforts more efficient and the projects more successful. To ensure that DFSS is institutionalized, a successful organization will need to plan, track, and monitor all of its DFSS projects. The goal would be to have every project use DFSS techniques every time, especially new product intro-ductions (NPI). By contrast, making DFSS "just how we do things" should not mean that those who put out these extra efforts should lose out on recognition for their efforts and results. Some future directions for project management will lead to creating a project portfolio management system that allows for optimized proj-ect selection and resource allocation. This will require creating an infrastructure and set of rules to ensure that projects align with the business strategic goals. The DFSS for project portfolio management optimizes the implementation of project selection, feature management, schedule management, and resource allocation while maximizing profit.

5. Integration of DFSS best practices with systems engineering: this book goes a long way toward achieving this integration, and there have been discussions about using this approach as part of university courses for systems engineering associated with engineering curricula at the graduate level. Part of the power would come from starting from the perspective of enabling an excellent customer experience, and driv-ing from there to requirements for the software and hardware subsystems to support customer experiences.

6. Innovation is a recurring theme and point of discussion with regard to Six Sigma in general. Chapter 8 provided a step-by-step process for developing new concepts, involving several proven methods, and—interestingly—in the majority of classes where this step-by-step process has been used, participants claimed to have devel-oped at least one patentable idea that they intended to pursue.

The very nature of product development involves innovation and innovative thinking, and Design for Six Sigma provides a key framework within which innovation should be encouraged. Moreover, there should be a goal and methods to involve as much of the development team as possible in the innovative thinking. Although innovation apparently is a skill that can be acquired or taught to some extent, it is possible that some people are more innovative or creative than others. However, the bigger determinant seems to be the will and stamina to pursue the innovation and drive it into practice.

Bluetooth headsets look like little insects affixed to people's ears as people walk through airports, seemingly talking to themselves. The frequency hopping spread spectrum approach used in Bluetooth is derived from a patent from Hedy Lamarr, a well-known actress who had met Hitler and Mussolini at parties in Europe. She was quietly appalled by the path they were taking, and she headed to the United States. Later, Hedy Lamarr came up with the concept of frequency hopping while singing at various pitches corresponding to keys on the piano, after having watched a newsreel on Nazi soldiers listening in on the communications between American soldiers using handy talkies. She filed for the patent without intending to profit, certainly not compared to her income as a movie star, but to help defeat the scourge of Nazi philosophy that she knew firsthand.

If an actress with little or no technical training can develop a key innovation, based on her interest and her drive and her receptiveness to the "ah-ha moment," generalizing an idea or an experience into an opportunity to solve real problems, then it should be easy to energize a team and deputize the team members as innovators—if not innovating like Thomas Edison,[3] then perhaps like Hedy Lamarr.

7. Agile approaches provide agility in handling changes in requirements; the approach should be applied to systems, both hardware and software. Some purists seem to prefer to keep Agile methods confined to software, and confined to manual methods—but the power of combining modeling tools, linked tools, automated tools, and the ability to directly drive test software offer a powerful set of methods for Agile new product development.

8. As mentioned in the two chapters on predictive engineering, for discrete pass/fail requirements, the combination of design of experiments (DOE) and possibly response surface methodology (RSM) with binary logistic regression could provide an excellent approach for minimizing defects. Interestingly, the value of this combination has already been proven in marketing but does not yet seem to have been tried very often in engineering.

---

3. Michael J. Gelb and Sarah Miller Caldicott, *Innovate Like Edison,* Dutton, 2007.

9. Dr. Kevin Otto has suggested that the "V" model for critical parameter management flow-down and flow-up be replaced with a "W" model that includes the use of modeling during evaluation of prototypes.[4] If a designed experiment was performed on prototypes and the same experiment on the model or simulator, a feedback loop could be established to reduce model error, the discrepancy between the model or simulator and reality, which would provide a means to improve the model and use the improved model to resolve issues and finalize the robust design.

10. Reliability modeling can be enhanced through using the physics of failure mechanisms—which are often fairly well understood—in combination with a carefully selected combination of acceleration conditions for those failure mechanisms that can provide a more accelerated life testing approach that is predictive of field reliability. Approaches such as Bayesian belief networks can be used to develop initial reliability models, which can then be instantiated with data from accelerated life testing results and correlations to factors from experimentation to provide a reliability model that learns and becomes more predictive.

In our experience, nothing will stand in the way of an intelligent person with a plan, with the right tools, and with the will to succeed. We hope to help enable you to do some amazing things, and look forward to benefitting from new medical systems, new telecommunication systems, new entertainment systems, new transportation systems—the new products that will change our lives and those of our children in ways that we might not be able to imagine.

---

4. Dr. Kevin Otto, "Deploying System Modeling and Analysis to Enable One-Pass Design," http://www.robuststrategy.com.

Index

Numbers

100-point rating method, for customer requirements, 125

A

Abstraction design heuristic, for software architecture, 228
Accelerated life testing, 328–330
Acceptability criteria, for success metric, 57
Acceptance testing, 359
Accuracy, MSA and, 307
Actors, in use case modeling, 295
Ad-hoc testing, 359
Affinity diagramming
  interviewing and, 114
  KJ analysis applying, 117
Affordances, in software mistake proofing, 300
Agile Alliance, 211–212
Agile development
  applying critical parameters to development of feature sets, 213–214
  data collection plan for Motorola ViewHome project, 219–220
  DFSS tools used with, 215–217
  handling changes in requirements, 389

iterative development and, 212–213
manifesto for, 212
measuring agile designs, 218–219
noise factors and, 217–218
overview of, 211–212
requirements gathering (VOC and VOB), 214
schedule risks and, 106
SDLC (software development lifecycle) and, 212
summary, 221
verification process, 218
Algebra, for deriving equations, 246
All-pairs or pairwise testing, 356–357
Alpha testing, 359
Altshuller, Dr. Genrich, 141
Alvarez, Antonio, 2–3
Analysis aspect, of risk management, 60
Analyzing concerns and issues, 34–39
  job titles and responsibilities and, 38–39
  overview of, 34
  resource requirements and, 36–38
  time required for DFSS, 35–36
  waiting out corporate initiatives, 36
  when product development process is already in place, 34–35

Anderson-Darling test, 315
Application framework (layer 3), software architecture design, 224
Application support services (layer 2), software architecture design, 224
Applications (layer 4), software architecture design, 224
Architecture
    alternate architecture generation for hardware and software, 143–146
    approaches to selecting, 49
    risks of changing legacy architecture, 230
    software. *See* Software architecture
Architecture phase, RADIOV
    in CDMA cellular phone example, 17–18
    overview of, 75, 77–78
    tools for, 77
Assembly, DFMA (Design for manufacturability and assembly), 366–369
Availability
    critical parameters and, 172–174
    flow-down, 321–322
    measuring, 320
    modeling case study, 342–346
    software architecture design tactics for, 229
    verifying. *See* Verification of reliability and availability
Axiomatic design, 145

B

Barriers, to acceptance of DFSS
    analyzing potential, 34
    existing processes as, 34–35
    job titles and responsibilities as, 38–39
    removing, 42–43
    resource requirement as, 36–38
    schedule risk as, 95
    time requirements as, 35–36
Baselines, success metrics and, 41–42
Bathtub curve
    early life failures/infant mortality and, 326
    reliability and availability and, 322–325

useful life/constant failure rates and, 326–327
    wear out mechanisms and, 327
Bayes theorem, in WeiBayes approach to failures, 330–331
Behavioral design patterns, 354–355
Benchmarking, applying Six Sigma to process goals, 5
Beta testing, 359
Bill of materials (BOM), 155–156
Binary logistic regression, for defect reduction, 389
Black belts
    certification of, 62–63
    risk assessment role of, 352
Black box tests
    definition of, 359
    software testing, 347
    system testing, 348
BOM (bill of materials), 155–156
Box-Cox transformation, 315
Brainstorming
    for concept generation, 140–141
    feasibility screening and, 148
    as method to flow-down requirements, 191
    for VOC-responsive features, 214
Branches/conditions coverage, in end-to-end testing, 349
Buffer overflows, preventing in software mistake proofing, 303
Buffers, in critical chain project management
    personal, 104–105
    project, 105
"Burning platform"
    articulating, 33
    as means of overcoming resistance to change, 30–32
    supporting DFSS deployment, 29–30
Business case risks
    adjustments based on commitments and strategic direction, 92–94
    analyzing projects already underway, 93–94

goals, constraints, considerations, and distractions, 91–92
metrics for, 59, 85–89
overview of, 83–84
portfolio decision making as optimization process, 84–85
resource constraints and, 89–91
summarizing address to, 94
Buy-in, in concept selection, 149

C

Calculus, for deriving equations, 246
Calibration standard, MSA and, 309
Capability
   design capability. *See* Design capability
   predicting software release capability, 304–305
   process capability. *See* Process capability
Causes, of failure
   controls for, 162–163
   listing, 160–161
   Pareto analysis of cause categories, 171
CBAM (cost-benefit analysis method), 232–234
CCRs (critical customer requirements), 119
CCD (central composite design), 256–257
CDMA base station example, 14–26
   Architecture phase, 17–18
   Integration and Optimization phases, 18–21
   IP-BSC (IP-based base station controller) and, 14–16
   Requirements phase, 15, 17
   Verification phase, 21–24
CDOV (Concept, Design, Optimize, Verify)
   critical parameter flow-up and, 263–265
   DFSS process nomenclature and, 69–70
   DFSS steps, tools, and methods associated with, 12–13, 71–72
   DOE (design of experiments) and, 251
   identification of critical parameters, 153
   as phase of DFSS project, 9
   RADIOV phases corresponding to, 15
   requirements flow-down and, 187
   schedule risks and, 95

software verification testing and, 350
   verification of design capability and, 307
   verification of reliability and, 319
   verification of supply chain readiness and, 364
Cellular phones
   as example of new product development, 385
Center for Quality Management, 109
Central composite design (CDD), 256–257
Certification, 62–64
Champions. *See also* Leadership
   certification of, 64
   obtaining for DFSS projects, 46
   removing roadblocks and impediments, 42
   supportive project reviews and, 54–55
Change agents, certification of, 62–63
Change management
   DFSS deployment and, 44
   handling changes in requirements, 389
   overcoming resistance to change, 30–32
Coalition, role of guiding coalition, 32
Code
   end-to-end testing, 349
   libraries, 224
   unit testing, 347
Cognition Cockpit, 203
Combinatorial design method, 356–358
Combinatorial optimization algorithms, 356–357
Commitments, adjusting decisions based on, 92–94
Communication
   risk management and, 60
   of vision, 40–41
Comparison tests, 359
Compatibility testing, 359
Competitive analysis, building House of Quality and, 129
Component level, concept generation at, 139
Comprehensiveness criteria, for success metrics, 57
Compression design heuristic, for software architecture, 228

Concept, Design, Optimize, Verify. *See* CDOV (Concept, Design, Optimize, Verify)

Concept engineering, 109

Concept generation/selection
alternate architecture generation for hardware and software, 143–146
approaches to, 137–139
brainstorming and mind-mapping, 140–141
concept selection process, 49, 149–151
consideration of existing solutions, 147–148
developing feasible concepts to consistent levels, 148–149
feasibility screening, 148
flowchart for, 138
Kansei engineering approach, 152
position within DFSS flow, 137
robust design concepts, 146–147
summary, 152
TRIZ, 141–143

Concerns, analyzing. *See* Analyzing concerns and issues

Conflict resolution, 55

Consistency, concept generation and, 148–149

Consolidating gains, in DFSS deployment, 44–45

Constant failure rates, reliability and availability and, 325–326

Constraints. *See also* critical chain (theory of constraints) project management
critical parameters and, 156
vs. functions, 155–156
portfolio decision making and, 91–92
portfolio optimization and, 84–85
resource constraints, 89–91
software mistake proofing, 301–303

Context-driven testing, 359

Continuous (and ordinal) critical parameters
vs. discrete critical parameters, 238–241
methods for deriving transfer functions from, 244–245
MSA (measurement system analysis) and, 307

Contractual obligations, complications impacting prioritization, 93

Controls, for potential causes of failure in FMEA, 162–163

Cost-benefit analysis method (CBAM), 232–234

Counterproductive detractors, in portfolio decision making, 91

Cp/Cpk indices. *See also* Design capability; Process capability
applying Six Sigma to process goals, 5
calculating values for, 267–269
capability analysis and, 202–203
cooptimizing Cpk's, 282–283
determining process capability with, 315–316
difference between Cp and Cpk indices, 271–273
forecasting/predictive use of, 265, 270
M/PCpS (Machine/Process Capability Study), 6
variance reduction and, 273–274

CPM, planning deployment and, 39

Creational design patterns, 354–355

Critical chain (theory of constraints) project management
critical paths compared with, 98–102
overview of, 103–105

Critical customer requirements (CCRs), 119

Critical parameter flow-down. *See also* Requirements flow-down
examples, 206–208
model for managing, 390
overview of, 203–206

Critical parameter flow-up
model for managing, 390
Monte Carlo simulation of, 266–267
overview of, 263–266

Critical Parameter Management
DFSS goals and, 49
formal gate reviews and, 56

Critical parameter scorecard, 269–270

Critical parameters
constraints and, 156
decision-making for software architecture and, 225
definition of, 153–155

discrete vs. continuous, 238–241
examples of, 174–176
feature sets based on, 213–214
flow-down, 203–206
flow-down examples, 206–208
flow-up, 263–266
identifying, 49
models for managing flow-down and
    flow-up, 390
Monte Carlo simulation of flow-up,
    266–267
position within DFSS flow, 153
predictive engineering and, 237
prioritization and selection of, 157–160
project schedule treated as, 95–96
RADIOV Architecture phase and, 75, 78
reliability and availability and, 172–174
requirements flow-down. *See* Requirements
    flow-down
target values and specification limits for,
    197–198
VOB (voice of business) considerations,
    155–156
Critical paths
  critical chains compared with, 98–102
  product development delays due to
      wandering critical path, 96
Crystal Ball
  Monte Carlo simulation with, 84
  portfolio decision making and, 89–90
Current Reality Tree, project selection and
    prioritization and, 37
Customer requirements
  100-point rating method, 125
  building House of Quality, 128–129
  CCRs (critical customer requirements), 119
  identifying challenging, 120–121
  Kano model of, 122
  translating into system requirements, 124–128
  validation and prioritization of, 124
Customers. *See also* VOC (voice of customer)
  critical parameters for optimization based
      on expectations of, 270

critical parameters impact on satisfaction of,
    153
interview guide for, 113–115
interview process, 116
planning visits and interviews, 115–116
profile matrices, 111–112
reasons for failures experienced by, 313
reliability and availability expectations of,
    172–174
reliability and availability perspective of,
    319–321
retention impacting prioritization, 93
VOC gathering and, 111–112
voices and images in interview process,
    112–113

D

DACE (Design and Analysis of Computer
    Experiments)
  DOE compared with, 259–260
  steps in, 260
Data collection plan, for Motorola ViewHome
    project, 219–220
Decision making
  product portfolio. *See* Portfolio decision
      making
  software architecture design, 224–227
Decomposition, 203–206. *See also*
    Flow-down/flow-up
Decomposition design heuristic, for software
    architecture, 228
Defaults, software mistake proofing and, 303
Defect discovery rate collection plan,
    303–305
Defects
  applying Six Sigma and, 5
  binary logistic regression combined with
      DOE or RSM to minimize, 389
  financial benefits of early detection, 10–11
  software FMEA for detection of, 168
  software stability and, 303–305
  software testing for reducing, 347
  testing as means of locating, 350

Define, Measure, Analyze, Design, Optimize, and Verify. *See* DMADOV (Define, Measure, Analyze, Design, Optimize, and Verify)

Define, Measure, Analyze, Improve, and Control. *See* DMAIC (Define, Measure, Analyze, Improve, and Control)

Delays
allocation of tolerances, 209
product development and, 96–98

Delighter's
assessing importance of, 157
in Kano model of customer requirements, 122–123

Deliverables
formal gate reviews and, 56
verification of supply chain readiness and, 363–364

Delivery. *See* On-time delivery

Demand uncertainty, supply chain decisions and, 372–373

Deploying DFSS
goals for DFSS and, 48–49
ideal scenario for, 29–30
overview of, 29
single project approach, 45–47
step 1: "burning platform", 30–32
step 2: guiding coalition, 32
step 3: defining the vision, 33–34
step 4: analyzing issues, 34–39
step 5: planning deployment, 39–40
step 6: communicating the vision, 40–41
step 7: executing deployment campaign, 41–42
step 8: removing impediments, 42–43
step 9: generating short-term wins, 43–44
step 10: consolidating gains, 44–45
success of, 50
summary, 50–51
tool set for, 47–48

Deployment experts, 29

Design
axiomatic, 145

of experiments. *See* DOE (design of experiments)
fractional factorial, 254–257
making insensitive to noise, 147
measuring agile, 218–219
robust concepts, 146–147
software architecture, 227–228, 234–235, 354
system, 187–190

Design and Analysis of Computer Experiments (DACE)
DOE compared with, 259–260
steps in, 260

Design capability. *See also* Cp/Cpk indices
assessment of, 202–203
calculating values for, 267–269
forecasting and, 270
predictive use of capability indices, 265
verification of. *See* Verification of design capability

Design FMEA. *See* DFMEA (Design FMEA)

Design for manufacturability and assembly. *See* DFMA (Design for manufacturability and assembly)

Design for Six Sigma. *See* DFSS (Design for Six Sigma) overview

Design heuristics, for software architecture, 227–228

Design patterns
benefits of, 355
example applications of, 355–356
GoF (Gang of Four), 354–355
software design and, 234–235, 354

Design phase, RADIOV, 78, 79

Detection ratings, for potential causes in FEMA, 162

Development manager (DM), risk assessment role of, 61, 352

DFMA (design for manufacturability and assembly), 366–369
best practices and benefits of, 369
list of key aspects and practices, 367–368
overview of, 366

DFMEA (Design FMEA). *See also* FMEA
   (Failure Modes and Effects Analysis)
  for anticipation of problems, 194
  benefits of, 196–197
  in deployment planning, 39
  DFSS goals and, 49
  overview of, 195
  tool summary, 196
DFSS (Design for Six Sigma) overview
  Architecture phase, 17–18
  CDMA base station example, 14–26
  charter for IP BSC, 15–16
  deployment. *See* Deploying DFSS
  flowchart, 11, 364–365
  history of, 8–9
  Integration and Optimization phases,
    18–21
  key tools and methods, 12–13
  preliminary steps in DFSS projects, 15
  processes in, 9–11
  Requirements phase, 15, 17
  software DFSS. *See* SDFSS (software DFSS)
  Verification phase of, 21–24
DFSS project Manager (DPM), 60
Diagnostic criteria, for success metrics, 57
Direction of goodness, in building House of
  Quality, 131
Discount rates, metrics for portfolio decision
  making, 88
Discrete critical parameters
  vs. continuous critical parameters, 238–241
  logistic regression for, 242–244
  methods for deriving transfer functions
    from, 241–242
Distractions, portfolio decision making and,
  91–92
DM (development manager), risk assessment
  role of, 61, 352
DMADOV (Define, Measure, Analyze, Design,
  Optimize, and Verify)
  critical parameter flow-up and, 263–265
  DFSS process nomenclature and, 69–70

DFSS steps, tools, and methods associated
  with, 12–13, 71–72
DOE (design of experiments) and, 251
in GE's DFSS project, 9
identification of critical parameters and,
  153
RADIOV phases corresponding to, 15
requirements flow-down and, 187
schedule risks and, 95
software verification testing and, 350
verification of design capability and, 307
verification of reliability and, 319
DMAIC (Define, Measure, Analyze, Improve,
  and Control)
  capability analysis and, 202
  generating short-term wins, 43–44
  identification of critical parameters, 153
  overview of, 6–7
  for problem-solving aspect of Six Sigma, 69
DOE (design of experiments), 251–256
  applying to call processing failures in
    CDMA cell phone example, 22
  benefits of, 251–252
  DACE compared with, 259–260
  fractional factorial design, 254–256
  logistic regression combined with,
    244–245, 389
  measurement error and, 310
  sparcity of effects principle in, 253–254
  statistical methods for improving quality,
    1–2
DPM (DFSS project Manager), 60

E

Early life failures
  product tolerances, 4
  verification of reliability and availability
    and, 326
  Weibull distribution and, 323
Economic commercial value (ECV), 88–89
ECV (economic commercial value), 88–89
Electrical engineering equations, 246

Electrical tolerances, allocating, 209
Electronics products, applying DFSS to, 66–69
Empirical modeling
    using DOE, 251–256
    using historical data, 247–251
    using response surface methods, 256–259
End-to-end software testing
    definition of, 359–360
    overview of, 348–349
Entry/exit coverage, in end-to-end testing, 349
Equations
    for Cp and Cpk, 315
    electrical engineering and mechanical engineering, 246
    for modeling, 244–245
Errors
    preventing manufacturing and assembly errors, 366, 369
    preventing software mistakes and errors, 299–303
    sources of measurement error, 309–310
Event-driven reviews, 54
Excel, for modeling within spreadsheets, 246
Executing deployment campaign, steps in DFSS deployment, 41–42
Exploratory testing, 360

F
FACT TOPS Team, 9, 274, 283–288
Failover testing, 360
Failure modes. *See also* FMEA (Failure Modes and Effects Analysis)
    benefits of FMEA in anticipating, 195–197
    benefits of P-diagrams in anticipating, 195
    determining risk of failure, 315–316
    listing in FMEA, 160
    requirements flow-down process and, 193–194
Failure Modes and Effects Analysis. *See* FMEA (Failure Modes and Effects Analysis)
Failures
    acceleration factors, 329–330
    early life failures/infant mortality, 326
    list of common software failures, 353
    risk of failures despite verification, 331–332
    useful life/constant failure rates, 325–326
    wear out mechanisms, 326
    WeiBayes approach to, 330–331
Fault tree analysis. *See* FTA (fault tree analysis)
Feasibility screening, in concept generation/selection, 148–149
Feature development, iterative approach to, 213–214
Feedback
    governance as feedback mechanism, 53
    software mistake proofing and, 301
Feldbaumer, David, 9
Field testing, 360
Fiero, Janet, 3
Financial metrics, for portfolio decision making, 85–89
Flow-down/flow-up
    availability and reliability and, 321–322
    criteria for success metrics, 57
    critical parameter flow-down. *See* Critical parameter flow-down
    critical parameter flow-up. *See* Critical parameter flow-up
    requirements flow-down. *See* Requirements flow-down
FMEA (Failure Modes and Effects Analysis)
    design FMEA for anticipation of problems, 194–197
    design FMEA for deployment planning, 39
    DFSS goals and, 49
    evaluating system-level risks, 157–158
    formal gate reviews and, 56
    risk reduction and, 353
    software FMEA acronyms and definitions, 164
    software FMEA benefits, 168–169
    software FMEA cost savings and ROI, 167–168
    software FMEA implementation case study, 169–172

software FMEA process documentation, 176–185

software FMEA process phases, 165–167

software FMEA roles and responsibilities, 165

system FEMA, steps in, 160–163

system FMEA risk evaluation, 157–158

Forums, for communicating vision, 41

Fractional factorial design

CCD (central composite design) based on, 256–257

DOE (design of experiments) and, 254–255

FTA (fault tree analysis)

applying to CDMA example, 19

DFSS goals and, 49

of reliability, 173–174

Fullerton, Craig, 9

Functional modeling, for hardware, 144

Functional testing, 360

Functions/functionality

decision-making for software architecture and, 225

end-to-end testing and, 349

functions vs. constraints, 155–156

interface functionality, 369

listing functions in FEMA, 160

measurable requirements and, 156

software testing and, 347, 350

transfer functions. See Transfer functions

Future directions

Agile approach to change management, 389

DFSS integration with systems engineering, 388

innovation, 388–389

logistic regression for minimizing defects, 389

modeling flow-down and flow-up of critical parameters (Otto), 390

portfolio including services as well as projects, 388

project and program management, 388

reliability modeling, 390

risk management process (Mun), 387–388

strategies becoming tactics, metrics, and actions, 386–387

Future Reality Tree, 38

"Fuzzy front end", causing product development delays, 97–98

G

Galvin, Bob

Malcolm Baldrige National Quality Award, 7–8

role in development of Six Sigma, 5

Garvin, David, 108–109

Gate reviews, for governance, 55–56

Gauge repeatability and reproducibility. See GR&R (gauge repeatability and reproducibility) index

General Electric History of Six Sigma and, 2, 9

General linear model (GLM), 247–251

Generation of system moments

critical parameter scorecard, 269–270

predicting critical parameter values, 267–269

GLM (general linear model), 247–251

Goals

for DFSS deployment, 48–49

portfolio decision making and, 91–92

Governance, 53–56

formal gate reviews, 55–56

overview of, 53–54

supportive project reviews, 54–55

GR&R (gauge repeatability and reproducibility) index

improving inadequate measurement systems, 312

MSA and, 202, 309–310

Green belts

certification of, 62–63

risk assessment role of, 352

"Guard-banding" approach, to measurement error, 314

Guiding coalition

for DFSS deployment, 32

role in removing roadblocks and impediments, 42

H

Hardware
   alternate architecture generation for, 143–146
   applying DFSS to, 66–67
   RADIOV and, 11
   requirements flow-down for, 190–193
   tolerance expectations, 366
Harry, Dr. Mikel
   documentation of Six Sigma concepts, 6
   as head of SSRI, 8
Heuristics (rule of thumb)
   design heuristics for software architecture, 227–228
   for generating alternative concepts and architectures, 145–146
Higher-is-better critical parameters, 197
Historical data, empirical modeling from, 247–251
House of Quality
   competitive analysis and, 129
   constructing, 128
   critical parameter flow-down or decomposition, 203
   customer requirements, 128–129
   direction of goodness, 131
   as method to flow-down requirements, 191–192
   prioritization and, 129, 133
   relationship matrix, 131–132
   system-level critical parameters, 190–191
   system requirements, 129–131
   targets and units for system requirements, 133
   trade-offs among system requirements and, 132
   translating customer requirements into system requirements, 127–128

I

Iacocca, Lee, 108
IAR (integrated alternator regulator), predictive engineering case study, 288–290

ICs (integrated circuits), statistical methods for improving quality of, 2
Identification
   of challenging customer requirements, 120–121
   of critical parameters, 49, 153
   in risk management, 60
IDOV (Identify, Design, Optimize, Verify)
   DFSS process nomenclature, 69–70
   DFSS project phases, 9
   DFSS steps, tools, and methods associated with, 71–72
   RADIOV phases corresponding to, 15
Images
   KJ analysis and, 117–118
   VOC gathering and, 112–113
Impediments. See Barriers, to acceptance of DFSS
Incremental integration testing, 360
Indifferent's, in Kano model, 122–123
Infant mortality, reliability and availability and, 326
Initial tolerance allocation, in requirements flow-down, 208–210
Innovation, future directions for applying DFSS, 388–389
Instability, of requirements, 59
Install/uninstall testing, 360
Integer programming, portfolio decision making and, 89
Integrate phase, RADIOV, 78, 79
   in CDMA cellular phone example, 18–21
Integrated alternator regulator (IAR), predictive engineering case study, 288–290
Integrated circuits (ICs), statistical methods for improving quality of, 2
Inter-operability testing (IOT), 348
Interfaces
   appropriateness and acceptability of flows between, 369
   software mistake proofing and, 303

Interruptions, product development delays due to, 101
Interviewing customers
  guide for, 113–115
  planning, 115–116
  process of, 116
Inventory, in supply chain decision-making, 372, 375, 378
IOT (inter-operability testing), 348
IP-BSC (IP-based base station controller)
  aligning critical parameters with DFSS tools, 19
  benefits of DFSS project for, 25–26
  CDMA base station example and, 14–16
Iridium project, Motorola, 107
Issues, analyzing. See Analyzing concerns and issues
Iterative development
  Agile development and, 212–213
  DFSS tools used with, 215–217
  schedule risks and, 105–106

J

Japan Society of Kansei Engineering (JSKE), 152
Job titles and responsibilities, as impediment to acceptance of DFSS, 38–39
JSKE (Japan Society of Kansei Engineering), 152

K

"k" factor, in Cp and Cpk capability indices, 271–273
Kano, Dr. Noriaki, 122
Kano model, 122–123
Kansei engineering, 152
Kawakita, Jiro, 117
Key performance indicators (KPIs), 294–295, 297
KJ (Jiro Kawakita) analysis
  affinity diagramming compared with, 114
  overview of, 117–120
  risk management and, 48

"Knapsack problem", portfolio optimization and, 84–85
Kotter, Dr. John, 30
Kougaku, Kansei, 152
KPIs (key performance indicators), 294–295, 297

L

Launch schedules, products, 369–370
Lawson, Dr. J. Ronald
  documentation of Six Sigma concepts, 6
  in history of Six Sigma and DFSS, 2
Layers, software architecture, 223–224
Lead time
  supply chain readiness and, 370–372
  trade-offs between on-time delivery, lead time, and inventory levels, 372, 380
Leadership
  consolidating gains and, 44
  by example, 41
  ideal scenario for DFSS deployment, 30
  role in removing roadblocks and impediments, 43
Leading Change (Kotter), 30
  Leading indicators, criteria for success metrics, 57
  Lean development
  Agile development compared with, 211
  early example of, 2
Legacy architecture, risks of changing, 230
Linear regression, empirical modeling using historical data, 247–251
Linear Satisfier's, in Kano model, 122–123
Load tests
  definition of, 360
  software testing and, 347
Logistic regression
  for discrete critical parameters, 242–244
  DOE combined with, 244–245
  minimizing defects, 389
"Loss leaders", financial metrics for portfolio decision making, 85
Lower-is-better critical parameters, 197

M

M/PCpS (Machine/Process Capability Study), 6
Maass, Eric
  FACT TOPS Team, 9, 274, 283-288
  in history of Six Sigma and DFSS, 2
Machine/Process Capability Study
  (M/PCpS), 6
Malcolm Baldrige National Quality Award, 7
Management
  DFSS goals and, 48
  improving project and program management,
    388
  role in removing roadblocks and
    impediments, 43
Manufacturability
  DFMA (design for manufacturability and
    assembly), 366–369
  tools for reviewing, 49
Market penetration, complications impacting
  prioritization, 93
Marketing
  customer profile matrices and, 111–112
  VOC gathering and, 107–108
Mathematical modeling software, 246
MBB (Master Black Belt)
  certification by, 62–63
  risk management and, 61
MCM (multichip module), supply chain
  readiness case study, 380–382
Mean
  generation of system moments for predicting
    values of, 267
  in optimization of critical parameters,
    271–273
Mean time between failures (MTBF), 320–321
Mean-time-to-failure (MTTF)
  critical parameters and, 197
  reliability and availability and, 320–321
Measurable requirements
  Agile development and, 218–219
  critical parameters as, 153
  data collection plan for ViewHome project,
    219–220

functionality and, 156
Measurement error
  "guard-banding" approach to, 314
  MSA and, 199
  sources of, 309–310
Measurement phase, DFSS, 218
Measurement system analysis. *See* MSA
  (measurement system analysis)
Measurement systems
  averaging measurements to improve, 312
  improving inadequate, 310–311
Mechanical engineering equations, 246
Mechanical tolerances, allocating, 209
Media, for communicating organizational
  vision, 41
Metrics. *See also* Success metrics
  applying Six Sigma to process goals, 5
  defining, 41–42
  key performance, 297
  portfolio decision making and, 85–89
  risks and, 59
MICARL (Motorola Integrated Circuits
  Applications Research Laboratory), 1
Microsoft Project, 104
Middleware, in software architecture, 224
Mind-mapping, for concept generation,
  140–141
Minitab reliability tools, 330–331
Modeling
  availability, 342–346
  critical parameter flow-down/flow-up, 390
  DACE and, 259–260
  empirical modeling using DOE, 251–256
  empirical modeling using historical data,
    247–251
  empirical modeling using RSM, 256–259
  evaluating software optimization models,
    298–299
  existing or derived equations for, 245–246
  reliability, 390
  schedule risks, 95–96
  software architecture design, 234–235
  software options for, 246–247

supply chain decisions in optoelectronic multichip module, 380–382
supply chain decisions in semiconductor manufacturing, 372–379
Modifiability tactic, in software architecture design, 229
Modules
concept generation at module level, 139
interface flows and, 369
Monte Carlo simulation
for cooptimizing Cpk's, 283
of critical parameter flow-up, 266–267
critical parameter scorecard and, 269–270
for DFSS goals, 49
history of development of, 266
for optimizing guard-bands, 314–315
for overcoming resistance to change, 32
for product or service evaluation, 83–84
for schedule estimation, 99–100
of selling price resulting in profit or losses, 86–87
showing impact of Parkinson's Law on project scheduling, 102
software options for modeling, 247
of system availability, 21, 23
use case modeling and, 298
MotoOATSGen tool, 356
Motorola
history of Six Sigma and DFSS, 1–2
Iridium project, 107
Malcolm Baldrige National Quality Award, 7–8
TCS (total customer satisfaction) competition, 8
Motorola Integrated Circuits Applications Research Laboratory (MICARL), 1
Motorola Training and Education Center (MTEC), 3
MSA (measurement system analysis), 198–202
linking to test and verification phase, 201–202
measurement error and, 199
overview of, 198–199

performing on measurable requirements, 75
variance and, 200–201, 272
verifying design capability, 307–310
MTBF (mean time between failures), 320–321
MTEC (Motorola Training and Education Center), 3
MTTF (mean-time-to-failure)
critical parameters and, 197
reliability and availability and, 320–321
Multichip module (MCM), supply chain readiness case study, 380–382
Multiple response optimization
overview of, 280–282
software performance and, 293–294
YSM (Yield Surface Modeling) and, 283–288
Multitasking
critical chain project management and, 100–102
minimizing, 104
product development delays due to, 96
Mun, Dr. Johnathan, 387–388
Must-Be's, in Kano model, 122–123
Mutation testing, 360

N

National Institute of Standards and Technology (NIST), 309
Natural mapping, software mistake proofing, 299–300
Net present value (NPV) metric, for portfolio decision making, 88–89
New product introduction (NPI), 58–59
New-unique-difficult (NUDs), customer requirements, 120–121
NIH ("Not Invented Here") syndrome, 32
NIST (National Institute of Standards and Technology), 309
Noise
Agile development and, 217–218
making design insensitive to, 147
"Not Invented Here" (NIH) syndrome, 32
NPI (new product introdution), 58–59

NPV (net present value) metric, for portfolio decision making, 88–89

NUDs (new-unique-difficult), customer requirements, 120–121

Numbers, software mistake proofing, 303

O

OATS (Orthogonal-array based testing), 356–358, 360

Objectives, of VOC gathering, 110

Occurrence ratings, assigning to causes of failure, 161–162

On-time delivery
  supply chain readiness and, 370–372
  trade-offs between on-time delivery, lead time, and inventory levels, 372

Ooi, C.C., 283

Open-ended questions, in interviewing, 114

Operating systems (OSs), 223–224

Optimization
  combinatorial optimization algorithms, 356–357
  cooptimizing Cpk's, 282–283
  mean and/or variance in, 271–273
  multiple response optimization, 280–282
  portfolio decision making as optimization process, 84–85
  robustness achieved through variance reduction, 273–280
  selecting critical parameters for, 270
  software optimization. *See* Software optimization

Optimize phase, RADIOV
  in CDMA cellular phone example, 18–21
  overview of, 78–80
  tools, 80

Optobus module, 380

Optoelectronic multichip module (MCM), supply chain readiness case study, 380–382

OptQuest
  cooptimizing Cpk's, 283
  portfolio decision making and, 89–91

Oracle, Crystal Ball utility, 84

Ordinal critical parameters. *See* Continuous (ordinal) critical parameters

Orthogonal-array based testing (OATS), 356–358, 360

Orthogonality criteria, for success metrics, 57

OSs (operating systems), 223–224

Otto, Dr. Kevin, 390

P

P-diagrams
  anticipation of potential problems, 195
  applying DFSS tools to Agile development, 216–217
  DFSS goals and, 49
  for requirements flow-down, 191

P/T (precision-to-tolerance) ratio
  improving inadequate measurement systems, 312
  MSA and, 202, 309–310
  system analysis and, 197

P × I × T (Probability × Impact × Time frame), risk formula, 61–62

Parameters, critical. *See* Critical parameters

Pareto analysis, of cause categories, 171

Parkinson's Law
  critical chain project management and, 100–101
  Monte Carlo simulation showing, 102
  product development delays due to, 96

Path coverage, end-to-end testing and, 349

Payback time metric, for portfolio decision making, 85–89

PDMA (Product Development and Management Association), 107

Perez-Wilson, Mario, 6

Performance
  metrics for, 294–295, 297
  as quality attribute, 227
  software architecture design tactics, 229
  software systems and, 293–294
  software testing and, 347
  tools for measuring robustness of, 49

Performance testing, 347, 360
Personal buffers, in critical chain project management, 104
Phases, DFSS, 64
Pilot runs, manufacturability and, 49
Planning
  customer visits and interviews, 115–116
  DFSS deployment, 39–40, 47
  product launch, 370
  risk management and, 60, 351–352
  RPN reduction, 163
Planning backwards, in critical chain project management, 103
PM (Project manager), 60, 352–353
Poka Yoke, 369
Poppendieck, Mary, 211
Poppendieck, Tom, 211
Portfolio decision making
  business case risk and, 94
  financial metrics and, 85–89
  goals, constraints, considerations, distractions, 91–92
  impact of commitments and strategic direction on, 92–94
  as optimization process, 84–85
  overview of, 83–84
  resource constraints and, 89–91
  steps in analyzing project already underway, 93–94
Pp/Ppk. See Cp/Cpk indices
Precision
  measurement variability and, 310
  MSA and, 309
Precision-to-tolerance ratio. See P/T (precision-to-tolerance) ratio
Prediction criteria, for success metrics, 57
Prediction indices, for capability, 265
Predictive engineering
  cooptimizing Cpk's, 282–283
  critical parameter flow-up and, 263–266
  critical parameter scorecard, 269–270
  DACE and, 259–260

deriving transfer functions for continuous critical parameters, 244–245
deriving transfer functions for discrete critical parameters, 241–242
discrete vs. continuous critical parameters, 238–241
empirical modeling using DOE, 251–256
empirical modeling using historical data, 247–251
empirical modeling using response surface methods, 256–259
existing or derived equations for modeling, 245–246
future directions for applying DFSS, 389
generation of system moments, 267–269
IAR (integrated alternator regulator) case study, 288–290
logistic regression for discrete parameters, 242–244
mean and/or variance in optimization of critical parameters, 271–273
Monte Carlo simulation of critical parameter flow-up, 266–267
multiple response optimization, 280–282
optimizing robustness through variance reduction, 273–280
overview of, 237–238
selecting critical parameters for optimization, 270
software optimization. See Software optimization
software options for modeling, 246–247
summary, 261, 290–291
YSM and, 283–288
Prioritization
  building House of Quality, 129
  complications in, 92–93
  of critical parameters, 157–160
  of customer requirements, 124
  of resources, 36–38
  of system requirements, 133
  trade-off analysis with Prioritization matrix, 232

Privileges, software mistake proofing and, 303

Probability × Impact × Time frame (P × I × T), risk formula, 61–62

Problem anticipation. *See* Failure modes

Process capability. *See also* Cp/Cpk indices
applying Six Sigma to process goals, 5
determining, 315–316
M/PCpS (Machine/Process Capability Study), 6

Product development
applying DFSS to, 65
benefits of, 385–386
existing process as impediment to acceptance of DFSS, 34–35
factors in product development time, 96
innovation and, 389
reasons for cancellation or failure, 107
risks, 58–60
schedule risks. *See* Schedule risks

Product Development and Management Association (PDMA), 107

Product launch schedules, 369–370

Products
applying VOC to, 48
assembly. *See* DFMA (design for manufacturability and assembly)
confidence in meeting on-time delivery, 370
metrics for delivery risks, 59
portfolio decision making. *See* Portfolio decision making
roadmaps for product portfolio, 387

Profit metric, for portfolio decision making, 85–89

Programs, improving management of, 388

Project buffers, in critical chain project management, 105

Project manager (PM), 60, 352–353

Project reviews, for governance, 54–55

Projects
improving management of, 388
removing doomed, 84

schedule risks. *See* Schedule risks

single project approach to DFSS deployment, 45–47

Proofing, for software mistakes and errors, 299–303

Prototyping, in software architecture design, 234–235

Pugh concept selection process
DFSS goals and, 49
overview of, 149–151

Pugh, Dr. Stuart, 149–151

Pugh matrix, trade-off analysis with, 231

Q

Q × A = E (quality × acceptance = effectiveness), 40

QFD (quality function deployment)
concept selection process and, 110
planning deployment and, 39

Qualifications, iterative development and, 105–106

Quality attributes, for software architecture systems, 225–227

Quality function deployment (QFD)
concept selection process and, 110
planning deployment and, 39

Quality × acceptance = effectiveness (Q × A = E), 40

R

RADIOV (Requirements, Architecture, Design, Integration, Optimization, and Verification)
alignment with DFSS flow, 75
Architecture phase, 17–18, 75, 77–78
capability analysis and, 202
critical parameter flow-up and, 263–265
Design phase, 78
DFSS process nomenclature and, 69–70
DFSS steps, tools, and methods associated with, 12–13, 71–72
DOE and, 251

Integrate phase, 18–21, 78
merging DFSS hardware and software, 11
Optimize phase, 18–21, 78–80
requirements flow-down and, 187
Requirements phase, 15, 17, 73–74, 76
SDFSS methods and, 215
software verification testing and, 350
TDD and, 212
trade-off analysis tools, 231
verification of design capability and, 307
verification of reliability and, 319
Verify phase, 21–24, 80–81
Rayleigh model, 304–305
Recognition, consolidating gains in DFSS deployment, 44
Recovery testing, 360
Red X effect, 274
Regression analysis, 247–251
Regression testing
  binary logistic regression, 389
  definition of, 360
  software testing and, 348
Relationship matrix, in building House of Quality, 131–132
Release capability, predicting for software, 304–305
Release to product, schedule risks and, 105–106
Reliability
  bathtub curve for interval aspect of, 323
  critical parameters and, 172–174
  customer perspective on, 320
  definition of, 322
  flow-down, 321–322
  metrics for reliability risks, 59
  modeling, 49, 390
  of software, 325
  software reliability case study, 333–341
  verifying. See Verification of reliability and availability
Removing impediments, steps in DFSS deployment, 42–43

Replication design heuristic, for software architecture, 228
Reproducibility, measurement variability and, 311
Requirements
  Agile handling of changes in, 389
  business requirements, 110
  customer requirements. See Customer requirements
  gathering (VOC and VOB), 214
  impact of changing requirements on scheduling, 97–98
  instability of, 59
  system requirements. See System requirements
  tolerance expectations for hardware requirements, 366
Requirements, Architecture, Design, Integration, Optimization, and Verification. See RADIOV (Requirements, Architecture, Design, Integration, Optimization, and Verification)
Requirements flow-down
  anticipation of potential problems, 193–194
  benefits of FMEA in anticipation of problems, 195–197
  benefits of P-diagrams in anticipation of problems, 195
  capability analysis and, 202–203
  critical parameter flow-down or decomposition, 203–206
  flow-down examples, 206–208
  for hardware and software systems, 190–193
  initial tolerance allocations and, 208–210
  MSA and, 198–202
  position within DFSS flow, 187–190
  summary, 210
  target values and specification limits for critical parameters, 197–198
Requirements phase, RADIOV
  in CDMA cellular phone example, 15, 17
  overview of, 73–74, 76

Resolution, risk management and, 60
Resource sharing design heuristic, for
    software architecture, 228
Resources
    critical chain (theory of constraints) project
        management and, 104–105
    dependencies in critical chain project
        management, 101
    ideal scenario for DFSS deployment, 30
    impact of constraints on portfolio decision
        making, 89–91
    resource requirement as impediment to
        acceptance of DFSS, 36–38
Response surface methods. See RSM
    (response surface methods)
Responsibilities
    impediments to acceptance of DFSS, 38–39
    in software FMEA, 165
Return on investment. See ROI (return on
    investment)
Revenue metric, for portfolio decision
    making, 85–86
Review process
    formal gate reviews, 55–56
    Rigorous Gate Reviews, 48
    supportive project reviews, 54–55
Rewards, consolidating gains and, 44
Rigorous Gate Reviews, 48
Risk management
    DFSS goals and, 48
    Mun proposal for, 387–388
    planning, 351–352
Risk management roles
    development manager, 61, 352
    overview of, 60
    project manager, 60, 352–353
    risk owner, 61–62, 352–353
Risk of failures
    despite verification, 331–332
    verification of design capability and,
        313–315

Risk owner, 61–62, 352–353
Risk priority numbers. See RPNs (risk priority
    numbers)
Risks
    business case. See Business case risks
    existing solutions and, 147
    FMEA for evaluating system-level,
        157–158
    plans for reducing RPNs, 163
    prioritization and selection of critical
        parameters and, 157
    product development, 58–60
    schedule. See Schedule risks
    stakeholders assessing, 159–160
    steps in assessing, 352
    success metrics and, 155–156
    trade-off analysis for assessing, 230
Roadblocks, analyzing potential, 34
Robust design
    concepts, 146–147
    DFMA and, 366–369
    software verification testing and, 350
    variance reduction and, 273–280
ROI (return on investment)
    CBAM (cost-benefit analysis method) and,
        233–234
    financial metrics for portfolio decision
        making, 85–89
    software FMEA and, 167–168
Roles
    development manager, 61, 352
    project manager, 60, 352–353
    risk owner, 61–62, 352–353
    software FMEA, 165
    use case modeling and, 295
RPNs (risk priority numbers)
    calculating and applying in FEMA,
        162–163
    formal gate reviews and, 56
    formula for determining, 61–62
    system availability and, 18

RSM (response surface methods)
benefits of combining with binary logistic regression, 389
cooptimizing Cpk's, 283
empirical modeling, 256–259
multiple response optimization, 280–282
YSM (Yield Surface Modeling) compared with, 283–284
Rule of thumb. *See* Heuristics (rule of thumb)

S

Sanity testing, 361
Satisfaction, in Kano model, 122
Schedule risks
allocation of tolerances, 209
changing requirements and, 97–98
critical chain theory of constraints project management and, 103–105
critical paths vs. critical chains in determining durations, 98–102
iterations, qualification, and release to product, 105–106
metrics for, 59
model for, 95–96
position within DFSS flow, 95
summary of, 106
Schedules, product launch, 369–370
Scripting languages, 224
SDFSS (software DFSS). *See also* Agile development
applying DFSS to software development, 67
combining with Agile development, 211
DFSS tools used with Agile development, 215–217
in Motorola ViewHome project, 219–220
SDLC (software development lifecycle)
Agile development and, 212
DFSS supporting, 214
SDM (systems development manager), 61
Security tactics, in software architecture design, 229–230

Security testing, 348, 361
Semiconductor manufacturing, product delivery example, 372–379
"Sense engineering", 152
Sensitivity studies, model evaluation and, 298
Separation design heuristic, for software architecture, 228
Services
application support services in software architecture design, 224
applying DFSS to, 65–66
applying VOC to, 48
portfolio of projects including, 388
system services in software architecture design, 223–224
Severity ratings, assigning to potential failure modes, 160–161
Shainin, Dorian, 3–4, 274
Short-term wins, generating in DFSS deployment, 43–44
Simulation
DACE and, 259–260
modeling with simulation software, 246
Monte Carlo simulation. *See* Monte Carlo simulation
software architecture design and, 234–235
SPICE simulation, 285–286, 289
SUPREM for process simulation, 2
YSM and, 283
Single project approach, to DFSS deployment, 45–47
Six Sigma
background of, 1–3
Bill Smith's role in birth of, 3–5
Cp and Cpk indices for, 315
preference for continuous parameters in, 238
six steps to, 6–7
verification test strategy using, 350–354
Six Sigma Design Methodology (SSDM), 9
Six Sigma Research Institute (SSRI), 8

Smith, Bill, 3–5
Smoke testing, 361
Software
    development lifecycle. See SDLC (software development lifecycle)
    execution models, 298
    for modeling, 246–247
    RADIOV and, 11, 78
    reliability, 325
    reliability case study, 333–341
    requirements flow-down, 190–193
Software architecture
    alternate architecture generation for, 143–146
    decision-making process for, 224–227
    design heuristics for, 227–228
    design patterns, simulation, modeling, and prototyping, 234–235
    flowchart for decision-making, 226
    layers, 223–224
    overview of, 223
    summary, 235
    tactics, 228–230
    trade-off analysis, 230–234
Software DFSS. See SDFSS (software DFSS)
Software FMEA. See also FMEA (Failure Modes and Effects Analysis)
    acronyms and definitions, 164
    benefits of, 168–169
    cost savings from, 167–168
    implementation case study, 169–172
    presentation template, 168
    process documentation, 176–185
    process phases, 165–167
    roles in, 165
    tracker, 167
Software optimization
    model evaluation, 298–299
    multiple response optimization, 293–294
    overview of, 293
    proofing for mistakes and errors, 299–303
    stability, 303–305

    summary, 305
    use case modeling, 294–298
Software testing
    list of common software failures, 353
    overview of, 347–350
    summary, 358–359
    terminology related to, 359–361
    test case development with design patterns, 354–356
    types of, 359–361
    verification test strategy using Six Sigma, 350–354
    verification testing using combinatorial design method, 356–358
Sources of variability (SOV), 2, 272
SOV (sources of variability), 2, 272
Sparcity of effects principle, in statistics, 253–254
Special characters, software mistake proofing, 303
Specification limits, for critical parameters, 197–198
SPICE simulation
    IAR (integrated alternator regulator) case study and, 289
    YSM and, 285–286
Spreadsheets, modeling within, 246
Sprints, in Agile development, 213
SSDM (Six Sigma Design Methodology), 9
SSRI (Six Sigma Research Institute), 8
Stability
    instability of requirements, 59
    software optimization and, 303–305
Stakeholders
    certification of, 63
    engaging in DFSS deployment, 32
    numerical assessment of risks by, 159–160
    role in defining the vision, 33
    single project approach to DFSS deployment, 46–47
Standard deviation, 267, 271–273

Stanford University Process Engineering
    Model (SUPREM), 2
Statistics
    methods for improving quality, 1–3
    sparcity of effects principle in, 253–254
Strategies
    adjusting decisions based on, 92–94
    portfolio decision making and, 91
    tactics derived from, 386
Stress testing
    interface functionality, 369
    overview of, 361
    software, 347
Strings, software mistake proofing, 303
Structural design patterns, 354–355
Subsystem level
    concept generation at, 139
    interface flows and, 369
    system design and, 189–190
Success metrics
    additive and multiplicative models for,
        57–58
    criteria for, 57
    defining, 41–42
    formal gate reviews and, 56
    program risks and, 155–156
Supply chain readiness
    interface flows and, 369
    on-time delivery and lead time
        commitments and, 370–372
    optoelectronic multichip module case study,
        380–382
    position within DFSS flow, 363–365
    product assembly robustness and,
        366–369
    product launch schedules and, 369–370
    semiconductor manufacturing example
        related to on-time delivery, 372–379
    summary, 382–383
    tolerance expectations and, 366
    trade-offs between on-time delivery, lead
        time, and inventory levels, 372

SUPREM (Stanford University Process
    Engineering Model), 2
Symbols, in use case modeling, 295
System engineering, incorporating DFSS with,
    388
System FMEA. See also FMEA (Failure Modes
    and Effects Analysis)
    evaluating system-level risks, 157–158
    steps in system/subsystem-level FEMA,
        160–163
System level
    concept generation at, 139
    design process, 187–189
    quality attributes for software architecture
        systems, 225–227
    question to ask for system verification, 351
System requirements. See also Requirements
    flow-down
    building House of Quality, 129–131
    prioritization of, 133
    process of system design and, 187–189
    setting roof on trade-offs among, 132
    targets and units for, 133
    translating customer requirements into,
        124–128
System services (layer 1), software architecture
    design, 223–224
System testing, 348, 361

T
Tactics
    software architecture design, 228–230
    strategies becoming, 386
Target-is-best critical parameters, 198
Target values, for critical parameters,
    197–198
Taylor series expansion, 267–268
TCS (total customer satisfaction), Motorola
    competition, 8
TDD (test-driven development)
    Agile development and, 218
    overview of, 212

TDFSS (Technology Development for Six Sigma), 65

Team
  role in defining the vision, 33–34
  scenario for DFSS deployment, 30
  stakeholder engagement in DFSS deployment, 32
  support for, 44
  VOC gathering, 110–111

Technical experts, compared with deployment experts, 29

Technical lead, 63–64

Technical requirements, 154, 157

Technical risks, metrics for, 59

Technology Development for Six Sigma (TDFSS), 65

Teoriya Resheniya Izobreatatelskikh Zadatch (TRIZ), 141–143

Test cases
  design patterns for developing, 354–356
  principals for developing, 349–350

Test-driven development (TDD)
  Agile development and, 218
  overview of, 212

Test escapes
  reliability and, 331–332
  verification of design capability and, 313–315

Testability tactics, in software architecture design, 229

Tests
  accelerated life testing, 328–330
  software. See Software testing
  types of, 359–361

Theory of Innovative Problem Solving (TIPS). See TRIZ (Teoriya Resheniya Izobreatatelskikh Zadatch)

Thread-based testing, 361

Time requirements, as impediment to acceptance of DFS, 35–36

"Time trap", in Lean Six Sigma, 380

Timing/delay, allocation of tolerances, 209

TIPS (Theory of Innovative Problem Solving). See TRIZ (Teoriya Resheniya Izobreatatelskikh Zadatch)

Tolerances
  initial tolerance allocation, 208–210
  P/T (precision-to-tolerance) ratio. See P/T (precision-to-tolerance) ratio
  in supply chain readiness, 366

Tools
  associated with DFSS steps, 71–72
  DFSS tools used with Agile development, 215–217
  minimum toolset for deploying DFSS, 47–48
  RADIOV, 77, 79–81
  single project approach to DFSS deployment, 46
  for trade-off analysis, 231

Total customer satisfaction (TCS), Motorola competition, 8

Trade-off analysis
  CBAM (cost-benefit analysis method), 232–234
  decision-making and, 224–225
  prioritization matrix, 232
  Pugh matrix, 231
  software architecture design decisions, 230

Training
  certification and, 62–63
  DFSS deployment and, 39, 46

Transfer functions
  in conjunction with estimated distributions, 264–265
  definition of, 237
  deriving from continuous critical parameters, 244–245
  deriving from discrete critical parameters, 241–242
  determining need for, 237–238
  Monte Carlo simulation used with, 267
  predicting performance with, 263

TRIZ Contradiction Matrix, 142–143

TRIZ (Teoriya Resheniya Izobreatatelskikh Zadatch), 141–143

U

UML (unified modeling language)
  software development and, 144
  use case modeling and, 296
Uncertainty, supply chain decisions and, 372
Unified modeling language (UML)
  software development and, 144
  use case modeling and, 296
Unit tests
  definition of, 361
  in software testing, 347
Usability tactics, in software architecture design, 229
Usability testing, 361
Use-based (cluster) testing, 361
Use case modeling, 294–298
Useful life, in verifying reliability and availability, 325–326
User acceptance testing, 361
User interface (layer 3), software architecture design, 224
Users
  interactions via software applications, 224
  interface flows and, 369

V

"V" model, 390
Validation
  of customer requirements, 124
  of inputs, in software mistake proofing, 302
  question to ask for system validation planning, 351
Validation testing, 361
Variance
  MSA and, 200–201
  in optimization of critical parameters, 271–273
  red X effect and, 274
  reduction methods, 273–280

Six Sigma focus on variance reduction, 3
Verification
  Agile development and, 218
  data collection plan for ViewHome project, 220
  linking MSA to, 201–202
  software testing. *See* Software testing
  of supply chain readiness. *See* Supply chain readiness
Verification of design capability
  assessing capability, 315–316
  improving measurement systems, 310–313
  MSA and, 307–310
  position within DFSS flow, 307
  risk of failures despite, 313–315
  summary, 316–317
Verification of reliability and availability
  accelerated life testing, 328–330
  availability and reliability flow-down, 321–322
  bathtub curve and Weibull distribution and, 322–325
  customer perspective on, 319–321
  early life failures/infant mortality, 326
  flowchart for, 327–328
  methods for improving, 332
  modeling availability case study, 342–346
  risk of failures despite verification, 331–332
  software reliability, 325
  software reliability case study, 333–341
  summary, 333
  useful life/constant failure rates, 325–326
  wear out mechanisms, 326
  WeiBayes approach, 330–331
Verification testing
  combinatorial design method for, 356–358
  using Six Sigma, 350–354
Verify phase, DFSS, 307, 363
Verify phase, RADIOV
  in CDMA base station example, 21–24
  overview of, 80–81
  tools, 81

ViewHome project (Motorola), 219–220
Visibility, software mistake proofing, 300
Vision (organizational)
    communicating, 40–41
    defining, 33–34
VOB (voice of business)
    business requirements, 110
    critical parameters and, 155–156
    requirements gathering, 214
VOC (voice of customer)
    applying to new products or services,
        48–49
    critical parameters as response to, 153–155
    customer expectations regarding reliability
        and availability, 319–320
    customer interview guide, 113–115
    customer selection for, 111–112
    developing services in response to, 388
    DFSS aligned with, 9–10
    House of Quality and, 128–133
    identifying challenging customer
        requirements, 120–121
    implicit VOC, 155
    importance of, 107–108
    interviewing customers, 116
    Kano model, 122–123
    KJ analysis, 117–120
    planning customer visits and interviews,
        115–116
    position in DFSS flow, 108–110
    purpose and objectives of, 110
    RADIOV requirements phase and, 73
    requirements gathering, 214

Six Sigma steps related to, 6
    summary of, 134–135
    team for VOC gathering, 110–111
    translating customer requirements into
        system requirements, 124–128
    validation and prioritization of customer
        requirements, 124
    voices and images in interview process,
        112–113
Voices
    KJ analysis and, 117–118
    VOC gathering and, 112–113

W

"W" model (Otto), 390
Waterfall model, 105
Wear out mechanisms, 325, 326
WeiBayes approach, 330–331
Weibull distribution, 322–325
    early life failures/infant mortality, 326
    shape and scale parameters, 323
    useful life/constant failure rates, 326–327
    wear out mechanisms, 327
    WeiBayes approach to failures, 330–331
White box tests
    definition of, 361
    in software testing, 347
Who Moved My Cheese? (Johnson and
    Blanchard), 30–31

Y

YSM (Yield Surface Modeling), 283–288